Lecture Notes in Computational Science and Engineering **137**

Editors:

More information about this series at http://www.springer.com/series/3527

Marta D'Elia • Max Gunzburger • Gianluigi Rozza
Editors

Quantification of Uncertainty: Improving Efficiency and Technology

QUIET selected contributions

Editors
Marta D'Elia
Computational Science and Analysis
Sandia National Laboratories
Livermore, CA, USA

Max Gunzburger
Dept of Scientific Computing
Florida State Univ
Tallahassee, FL, USA

Gianluigi Rozza
SISSA mathLab, Mathematics Area
International School for Advanced Studies
Trieste, Italy

ISSN 1439-7358 ISSN 2197-7100 (electronic)
Lecture Notes in Computational Science and Engineering
ISBN 978-3-030-48723-2 ISBN 978-3-030-48721-8 (eBook)
https://doi.org/10.1007/978-3-030-48721-8

Cover illustration: Two face field fluid flow and a geometrically parameterized evolutionary 4th order Cahn-Hilliard system with physical boundary conditions: Background geometry and five reduced basis components, based on a full order cut finite element method. Courtesy Dr Efthymios Karatzas (SISSA mathLab).

This Springer imprint is published by the registered company Springer Nature Switzerland AG.
The registered company address is: Gewerbestrasse 11, 6330 Cham, Switzerland

Preface

The international workshop "Quantification of Uncertainty: Improving Efficiency and Technology" (QUIET) was held in July 2017 at SISSA (International School for Advanced Studies) in Trieste, Italy. The workshop focused on promising approaches for near-future improvements in the way uncertainty quantification problems in the partial differential equation setting are solved. Its central objectives were to bring together both senior and junior researchers: (1) to share and compare their latest findings; (2) to provide guidelines for the most promising approaches for near-future research; and (3) to provide junior researchers with new problems that they can incorporate into their research plans.

Group picture at QUIET workshop, SISSA Trieste, summer 2017

One particular focus in the workshop's planning was on problems with a high number of random parameters and on four specific avenues that have recently shown considerable promise, namely: reduced order modeling, more efficient solvers, high-dimensional approximation, and applications.

Through the formal presentations and daily discussion sessions, the outcomes of the workshop included: (1) synergistic exchanges across topics facilitated by the commonality of algorithms used for more than one topic; (2) the exchange, among participants in the four focus topics, of recent and often unpublished progress and results; (3) the exposure of a sizable group of junior researchers already active in uncertainty quantification research to new problem areas and new directions for their own research; and (4) extensive discussions on identifying the most promising future research directions.

This volume consists of 11 peer-reviewed chapters based on workshop presentations. Although these selected contributions are only a sample of the talks given at the workshop, they are eminently representative of the presentations' broad and deep coverage of algorithms and applications, and give ample illustrations of the aforementioned outcomes of the workshop.

QUIET 2017 was supported by SISSA, International School for Advanced Studies, in Trieste, Italy; the US National Science Foundation, Division of Mathematical Sciences; US Air Force Office of Scientific Research, Computational Mathematics Program; and Florida State University, Department of Scientific Computing, Tallahassee, FL.

More information on the workshop can be found at indico.sissa.it/event/8.

We would like to thank the SISSA mathLab organizing team, the anonymous reviewers for their diligent work, as well as Springer Nature, especially Francesca Bonadei and the LNCSE series editorial board, for their valued support.

Livermore, CA, USA — Marta D'Elia
Denver, CO, USA — Max Gunzburger
Trieste, Italy — Gianluigi Rozza
March 2020

Contents

About the Editors

Dr. Marta D'Elia, born in 1983, is a Senior member of the technical staff at Sandia National Laboratories, Center for Computing Research. She graduated with honors in Mathematical Engineering at Politecnico di Milano (Bachelor 2005, Master 2007), and she has a PhD in Applied Mathematics from the Emory University, Atlanta, GA (2011). She was a postdoctoral appointee at Florida State University, Tallahassee, FL, from 2012 to 2014. Her research is focused on computational science and engineering, mostly in the field of nonlocal and fractional problems. She authored more than 30 research papers and she is the principal investigator of the Laboratory Directed Research and Development grant MATNIP. She is also associate editor of the SIAM Journal on Scientific Computing. She organized several international workshops and she has been part of conference scientific committees.

Max Gunzburger is the Robert O. Lawton Distinguished Professor and Frances Eppes Eminent Professor and Founding Chair of the Department of Scientific Computing at Florida State University. He has received numerous awards and honors including the W. T. and Idelia Reid Prize in Mathematics from the Society for Industrial and Applied Mathematics (SIAM), being a member of the initial set of fellows of SIAM, and a NASA Innovator's Prize for Inventions and Contributions. He received the Rostchild Visiting Fellow and Simons Fellowship awards from the Isaac Newton Institute located at Cambridge University and the OCCAM Visiting Fellow award from Oxford University. He has held a Distinguished Professorship at Yonsei University in South Korea, a Guest Professorship at Peking University, Distinguished Visiting Professorship Hong Kong Polytechnic University, and has also served an an External Scientific Member of the Basque Center for Applied Mathematics. He has delivered distinguished lectures at several universities, was a plenary speaker at many major conferences, including four SIAM conferences and at an International Congress of Mathematicians, and delivered eighteen short courses at venues in North and South America and Europe.

Max Gunzburger received his B.S., M.S., and PhD from New York University. He held a National Research Council Postdoctorate Associateship at the Naval Ordnance Laboratory and was among the first group or researchers at the Institute for Computational Science and Engineering (ICASE) at NASA Langley Research Center. Subsequently, he has served as a Distinguished Professor of Mathematics and Chair of the Mathematics Department at Iowa State University and as Professor of Mathematics at Virginia Tech, Carnegie Mellon University, and the University of Tennessee. He has served on the editorial board of numerous journals and book series, including three SIAM journals, and was Editor in Chief of the SIAM Journal on Numerical Analysis and is a Founding Editor and Senior Editor of the SIAM/ASA Journal on Uncertainty Quantification. He has served on numerous SIAM committees and was Chairman of the Board of Trustees of SIAM. He has served as a consultant to three US Department of Energy Laboratories and two NASA laboratories as well as several industrial and commercial organizations. He maintains active collaborations with researchers in the United States and overseas, including currently with three Department of Energy laboratories.

Max Gunzburger has served as the major professor for 46 PhD students and 36 postdoctoral researchers, almost all of whom are still active in research, and many of whom have had significant research careers in their own right. He has authored or co-authored close to 300 journal articles as well as may proceedings and other publications. His research has been funded, at one time or another, by the several US agencies, including the Air Force Office of Scientific Research, Army Research Office, Department of Commerce, Department of Energy, National Aeronautics and Space Administration, National Science Foundation, Office of Naval Research and also by the North Atlantic Treaty Organization, several government laboratories and companies. His research interests span the areas of numerical analysis, scientific computing, optimization and control, computational geometry, grid generation, and partial differential and integral equations with applications in diverse areas including fluid and solid mechanics, anomalous diffusion processes, climate, superconductivity, subsurface flows, fracture mechanics, materials, image processing, acoustics, electromagnetics, finance, biology, etc.

Prof. Gianluigi Rozza, born in 1977, is Full Professor in Numerical Analysis and Scientific Computing at SISSA, International School for Advanced Studies, Trieste, Italy. Lauream cum Laude in Aerospace Engineering at Politecnico di Milano (2002), PhD in Applied Mathematics at Ecole Polytechnique Federale de Lausanne (2005). Post-doc research associate at MIT, Massachusetts Institute of Technology, Center for Computational Engineering (2006–2008). Researcher and Lecturer at EPFL (2008–2012). Excellence research grant (NOFYSAS) at SISSA (2012–2014). He is a member of the recently created SISSA mathLab laboratory. He is lecturer in the SISSA doctoral program in Mathematical Analysis, Modelling and Applications, in the SISSA-ICTP master in High Performance Computing and at the master degrees in Mathematics and in Data Science and Scientific Computing. He is coordinator for the doctoral programme in Mathematical Analysis, Modelling and Applications. At SISSA he is director delegate for innovation, valorisation of research and technology transfer.

His research is mostly focused in numerical analysis and scientific computing, developing reduced order methods. Author of more than 100 scientific publications (editor of six books and author of two books). Co-advisor of 25 master thesis, co-director/director of 16 PhD theses since 2009. Principal Investigator of the European Research Council Consolidator Grant (H2020) AROMA-CFD and for the project FARE-AROMA-CFD and PRIN NA-FROM-PDEs funded by Italian Government. Within SISSA mathLab he is responsible of UBE (Under Water Blue Efficiency project), SOPHYA (Seakeeping of Planing Hull for YAchts) and PRELICA projects, within the regional maritime technological cluster MARE FVG, and coordinator of industrial projects with several companies, such as Danieli, Electrolux and Fincantieri. He is associate editor of SIAM SINUM, SIAM/ASA JUQ, Computing and Visualisation in Science. He is member of the Applied Mathematics Committee of European Mathematical Society.

Winner of the 2004 Bill Morton CFD Prize (Oxford University); ECCOMAS Award 2006 for the best PhD. Thesis in applied sciences and engineering; Springer Computational Science and Engineering prize in 2009; and ECCOMAS Jacques Louis Lions Award in 2014. He has organized national and international workshops and served in the scientific and organizing committee of several conferences.

Effect of Load Path on Parameter Identification for Plasticity Models Using Bayesian Methods

Ehsan Adeli, Bojana Rosić, Hermann G. Matthies, and Sven Reinstädler

Abstract To evaluate the cyclic behavior under different loading conditions using the kinematic and isotropic hardening theory of steel, a Chaboche viscoplastic material model is employed. The parameters of a constitutive model are usually identified by minimization of the distance between model response and experimental data. However, measurement errors and differences in the specimens lead to deviations in the determined parameters. In this article the Chaboche model is used and a stochastic simulation technique is applied to generate artificial data which exhibit the same stochastic behavior as experimental data. Then the model parameters are identified by applying an estimation using Bayes's theorem. The Gauss–Markov–Kalman filter using functional approximation is introduced and employed to estimate the model parameters in the Bayesian setting. Identified parameters are compared with the true parameters in the simulation, and the efficiency of the identification method is discussed. At the end, the effect of the load path on the parameter identification is investigated.

1 Introduction

In order to predict the behavior of loaded metallic materials, constitutive models are applied, which present a mathematical frame for the description of elastic and inelastic deformation. The models by Miller, Krempl, Korhonen, Aubertin, Chan, and Bodner are well-known constitutive models for isotropic materials [1–5]. In

E. Adeli (✉) · B. Rosić · H. G. Matthies
Institute of Scientific Computing, Technische Universität Braunschweig, Braunschweig, Germany
e-mail: e.adeli@tu-braunschweig.de

S. Reinstädler
Institute of Structural Analysis, Technische Universität Braunschweig, Braunschweig, Germany
e-mail: statik@tu-bs.de

M. D'Elia et al. (eds.), *Quantification of Uncertainty: Improving Efficiency and Technology*, Lecture Notes in Computational Science and Engineering 137,
https://doi.org/10.1007/978-3-030-48721-8_1

1983, Chaboche [6, 7] put forward what has become known as the unified Chaboche viscoplasticity constitutive model, which has been widely accepted.

All inelastic constitutive models contain parameters which have to be identified for a given material from experiments. In the literature only few investigations can be found dealing with identification problems using stochastic approaches. Klosowski and Mleczek have applied the least-squares method in the Marquardt–Levenberg variant to estimate the parameters of an inelastic model [8]. Gong et al. have also used some modification of the least-squares method to identify the parameters [9]. Harth and Lehn identified the model parameters of a model by employing some generated artificial data instead of experimental data using a stochastic technique [10]. A similar study by Harth and Lehn has been done for other constitutive models like Lindholm and Chan [11].

In this paper, a viscoplastic model of Chaboche is studied. The model contains five material parameters which have to be determined from experimental data. It should be noted that here virtual data are employed instead of real experimental data. A cyclic tension-compression test is applied in order to extract the virtual data.

The model is described in Sect. 2, whereas Sect. 3 explains how to propagate the uncertainty in the model and how to perform the update. The probabilistic model is reformulated from the deterministic model, and once the forward model is provided, the model parameters are updated using a Bayesian approach.

In Sect. 4 the desired parameters are identified from the measured data. In fact, the parameters which have been considered as uncertain parameters are updated and the uncertainties of the them are reduced while the random variables representing the uncertain parameters are updated during the process. The results are thoroughly studied and the identified parameters as well as the corresponding model responses are analyzed. Finally the prediction of the models is compared with the measured data for different applied load paths. It is also explained why different load paths cause different identification of model parameters.

2 Model Problem

The mathematical description of metals under cyclic loading beyond the yield limit that includes viscoplastic material behavior as well as the characterization of compulsory isotropic-kinematic hardening is here given in terms of a modified Chaboche model introduced in [12]. As we consider classical infinitesimal strains, we assume an additive strain decomposition. The material behavior is described for the elastic part by isotropic homogeneous elasticity, and for viscoplasticity the dissipation potential is given by

$$\phi(\sigma) = \frac{k}{n+1}\left\langle \frac{\sigma_{ex}}{k} \right\rangle^{n+1}, \tag{1}$$

with $\langle\cdot\rangle = \max(0, \cdot)$ and k and n as the material parameters. Here σ_{ex} is the over-stress, defined via the equivalent stress (σ_{eq}) which reads

$$\sigma_{eq} = \sqrt{\frac{3}{2}\text{tr}((\boldsymbol{\sigma} - \boldsymbol{\chi})_D.((\boldsymbol{\sigma} - \boldsymbol{\chi})_D)}, \tag{2}$$

where $(\cdot)_D$ denotes the deviatoric part and $\boldsymbol{\chi}$ is the back-stress of kinematic hardening. The over-stress σ_{ex} is given by

$$\sigma_{ex} = \sigma_{eq} - \sigma_y - R = \sqrt{\frac{3}{2}\text{tr}((\boldsymbol{\sigma} - \boldsymbol{\chi})_D.(\boldsymbol{\sigma} - \boldsymbol{\chi})} - \sigma_y - R, \tag{3}$$

where σ_y is the yield stress and R models the isotropic hardening which is introduced in the following. The partial derivative of the dissipation potential ϕ with respect to $\boldsymbol{\sigma}$ leads to the equation for the inelastic strain rate

$$\dot{\boldsymbol{\epsilon}}_{vp} = \frac{\partial\phi}{\partial\boldsymbol{\sigma}} = \left\langle\frac{\sigma_{ex}}{k}\right\rangle^n \frac{\partial\sigma_{ex}}{\partial\boldsymbol{\sigma}}. \tag{4}$$

The viscoplastic model allows for isotropic and kinematic hardening, which is considered in order to describe different specifications. Assuming $R(t)$ and $\boldsymbol{\chi}(t)$ with $R(0) = 0$ and $\boldsymbol{\chi}(0) = 0$ to describe isotropic and kinematic hardening respectively, the evaluation equations for these two are

$$\dot{R} = b_R(H_R - R)\dot{p} \tag{5}$$

and

$$\dot{\boldsymbol{\chi}} = b_\chi(\frac{2}{3}H_\chi\frac{\partial\sigma_{eq}}{\partial\boldsymbol{\sigma}} - \boldsymbol{\chi})\dot{p} \tag{6}$$

respectively. In the evaluation equations of the both hardening, $\dot{p}$ is the viscoplastic multiplier rate given as:

$$\dot{p} = \left\langle\frac{\sigma_{ex}}{k}\right\rangle^n, \tag{7}$$

which describes the rate of accumulated plastic strains. The parameter b_R indicates the speed of stabilization, whereas the value of the parameter H_R is an asymptotic value according to the evolution of the isotropic hardening. Similarly, the parameter b_χ denotes the speed of saturation and the parameter H_χ is the asymptotic value of the kinematic hardening variables. The complete model is stated in Table 1. Note that E represents the elasticity tensor.

By gathering all the desired material parameters to identify into the vector $\boldsymbol{q} = [\kappa\ G\ b_R\ b_\chi\ \sigma_y]$, where κ and G are bulk and shear modulus, respectively, which

Table 1 The constitutive model of Chaboche

Strain	$\boldsymbol{\epsilon}(t) = \boldsymbol{\epsilon}_e(t) + \boldsymbol{\epsilon}_{vp}(t)$
Hooke's law	$\boldsymbol{\sigma}(t) = E : \boldsymbol{\epsilon}_e(t)$
Flow rule	$\dot{\boldsymbol{\epsilon}}_{vp}(t) = \langle \frac{\sigma_{eq}(t) - \sigma_y - R(t)}{k} \rangle^n \frac{\partial \sigma_{ex}}{\partial \boldsymbol{\sigma}}$
Hardening	$\dot{R} = b_R(H_R - R)\dot{p}$
	$\dot{\boldsymbol{\chi}} = b_\chi(\frac{2}{3} H_\chi \frac{\partial \sigma_{eq}}{\partial \boldsymbol{\sigma}} - \boldsymbol{\chi})\dot{p}$
Initial conditions	$\boldsymbol{\epsilon}_{vp}(0) = 0,\ R(0) = 0,\ \boldsymbol{\chi}(0) = 0$
Parameters	σ_y (yield stress)
	k, n (flow rule)
	b_R, H_R, b_χ, H_χ (hardening)

determine the isotropic elasticity tensor, the goal is to estimate $\boldsymbol{q}$ given measurement displacement data, i.e.

$$u = Y(\boldsymbol{q}) + e, \tag{8}$$

in which $Y(\boldsymbol{q})$ represents the measurement operator and e the measurement (also possibly the model) error. Being an ill-posed problem, the estimation of $\boldsymbol{q}$ given u is not an easy task and usually requires regularization. This can be achieved either in a deterministic or a probabilistic setting. Here, the latter one is taken into consideration as further described in the text.

3 Bayesian Identification

By using additional (prior) knowledge on the parameter set next to the observation data, the probabilistic approach regularizes the problem of estimating $\boldsymbol{q}$ with the help of Bayes's theorem

$$\pi_{q|u}(\boldsymbol{q}|u) \propto L(u|\boldsymbol{q})\pi_q(\boldsymbol{q}), \tag{9}$$

in which the likelihood $L(u|\boldsymbol{q})$ describes how likely the measurement data are given prior knowledge $\pi_q(\boldsymbol{q})$. This in turn requires the reformulation of the deterministic model into a probabilistic one, and hence the propagation of material uncertainties through the model—the so-called forward problem—in order to obtain the likelihood [13, 14].

The main difficulty in using Eq. (9) lies in the computation of the likelihood. Various numerical algorithms can be applied, the most popular example of which are the Markov chain Monte Carlo methods. Being constructed on the fundamentals of ergodic Markov theory, these methods are characterized by very slow convergence. To avoid this, an approximate method based on Kolmogorov's definition of conditional expectation as already presented in [15] is considered here.

Let the material parameters $\boldsymbol{q}$ be modeled as random variables on a probability space $S := L_2(\Omega, \mathcal{B}, \mathbb{P})$. Here, Ω denotes the space of elementary events ω, $\mathcal{B}$ is the σ-algebra and $\mathbb{P}$ stands for the probability measure. This alternative formulation of Bayes's rule can be achieved by expressing the conditional probabilities in Eq. (9) in terms of conditional expectation. Following the mathematical derivation in [15–18], this approach boils down to a quadratic minimization problem by considering the forecast random variable q_f and the update of the forecast random variable q_a:

$$\bar{q}^{|z} = P_{\mathcal{Q}_{sn}} q_f = \underset{\eta \in \mathcal{Q}_{sn}}{\arg\min} \|q_f - \eta\|_{L_2}^2 = \Xi(u_f(\omega)), \tag{10}$$

where $P_{\mathcal{Q}_{sn}}$ is the orthogonal projection operator of q_f onto the space of the new information $\mathcal{Q}_{sn} := \mathcal{Q} \otimes S_n$ in which the space S_n is the space of random variables generated by the measurement $u = Y(\boldsymbol{q}) + e$. Due to the Doob–Dynkin lemma, $\bar{q}^{|z}$ is a function of the observation, where $u_f(\omega) = Y(q(\omega)) + e(\omega)$ is the forecast, and the assimilated value is $q_a(\omega) = q_f(\omega) + (\Xi(\hat{z}) - \Xi(u_f(\omega)))$.

Constraining the space of all functions Ξ to the subspace of linear maps, the minimization problem in Eq. (10) leads to a unique solution K. Note that the projection is performed over a smaller space than $\mathcal{Q}_{sn}$. An implication of this is that available information is not completely used in the process of updating, introducing an approximation error. This gives an affine approximation of Eq. (10)

$$q_a(\omega) = q_f(\omega) + K(\hat{z} - u_f(\omega)), \tag{11}$$

also known as a linear Bayesian posterior estimate or the so-called Gauss–Markov–Kalman filter (GMKF). Here, q_f represents the prior random variable, q_a is the posterior approximation, $u_f = Y(q_f(\omega)) + e(\omega)$ is the predicted measurement and K represents the very well-known Kalman gain

$$K := C_{q_f u_f} \left(C_{u_f} + C_\varepsilon \right)^{-1}, \tag{12}$$

which can be easily evaluated if the appropriate covariance matrices $C_{q_f u_f}$, C_{u_f} and C_ε are known.

An advantage of Eq. (11) compared to Eq. (9) is that the inference in Eq. (11) is given in terms of random variables instead of conditional densities. Namely, $q_a(\omega)$, $q_f(\omega)$, and $u_f(\omega)$ denote the random variables used to model the posterior, prior, observation, and predicted observation, respectively.

In this light the linear Bayesian procedure can be reduced to a simple algebraic method. Starting from the functional representation of the prior

$$\hat{q}_f(\omega) = \sum_\alpha q_f^{(\alpha)} \psi_\alpha(\omega), \tag{13}$$

where ψ_α are multivariate Hermite polynomials, and by considering the proxy in Eq. (13), one may discretize Eq. (11) as:

$$\boldsymbol{q}_a = \boldsymbol{q}_f + K(\hat{z} - \boldsymbol{u}_f), \tag{14}$$

where $\boldsymbol{q}_a = [\ldots, q_a^{(n)}, \ldots]$, etc. are the PCE coefficients. As the measurement is a deterministic value, $\hat{z} = [\hat{z}, 0, \ldots, 0]$ has only a zero-th order tensor. The covariances for the Kalman gain Eq. (12) are easily computed, e.g.

$$C_{\hat{q}_f, \hat{u}_f} = \sum_{\alpha>0} \alpha! \, q_f^{(\alpha)} (\boldsymbol{u}_f^{(\alpha)})^T. \tag{15}$$

4 Numerical Results

The identification of the material constants in the Chaboche unified viscoplasticity model is a reverse process, here based on virtual data. In case of the Chaboche model the best way of parameter identification is using the results of the cyclic tests, since more information can be obtained from cyclic test rather than creep and relaxation tests, specifically information regarding hardening parameters. The aim of the parameter identification is to find a parameter vector $\boldsymbol{q}$ introduced in the previous section. The bulk modulus (κ), the shear modulus (G), the isotropic hardening coefficient (b_R), the kinematic hardening coefficient (b_χ) and the yield stress (σ_y) are considered as the uncertain parameters of the constitutive model.

A preliminary study is on a regular cube, modeled with one 8 node element, completely restrained on the back face, and with normal traction on the opposite (front) face. Two cases are considered in order to compare the effect of applied force on identified parameters. For both cases the magnitude of the normal traction and a stress in the plane of the front face are plotted in Figs. 1 and 2, respectively. Blue and red colors represent the stress value in normal and in plane directions, respectively. As it is seen, the magnitude of the applied force for the case 1 is constant all time but for the case 2 the magnitude of the applied force grows gradually by time.

Considering the parameters listed in Table 2, the related $\sigma - \epsilon$ hysteretic graph obtained for the applied force cases 1 and 2 which can be seen in Figs. 3 and 4, respectively.

The displacements of one of the nodes on the front surface in normal and in plane directions are observed as the virtual data in this study. Applying the Gauss–Markov–Kalman filter with functional approximation as explained in the previous chapter and introducing measurement error in such a way that 15% of mean values are equal to the coefficient of variation for the related parameter, the probability density function (PDF) of prior and posterior of the identified parameters can be seen in Figs. 5 and 6 for the first and second case, respectively.

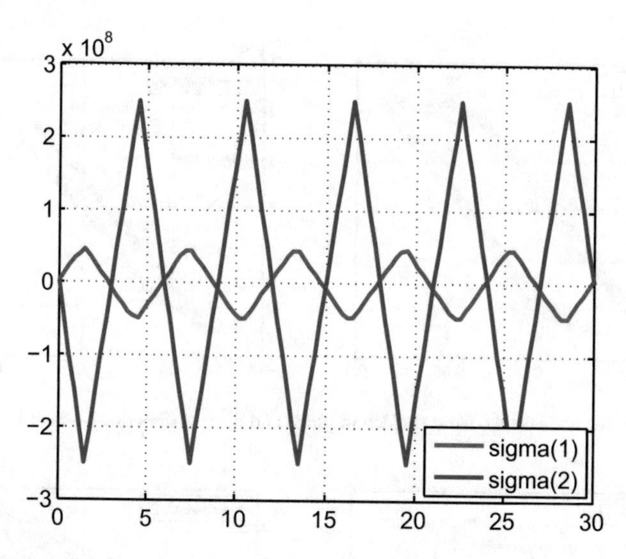

Fig. 1 Decomposed applied force on desired node according to time—case 1

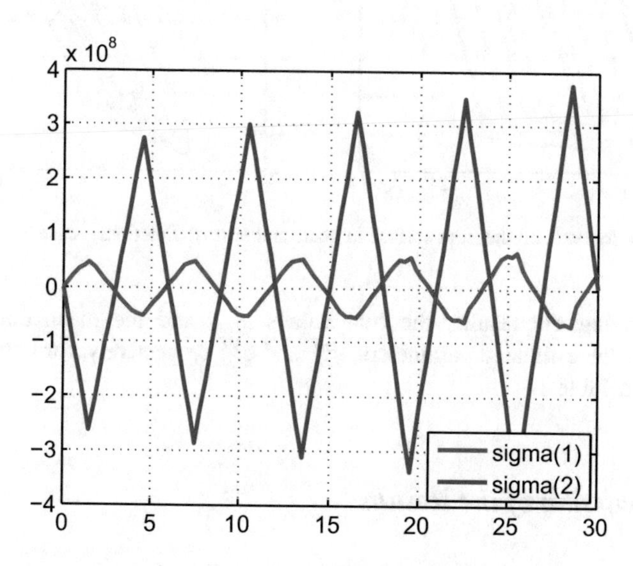

Fig. 2 Decomposed applied force on desired node according to time—case 2

Table 2 The model parameters

κ	G	σ_y	n	k	b_R	H_R	b_χ	H_χ
1.66e9	7.69e8	1.7e8	1.0	1.5e8	50	0.5e8	50	0.5e8

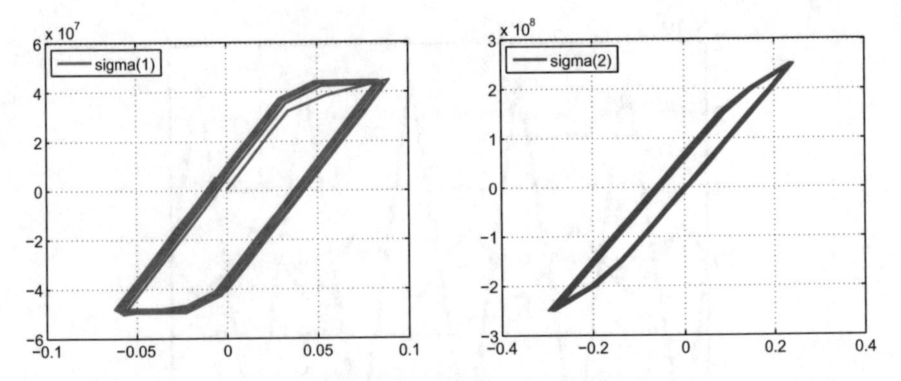

Fig. 3 $\sigma - \epsilon$ for node on the front surface in plane and normal directions—case 1

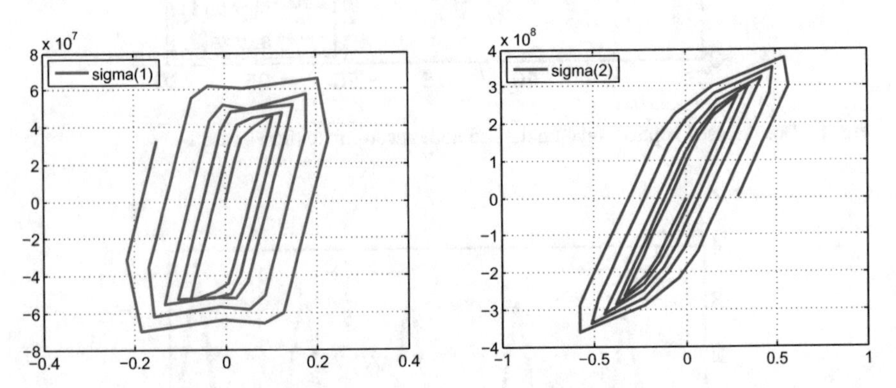

Fig. 4 $\sigma - \epsilon$ for node on the front surface in plane and normal directions—case 2

Summarizing the results, the true values q_{true} and the mean and standard deviation of the estimated parameters, $\boldsymbol{q}^m_{\text{est}}$ and $\boldsymbol{q}^{\text{std}}_{\text{est}}$ respectively, for both cases are compared in Table 3.

4.1 Discussion of the Results

From the sharpness of the posterior PDF of κ, G and σ_y, it can be concluded that enough information from virtual data is received and updating the parameters considering their uncertainty is done very properly for the both cases, as the standard deviation of the residual uncertainty is below 1% of the mean.

For the posterior PDF of b_R and b_χ, it can be inferred that better updating is done for the second case compared to the first case. Not only are the more accurate estimations of the exact hardening parameters, b_R and b_χ, predicted for the second case, but the uncertainty of the estimated hardening parameters are also reduced much more for the second case.

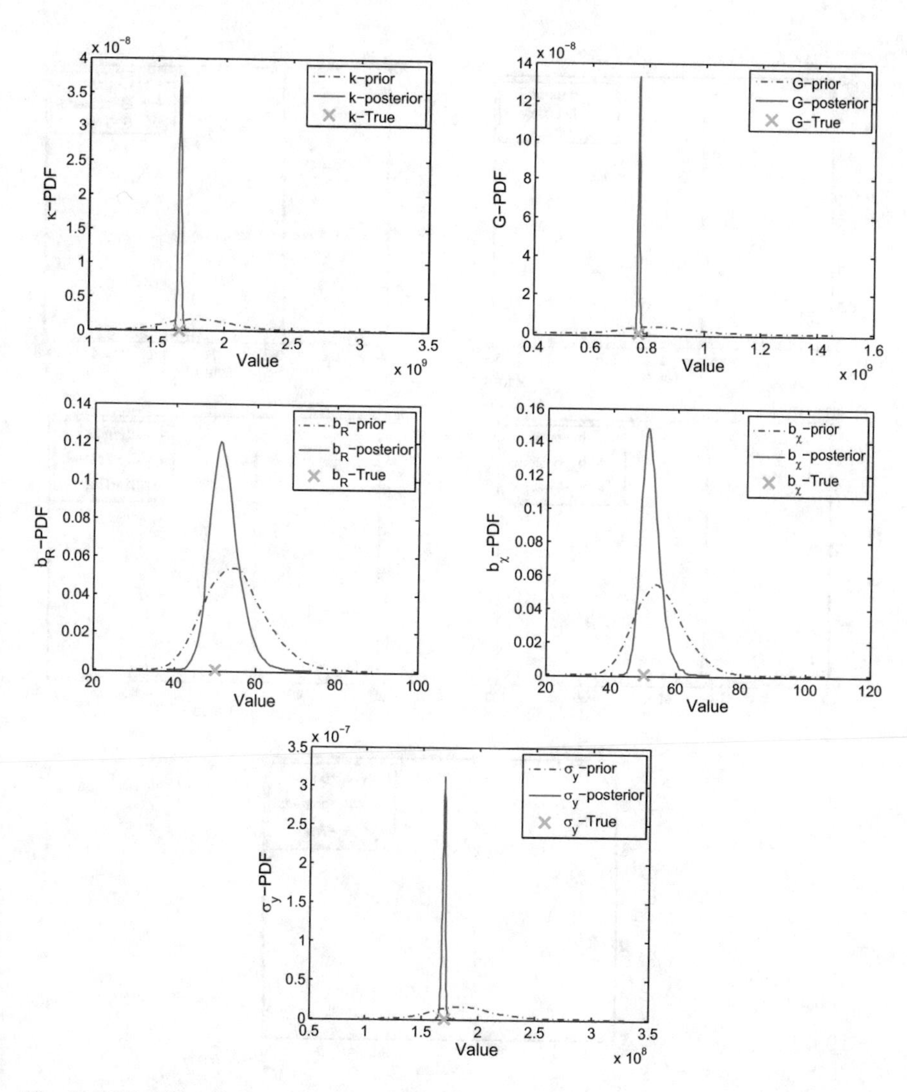

Fig. 5 PDF of identified parameters—case 1

One reason that can be mentioned is that the process is not always in the states that hardening equations are involved like the elastic states. Therefore less information from the whole simulation can be analyzed for estimating the hardening parameters and updating their parameters' uncertainties. Figures 7 and 8 prove this fact that since more states are out of the von Mises yield criterion for the second case compared to the first case, in which the hardening equations are involved only in these states, the better identification can be done for the second case, where a gradually varying increasing applied force is considered, for hardening parameters in comparison with the first case where a constant magnitude applied force is

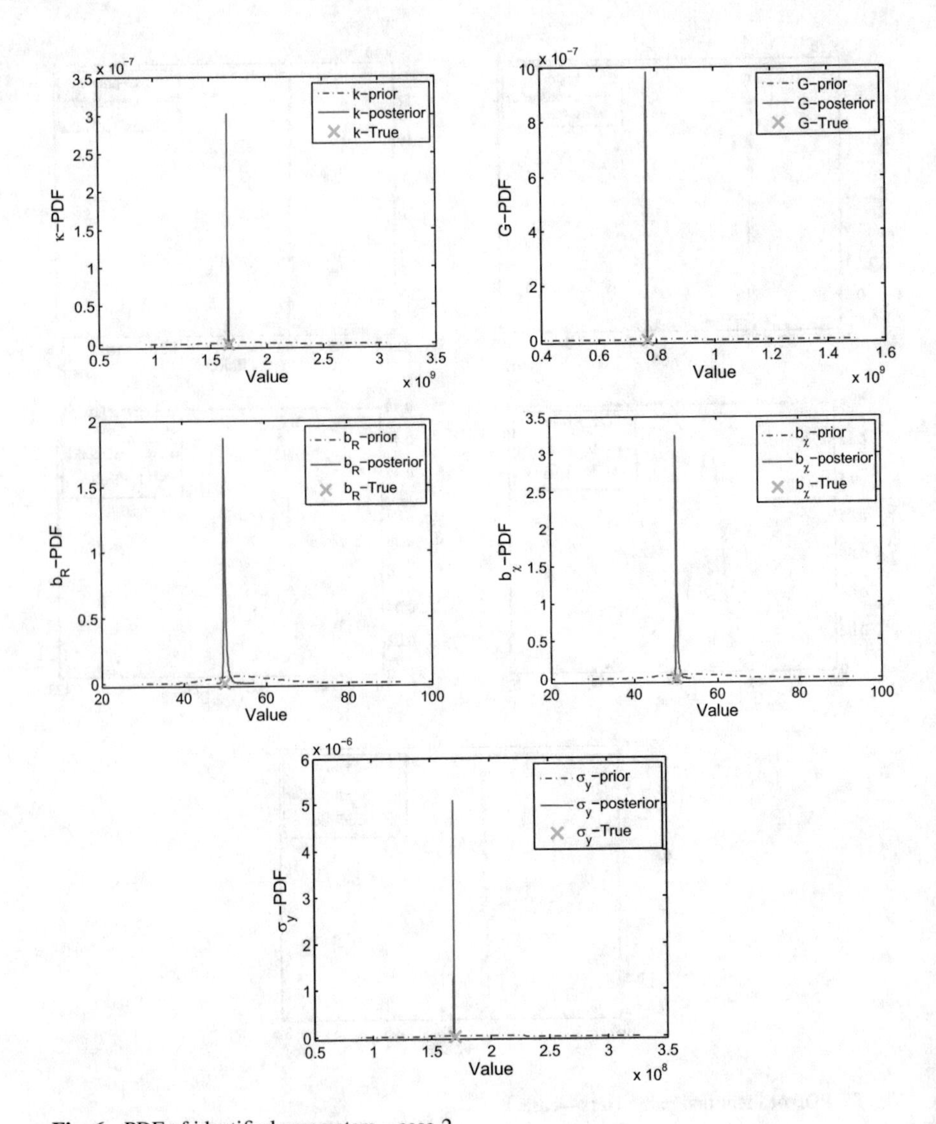

Fig. 6 PDF of identified parameters—case 2

Table 3 The identified model parameters

Parameters	$\boldsymbol{q}_{\text{true}}$	$\boldsymbol{q}^{m}_{\text{est-1}}$	$\boldsymbol{q}^{\text{std}}_{\text{est-1}}$	$\boldsymbol{q}^{m}_{\text{est-2}}$	$\boldsymbol{q}^{\text{std}}_{\text{est-2}}$
κ	1.66e9	1.66e9	1.13e7	1.66e9	2.59e6
G	7.69e8	7.68e8	3.47e6	7.68e8	6.39e5
b_R	50	52.36	3.71	50.27	0.29
b_χ	50	52.04	3.01	50.19	0.53
σ_y	1.7e8	1.69e8	1.35e6	1.69e8	1.52e5

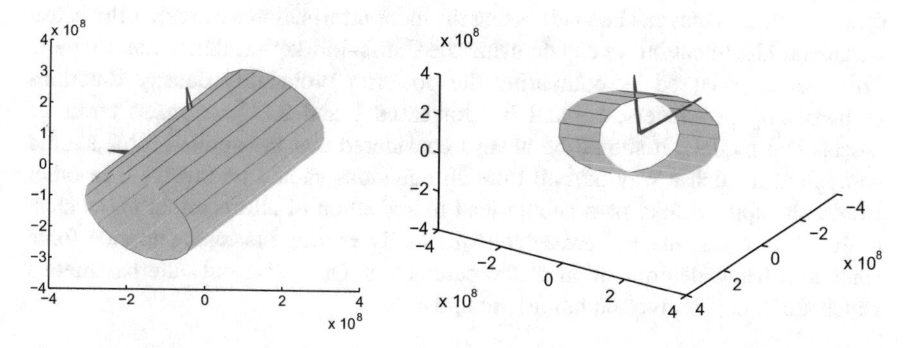

Fig. 7 Principal stresses of applied force in 3D considering the von Mises yield criterion—case 1

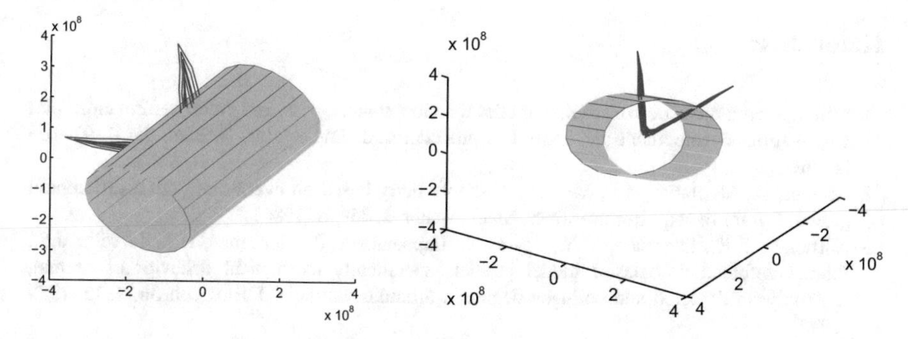

Fig. 8 Principal stresses of applied force in 3D considering the von Mises yield criterion—case 2

employed. In fact, the cyclic applied force in the second case causes the more activation of the desired parameters in the studied set of equations comparing to the first case and accordingly a better determination of the parameters can be carried out for the second case. It should be pointed out that the von Mises yield criterion is illustrated by the green cylinder in the mentioned figures i.e. inside and outside of the cylinder respectively refer to the elasticity and plasticity states, respectively. Also the blue color represents the principal stresses in these figures.

5 Summary

Using the Gauss–Markov–Kalman Filter method explained in Sect. 3 to identify the model parameters of the Chaboche model indicates that it is possible to identify the model parameters by employing this method using functional approximation. The parameters are well estimated and the uncertainty of the parameters is reduced while the random variables of the parameters are updated during the process. The other conclusion that can be made is that the more information we receive, the better parameter identification we can do using the Gauss–Markov–Kalman filter method. This fact is observed by comparing the posterior probability density functions of hardening parameters, b_R and b_χ, for cases 1 and 2. Therefore in terms of mechanical models, it should be always considered that the applied force should be applied in such a way that all time all equations should be involved. In other words, the applied load path should lead to activation of all uncertain parameters in the set of equations, as here a cyclic gradually varying increasing applied force leads to a better determination of the parameters. Otherwise only the parameters which are in the involved equations are updated.

Acknowledgement This work is partially supported by the DFG through GRK 2075.

References

1. Miller, A.: An inelastic constitutive model for monotonic, cyclic, and creep deformation: part I—equations development and analytical procedures. J. Eng. Mater. Technol. **98**(2), 97–105 (1976)
2. Krempl, E., McMahon, J.J., Yao, D.: Viscoplasticity based on overstress with a differential growth law for the equilibrium stress. Mech. Mater. **5**, 35–48 (1986)
3. Korhonen, R.K., Laasanen, M.S., Toyras, J., Lappalainen, R., Helminen, H.J., Jurvelin, J.S.: Fibril reinforced poroelastic model predicts specifically mechanical behavior of normal, proteoglycan depleted and collagen degraded articular cartilage. J Biomech **36**, 1373–1379 (2003)
4. Aubertin, M., Gill, D.E., Ladanyi, B.: A unified viscoplastic model for the inelastic flow of alkali halides. Mech. Mater. **11**, 63–82 (1991)
5. Chan, K.S., Bodner, S.R., Fossum, A.F., Munson, D.E.: A constitutive model for inelastic flow and damage evolution in solids under triaxial compression. Mech. Mater. **14**, 1–14 (1992)
6. Chaboche, J.L., Rousselier, G.: On the plastic and viscoplastic constitutive equations—part 1: rules developed with internal variable concept. J. Press. Vessel Technol. **105**, 153–158 (1983)
7. Chaboche, J.L., Rousselier, G.: On the plastic and viscoplastic constitutive equations—part 2: application of internal variable concepts to the 316 stainless steel. J. Press. Vessel Technol. **105**, 159–164 (1983)
8. Kłosowski, P., Mleczek, A.: Parameters' identification of Perzyna and Chaboche viscoplastic models for aluminum alloy at temperature of 120 °C. Eng. Trans. **62**(3), 291–305 (2014)
9. Gong, Y., Hyde, C., Sun, W., Hyde, T.: Determination of material properties in the Chaboche unified viscoplasticity model. J. Mater. Des. Appl. **224**(1), 19–29 (2010)
10. Harth, T., Lehn, J.: Identification of material parameters for inelastic constitutive models using stochastic methods. GAMM-Mitt. **30**(2), 409–429 (2007)
11. Chan, K.S., Bodner, S.R., Lindholm, U.S.: Phenomenological modelling of hardening and thermal recovery in metals. J. Eng. Mater. Technol. **110**, 1–8 (1988)
12. Velde, J.: 3D Nonlocal Damage Modeling for Steel Structures under Earthquake Loading. Department of Architecture, Civil Engineering and Environmental Sciences University of Braunschweig—Institute of Technology (2010)
13. Matthies, H.G.: Stochastic finite elements: computational approaches to stochastic partial differential equations. J. Appl. Math. Mech. **88**, 849–873 (2008)

14. Matthies, H.G.: Uncertainty quantification with stochastic finite elements. Encyclopedia of computational mechanics. In: Stein, E., de Borst, R., Hughes, T.R.J. (eds.) Wiley, Chichester (2007)
15. Matthies, H.G., Zander, E., Rosić, B.V., Litvinenko, A.: Parameter estimation via conditional expectation: a Bayesian inversion. J. Adv. Model. Simul. Eng. Sci. **3**(24), (2016)
16. Rosić, B.V., Litvinenko, A., Pajonk, O., Matthies, H.G.: Sampling-free linear Bayesian update of polynomial chaos representations. J. Comput. Phys. **231**(17), 5761–5787 (2012)
17. Matthies, H.G., Zander, E., Rosić, B.V., Litvinenko, A., Pajonk, O.: Inverse problems in a Bayesian setting. J. Comput. Methods Solids Fluids **41**, 245–286 (2016)
18. Rosić, B.V., Matthies, H.G.: Identification of properties of stochastic elastoplastic systems. In: Computational Methods in Stochastic Dynamics, pp. 237–253 (2013)

A Compressive Spectral Collocation Method for the Diffusion Equation Under the Restricted Isometry Property

Simone Brugiapaglia

Abstract We propose a compressive spectral collocation method for the numerical approximation of Partial Differential Equations (PDEs). The approach is based on a spectral Sturm-Liouville approximation of the solution and on the collocation of the PDE in strong form at randomized points, by taking advantage of the compressive sensing principle. The proposed approach makes use of a number of collocation points substantially less than the number of basis functions when the solution to recover is sparse or compressible. Focusing on the case of the diffusion equation, we prove that, under suitable assumptions on the diffusion coefficient, the matrix associated with the compressive spectral collocation approach satisfies the restricted isometry property of compressive sensing with high probability. Moreover, we demonstrate the ability of the proposed method to reduce the computational cost associated with the corresponding full spectral collocation approach while preserving good accuracy through numerical illustrations.

1 Introduction

Compressive Sensing (CS) is a mathematical principle introduced in 2006 that allows for the efficient measurement and reconstruction of sparse and compressible signals. Its success is now established in the signal processing community and its wide range of applications include medical imaging, computational biology, geophysical data analysis, compressive radar, remote sensing, and machine learning. More recently, CS has also started attracting more and more attention in scientific computing and numerical analysis, in particular, in the fields of numerical methods for Partial Differential Equations (PDEs), high-dimensional function approximation, and uncertainty quantification.

S. Brugiapaglia (✉)
Simon Fraser University, Burnaby, BC, Canada
e-mail: simone_brugiapaglia@sfu.ca

M. D'Elia et al. (eds.), *Quantification of Uncertainty: Improving Efficiency and Technology*, Lecture Notes in Computational Science and Engineering 137,
https://doi.org/10.1007/978-3-030-48721-8_2

In this paper, we present a novel technique for the numerical solution of PDEs based on CS. The proposed approach, called *compressive spectral collocation* takes advantage the CS principle in the context of spectral collocation methods. Its constitutive elements are: (1) Sturm-Liouville spectral approximation, (2) randomized collocation, and (3) greedy sparse recovery. In order to make the presentation easier and the theoretical analysis of the method accessible, we focus on the case of a stationary diffusion equation over a tensor product domain with homogeneous boundary conditions.

1.1 Main Contributions

We propose a novel numerical method for PDEs, called compressive spectral collocation, focusing on the case of a stationary diffusion equation over a tensor product domain with homogeneous boundary conditions. The approach leverages the CS principle by randomizing the choice of the collocation points and by promoting sparse solutions with respect to a Sturm-Liouville basis, which are recovered via the greedy algorithm orthogonal matching pursuit.

Our main contributions can be summarized as follows:

1. In Algorithm 1, we present a rigorous formulation of the compressive spectral collocation approach for the diffusion equation.
2. In Theorem 4, we prove that the matrix associated with the compressive spectral collocation approach satisfies the restricted isometry property of CS under suitable assumptions on the diffusion coefficient.
3. In Sect. 5, we demonstrate numerically that the compressive spectral collocation approach is able to recover sparse solutions with higher accuracy and lower computational cost than the corresponding "full" spectral collocation method. Moreover, in the case of compressible solutions, we show that the compressive approach is computationally less expensive than the full one while maintaining a good level of accuracy.

Before outlining the structure of the paper, we review the literature about CS-based methods in numerical analysis, placing particular emphasis on numerical methods for PDEs.

1.2 Literature Review

CS was proposed in 2006 by the pioneering works of Donoho [13], Candès et al. [9] and has triggered an impressive amount of work since then.

The very first attempt to apply CS to the numerical approximation of a PDEs can be found in [21]. The authors propose a Galerkin discretization of the Poisson problem, where the trial and test spaces are composed by piecewise linear finite elements. The technique is deterministic and relies on the successive refinement of the solution on different hierarchical levels and on a suitable error estimator. Recovery is based on ℓ^1-minimization.

The CS principle has then been applied to Petrov-Galerkin discretizations of advection-diffusion-reaction equations via the COmpRessed SolvING method (in short, CORSING), proposed in [7]. The method employs Fourier-type trial functions and wavelet-like test functions (or vice versa) and the dimensionality of the discretization is reduced by randomly subsampling the test space. The theoretical analysis of the method in the infinite-dimensional setting has been carried out in [8]. The CORSING method has also been applied to the two-dimensional Stokes' equation in [6].

Numerical methods for PDEs based on ℓ^1 minimization can be considered the ancestors of CS-based methods for PDEs. Lavery conducted some pioneering studies on finite differences for the inviscid Burgers' equation [23] and on finite volumes dicretizations for steady scalar conservation laws [24]. More recently, similar techniques have been analyzed for transport and Hamilton-Jacobi equations [17, 18]. Moreover, some works considered sparsity-promoting spectral schemes for time-dependent multiscale problems based on soft thresholding [25, 30] or on the sparse Fourier transform [12].

On a different but related note, there has been a lot of research activity around CS-based methods for the uncertainty quantification of PDEs with random inputs [5, 14, 26–28, 33]. In these works, the CS principle is combined with Polynomial Chaos in order to approximate a quantity of interest of the solution map of the PDE. Being very smooth for a wide family of operator equations, this map can be approximated by a sparse combination of orthogonal polynomials and the CS principle employed to lessen the curse of dimensionality.

Finally, it is worth mentioning recent works where CS is employed to learn the governing equations of a dynamical systems given time-varying measurements [32] and to solve inverse problems in PDEs [4].

1.3 Outline of the Paper

The paper is organized as follows.

In Sect. 2, we recall some of the main elements of CS, of particular interest in our context. We place more emphasis on greedy recovery via orthogonal matching pursuit and on recovery guarantees based on the restricted isometry property.

Equipped with the CS fundamentals, we present the compressive spectral collocation method in Sect. 3, focusing on the case of a homogeneous stationary diffusion equation.

Section 4 deals with the theoretical analysis of the method. We prove that the matrix associated with the compressive spectral collocation approach satisfies the restricted isometry property with high probability under suitable conditions on the diffusion coefficient. Moreover, we discuss the implications of the restricted isometry property for the recovery error analysis of the method.

In Sect. 5, we illustrate some numerical results for the two-dimensional diffusion equation. We assess the performance of the compressive spectral collocation approach when recovering sparse and compressible solutions. Moreover, we compare it with the corresponding "full" spectral collocation approach, demonstrating the computational advantages associated with the proposed strategy.

Conclusions and future directions are finally discussed in Sect. 6.

2 Elements of Compressive Sensing

We introduce some elements of CS that will be useful to define the compressive spectral collocation approach. Our presentation is based on a very special selection of topics. For a comprehensive introduction to CS, we refer the reader to [15].

CS deals with the problem of measuring a sparse or compressible signal by using the minimum amount of linear, nonadaptive observations, and of reconstructing it via convex optimization techniques (such as ℓ^1 minimization and its variants), greedy algorithms, or thresholding techniques.

Here, we focus on CS with greedy recovery via orthogonal matching pursuit. After introducing this setting in Sect. 2.1, we recall some theoretical results based on the restricted isometry property in Sect. 2.2.

2.1 Compressive Sensing and Greedy Recovery

Let us consider a vector $x \in \mathbb{R}^N$ (often called "signal"). We restrict the presentation to the real case, even though the theory can be generalized to the complex case. We collect m linear nonadaptive measurements of x into a vector $b \in \mathbb{R}^m$, i.e.

$$Ax = b. \tag{1}$$

The matrix $A \in \mathbb{R}^{m \times N}$ is called the sensing matrix and $m \ll N$. The problem of finding x given b is clearly ill-posed since the linear system (1) is highly underdetermined. In order to regularize this inverse problem, the a priori information assumed on x is *sparsity* or *compressibility*.

A vector is said to be s-sparse if it has at most s nonzero entries. More in general, x is said to be compressible if, for some $p \geq 1$, its best s-term approximation error (with respect to some ℓ^p norm) $\sigma_s(x)_p$ decays quickly in s, where $\sigma_s(x)_p$ is defined as

$$\sigma_s(x)_p := \inf\{\|x - v\|_p : v \in \mathbb{R}^N, \|v\|_0 \leq s\},$$

with

$$\|v\|_0 := |\operatorname{supp}(v)|, \quad \operatorname{supp}(v) := \{j \in [N] : v_j \neq 0\}, \quad \forall v \in \mathbb{R}^N,$$

and where we have employed the notation

$$[n] := \{1, \ldots, n\}, \quad \forall n \in \mathbb{N}.$$

Notice that if the signal x is s-sparse, then $\sigma_k(x)_p = 0$, for every $k \geq s$ and $p \geq 1$.

A plethora of recovery strategies is available in order to find sparse or compressible solutions to the linear system (1). In this paper, we focus on the greedy algorithm *Orthogonal Matching Pursuit* (OMP) (see [31] and references therein), outlined in Algorithm 1.

Algorithm 1 Orthogonal matching pursuit (OMP)

Inputs:

- $A \in \mathbb{R}^{m \times N}$: sensing matrix, with ℓ^2-normalized columns;
- $b \in \mathbb{R}^m$: vector of measurements;
- $K \in \mathbb{N}$: number of iterations.

Orthogonal Matching Pursuit:

1. Let $\hat{x}_0 = 0$ and $S_0 = \operatorname{supp}(\hat{x}_0) = \emptyset$;
2. For $k = 1, \ldots, K$, repeat the following steps:
   a. Find $j_k = \arg\max_{j \in [N]} |(A^T(A\hat{x}_{k-1} - b))_j|$;
   b. Define $S_k = S_{k-1} \cup \{j_k\}$;
   c. Compute $\hat{x}_k = \arg\min_{v \in \mathbb{R}^N} \|Av - b\|_2$ s.t. $\operatorname{supp}(v) \subseteq S_k$.

Output:

- $\hat{x}_K \in \mathbb{R}^N$: K-sparse approximate solution to (1).

OMP iteratively constructs a sequence of k-sparse vectors $\hat{x}_k$ that approximately solve (1), with $k = 1, \ldots, K$, by adding at most one new entry to the support at each iteration. During the k-th iteration, OMP seeks the column of A mostly correlated with the residual associated with the previous approximation $\hat{x}_{k-1}$. Then, the support is enlarged by adding the corresponding index, and the k-th approximation $\hat{x}_k$ is computed by solving an $m \times k$ least-squares problem. Observe that the least-square problem solved to compute $\hat{x}_k$ is overdetermined if $K \leq m$, which is usually the case in practice.

2.2 Recovery Guarantees Based on the Restricted Isometry Property

In order to quantify the approximation error associated with the OMP solution, we present some theoretical results based on the restricted isometry property, which has by now become a standard tool in CS.

Definition 1 A matrix $A \in \mathbb{R}^{m \times N}$ is said to satisfy the *restricted isometry property of order s and constant* $0 < \delta < 1$ if

$$(1-\delta)\|v\|_2^2 \leq \|Av\|_2^2 \leq (1+\delta)\|v\|_2^2, \quad \forall v \in \mathbb{R}^N, \ \|v\|_0 \leq s. \tag{2}$$

The smallest $0 < \delta < 1$ such that (2) holds is referred to as the *s-th restricted isometry constant* of A and it is denoted by $\delta_s(A)$.

Intuitively, the restricted isometry property requires the map $x \mapsto Ax$ to approximately preserve distances when its action is restricted to the set of s-sparse vectors, up to a distortion factor δ. Computing $\delta_s(A)$ given A is not computationally feasible in general since it implies a search over all the $\binom{N}{s}$ subsets of $[N]$ of cardinality s. However, what makes this tool extremely useful in practice is the fact that it is possible to show that certain classes of random matrices satisfy the restricted isometry property with high probability.

The following theorem gives sufficient conditions for a matrix $A \in \mathbb{R}^{m \times N}$ built by independently selecting m random rows according to a suitable probability density from a "tall" matrix $B \in \mathbb{R}^{M \times N}$ in order to satisfy the restricted isometry property (up to a suitable diagonal preconditioning). These conditions depend on the spectrum of the Gram matrix $B^T B$ and on the so-called *local coherence* of B, i.e., the vector whose entries are

$$\max_{j \in [N]} (B_{qj})^2, \quad \forall q \in [M].$$

The proof of this result can be found in [6, Theorem 1.21]. Let us note that this is an extension of the restricted isometry property analysis based on the local coherence for bounded orthonormal systems proposed in [22], where the orthonormality condition is relaxed.

Theorem 1 *Consider* $B \in \mathbb{R}^{M \times N}$, *with* $M \geq N$, *and suppose that there exist two constants* $0 < r \leq R < +\infty$ *such that the minimum and maximum eigenvalues of* $B^T B$ *satisfy*

$$0 < r \leq \lambda_{\min}(B^T B) \leq \lambda_{\max}(B^T B) \leq R.$$

Moreover, assume that there exists a vector $\nu \in \mathbb{R}^M$ such that the local coherence of B is bounded from above as follows:

$$\max_{j \in [N]} (B_{qj})^2 \leq \nu_q, \quad \forall q \in [M].$$

Then, for every $1 - \frac{r}{R} < \delta < 1$, there exists a universal constant $c > 0$ such that, provided

$$m \geq \widetilde{c}\, s \ln^3(s) \ln(N),$$

and $s \geq \widetilde{c} \ln(N)$, where

$$\widetilde{c} = c \max\left(\frac{\|\nu\|_1}{R}, 1\right)\left(\delta - \left(1 - \frac{r}{R}\right)\right)^{-2},$$

the following holds.

Let us draw $\tau_1, \ldots, \tau_m$ i.i.d. from $[M]$ distributed according to the probability density

$$p = \frac{\nu}{\|\nu\|_1} \in \mathbb{R}^M,$$

and define $A \in \mathbb{R}^{m \times N}$ and $D \in \mathbb{R}^{m \times m}$ as

$$A_{i,j} = B_{\tau_i, j}, \quad \forall i \in [m], \forall j \in [N], \quad D = \text{diag}\left(\left(\frac{1}{\sqrt{mRp_i}}\right)_{i \in [m]}\right), \tag{3}$$

where $\text{diag}(v)$ denotes the matrix having the entries of v on the main diagonal and zeros elsewhere. Then, the s-th restricted isometry constant of DA satisfies

$$\delta_s(DA) \leq \delta,$$

with probability at least $1 - N^{-\ln^3(s)}$.

The restricted isometry property is a sufficient condition to show that the vector $\hat{x}_K$ computed by K iterations of OMP is a good approximation to x. In particular, a suitable upper bound on the $26s$-th restricted isometry property constant is sufficient for OMP to reach the accuracy of the best s-term approximation error up to a universal multiplicative constant using $K = 24s$ iterations. The following theorem is a direct consequence of [15, Theorem 6.25]. The recovery error analysis of OMP based on the restricted isometry property was originally proposed in [34].

Theorem 2 *Let $A \in \mathbb{R}^{m \times N}$ with ℓ^2-normalized columns such that*

$$\delta_{26s}(A) < \frac{1}{6}. \tag{4}$$

Then, there exists a universal constant $C > 0$ such that for every $x \in \mathbb{R}^N$ and $b \in \mathbb{R}^m$ such that (1) *holds, the vector $\hat{x}_K$ computed by $K = 24s$ iterations of OMP (Algorithm 1) satisfies*

$$\|x - \hat{x}_K\|_2 \leq C \frac{\sigma_s(x)_1}{\sqrt{s}}. \tag{5}$$

This type of recovery error estimate is called "uniform" since it holds for every signal $x \in \mathbb{R}^N$. It is worth noticing that when x is s-sparse, it is recovered *exactly* since $\sigma_s(x)_1 = 0$. Moreover, let us observe that a more general version of this theorem holds in the case of noisy measurements, i.e., when $y = Ax + e$, where $e \in \mathbb{R}^m$ is a noise vector corrupting the measurements. In that case, an additive term directly proportional to $\|e\|_2$ appears on the left-hand side of (5) (see [15, Theorem 6.25]).

3 Compressive Spectral Collocation

We are now in a position to introduce the compressive spectral collocation method. Let us consider the following diffusion equation in strong form:

$$\begin{cases} -\nabla \cdot (\eta \nabla u) = F, & \text{in } \Omega, \\ u = 0, & \text{on } \partial\Omega, \end{cases} \tag{6}$$

where $\Omega = (0, 1)^d$ is the physical domain, $u \in C^2(\overline{\Omega})$ is the unknown solution, the function $\eta : \overline{\Omega} \to \mathbb{R}$, with $\eta \in C^1(\overline{\Omega})$ and $\eta(x) \geq \eta_{\min} > 0$ for every $x \in \overline{\Omega}$, is the diffusion coefficient, and $F \in C^0(\overline{\Omega})$ is the forcing term. We also consider the dimension d to be of moderate size.

We will define the compressive spectral collocation approach in two steps. First, we describe the Sturm-Liouville basis, the collocation grid employed, and the corresponding "full" spectral collocation method in Sect. 3.1. Then, in Sect. 3.2, we define the compressive approach, outlined in Algorithm 1.

3.1 The Spectral Basis and the Collocation Grid

We discretize Eq. (6) by using a spectral collocation method based on a Sturm-Liouville basis (for a comprehensive introduction to spectral methods, we refer the reader to [10, 16]). In particular, let us consider the functions

$$\xi_j(z) := \frac{2^{d/2}}{\pi^2 \|j\|_2^2} \cdot \prod_{k=1}^{d} \sin(\pi j_k z_k), \quad \forall x \in \overline{\Omega}, \quad \forall j \in \mathbb{N}^d. \tag{7}$$

The system $\{\xi_j\}_{j \in \mathbb{N}^d}$ is formed by eigenvectors of the Laplace operator with homogeneous Dirichlet boundary conditions, normalized such that

$$\|\Delta \xi_j\|_{L^2(\Omega)} = 1, \quad \forall j \in \mathbb{N}^d.$$

In fact, the system $\{\Delta \xi_j\}_{j \in \mathbb{N}^d}$ is an orthonormal basis for $L^2(\Omega)$ with respect to the standard inner product $\int_\Omega uv$. Expanding a function with respect to this basis (up to normalization) corresponds to the so-called "modified Fourier series expansion". The coefficients' decay rate of the modified Fourier series expansion of a function is related to its Sobolev regularity and to suitable boundary conditions involving its derivatives. Here, we will assume the solution u to be regular enough to guarantee the compressibility of its coefficients and, consequently, to enable the application of the CS principle. For more details on the approximation properties of univariate and multivariate modified Fourier series expansions and on their usage in spectral methods for PDEs, we refer the reader to [1, 2, 19, 20].

Let us now truncate the multi-index set $\mathbb{N}^d$ by using the tensor product multi-index space of order $n \in \mathbb{N}$, i.e.

$$[n]^d \subseteq \mathbb{N}^d,$$

of cardinality $N := n^d$. Given a truncation level n, we rescale of the basis functions and define

$$\psi_j(z) := \frac{\xi_j(z)}{(n+1)^{d/2}}, \quad \forall x \in \overline{\Omega}, \quad \forall j \in [n]^d. \tag{8}$$

(The normalizations chosen in (7) and in (8) will turn out to be crucial in order to guarantee the restricted isometry property.)

As a collocation grid, we consider a tensorial grid of uniform step $h = 1/(N+1)$ over Ω, defined as

$$t_q := \frac{q}{n+1}, \quad \forall q \in [n]^d.$$

Notice that we do not need collocation points on $\partial\Omega$ since the functions $\{\psi_j\}_{j\in[n]^d}$ already satisfy the homogeneous boundary conditions.

The resulting "full" spectral collocation discretization of (6) is given by

$$Bx^{\text{full}} = c, \tag{9}$$

where

$$B_{qj} = [\nabla \cdot (\eta \nabla \psi_j)](t_q), \quad c_q = F(t_q), \quad \forall q, j \in [n]^d, \tag{10}$$

and where we are implicitly assuming some ordering for multi-indices in $[n]^d$ (e.g., the lexicographic ordering). Given a solution x^{full} to the system (9), the full spectral approximation to u is defined as

$$u^{\text{full}} = \sum_{j\in[n]^d} x_j^{\text{full}} \psi_j. \tag{11}$$

3.2 The Compressive Approach

Before describing the compressive approach, let us explain the rationale behind the normalizations adopted in (7) and (8) by considering for a moment the simple case of the Poisson equation. The normalization chosen for the system $\{\psi_j\}_{j\in[n]^d}$ ensures that

$$\text{if} \quad \eta(x) = 1, \ \forall x \in \overline{\Omega}, \qquad \text{then} \quad B = \underbrace{S_n \otimes \cdots \otimes S_n}_{d \text{ times}},$$

where $\otimes$ denotes the matrix Kronecker product and where $S_n \in \mathbb{R}^{n\times n}$ is the matrix associated with the discrete sine transform, defined as

$$(S_n)_{ij} = \sqrt{\frac{2}{n+1}} \sin\left(\frac{ij\pi}{n+1}\right), \quad \forall i, j \in [n]. \tag{12}$$

In particular, the full spectral collocation matrix B is orthogonal, i.e. it satisfies

$$B^T B = I, \tag{13}$$

where I is the identity matrix, because S_n is orthogonal. Moreover, the local coherence of the matrix B satisfies the upper bound

$$\max_{j\in[n]^d} (B_{qj})^2 \leq \left(\frac{2}{(n+1)}\right)^d \leq \frac{2^d}{N}, \quad \forall q \in [n]^d. \tag{14}$$

Therefore, in view of Theorem 1, drawing m indices independently distributed according to the uniform measure over $[n]^d$, i.e.,

$$\tau_1, \dots, \tau_m \quad \text{i.i.d. with} \quad \mathbb{P}\{\tau_i = q\} = \frac{1}{N}, \quad \forall q \in [n]^d, \ \forall i \in [m],$$

it is natural to define the resulting compressive spectral collocation discretization as

$$Ax = b, \tag{15}$$

where the matrix $A \in \mathbb{R}^{m \times N}$ and $b \in \mathbb{R}^m$ are defined as

$$A_{ij} = \sqrt{\frac{N}{m}} B_{\tau_i, j}, \quad b_i = \sqrt{\frac{N}{m}} c_{\tau_i}, \quad \forall i \in [m], \ \forall j \in [n]^d. \tag{16}$$

The normalization by a factor $\sqrt{N/m}$ is done in order to ensure the restricted isometry property for A (see Theorem 1). In particular, we observe that since the probability density is uniform, we have $D = \sqrt{N/m} \cdot I$ in (3).

The compressive spectral collocation solution is then computed by applying OMP in order to find a sparse solution to (15), up to normalizing the columns of A with respect to the ℓ^2 norm. The proposed method is summarized in Algorithm 1.

Algorithm 1 Compressive spectral collocation

Inputs:

- $n \in \mathbb{N}$: order of the tensor product multi-index space $[n]^d$;
- $m \in \mathbb{N}$: number of randomized collocation points;
- $K \in \mathbb{K}$: number of OMP iterations.

Compressive spectral collocation:

1. Draw $\tau_1, \dots, \tau_m$ i.i.d. uniformly at random from $[n]^d$;
2. Build $A \in \mathbb{R}^{m \times N}$ and $b \in \mathbb{R}^m$ according to (16);
3. Build $M = \operatorname{diag}\left((\|a_j\|_2)_{j \in [n]^d}\right) \in \mathbb{R}^{N \times N}$, where a_j is the j-th column of A;
4. Define $\widetilde{A} = AM^{-1} \in \mathbb{R}^{m \times N}$;
5. Compute $\hat{x}_K \in \mathbb{R}^N$ using OMP (Algorithm 1) with inputs $\widetilde{A}$, b, and K;
6. Define $\hat{x} = M\hat{x}_K$;
7. Define $\hat{u} = \sum_{j \in [n]^d} \hat{x}_j \psi_j$, with $\{\psi_j\}_{j \in [n]^d}$ given by (8).

Output:

- $\hat{u} \in C^\infty(\overline{\Omega})$: Compressive spectral approximation to (6).

At least three questions naturally arise at this point:

(i) The compressive spectral collocation method looks tailored to the Poisson equation. Does this method work for nonconstant diffusion coefficients?
(ii) How to choose the input parameters n, m, and K in Algorithm 1?
(iii) What are the benefits (if any) of the compressive approach with respect to the full one?

Answering to these questions will be the objective of the next two sections. In particular, Sect. 4 will focus on questions (i) and (ii). Applying the theory of CS introduced in Sect. 2, we will give a sufficient condition on η that implies a positive answer to (i) and propose a recipe for (ii). In Sect. 5, we will deal with question (iii) by showing the benefits of the compressive approach with respect to the full one through a numerical illustration.

4 Theoretical Analysis

In the previous section, we have proposed the compressive spectral collocation method for the diffusion equation (6), summarized in Algorithm 1. In this section, we see that, given $n \in \mathbb{N}$, in order to recover the best s-term approximation error to x^{full} (up to a universal constant) with high probability, it is sufficient to choose the number of collocation points and the iterations of OMP such that

$$m \geq C 2^d s \ln^3(s) \ln(N) \quad \text{and} \quad K = 24s, \tag{17}$$

where $C > 0$ is a universal constant independent of s, n, and d. This shows that for $s \ll N$, the number of collocation points to employ is substantially less than the dimension of the approximation space N. In particular, it scales linearly with respect to the target sparsity s, up to logarithmic factors. The main ingredients of the theoretical analysis are Theorems 1 and 2.

4.1 Restricted Isometry Property

Let us first consider the case of the Poisson Problem, where $\eta(z) = 1$ for every $z \in \overline{\Omega}$. We have seen that in this case the full spectral collocation matrix B defined in (10) is orthogonal (recall (13)) and that its local coherence satisfies the upper bound (14). Therefore, a direct application of Theorem 1 with $r = R = 1$, yields the following restricted isometry result.

Theorem 3 *Let $d, s, N \in \mathbb{N}$, with $s \leq N$. Then, there exists a universal constant $c > 0$ such that the following holds. For the Poisson equation, the full spectral collocation matrix $B \in \mathbb{R}^{N \times N}$ defined by* (10) *is orthogonal and the corresponding* *compressive spectral collocation matrix* $A \in \mathbb{R}^{m \times N}$ *defined by* (16) *has the restricted isometry property of order* s *and constant* δ *with probability at least* $1 - N^{-\ln^s(s)}$, *provided that*

$$m \geq c\, 2^d \delta^{-2} s \ln^3(s) \ln(N), \tag{18}$$

and $s \geq c\, \delta^{-2} \ln(N)$.

Let us now consider the case of a nonconstant coefficient η. In this case, B is not necessarily orthogonal and, in order to apply Theorem 1, we need to estimate the minimum and maximum eigenvalue of the Gram matrix $B^T B$ and to find a suitable upper bound to the local coherence of B. Using this strategy, in the next theorem we give sufficient conditions on the diffusion coefficient η able to guarantee the restricted isometry property for the compressive spectral collocation matrix A with high probability.

Theorem 4 *Let* $d, s, N \in \mathbb{N}$ *with* $s \leq N$, *and* $\eta \in C^1(\overline{\Omega})$ *satisfying the following conditions:*

$$\eta_{\min} := \min_{z \in \overline{\Omega}} \eta(z) > 0, \tag{19}$$

$$\|\eta\|_{L^\infty(\Omega)} \sum_{k=1}^{d} \|(\nabla \eta)_k\|_{L^\infty(\Omega)} < \frac{\pi}{2} \eta_{\min}^2. \tag{20}$$

Then, the full spectral collocation matrix B *defined by* (10) *satisfies*

$$r \leq \lambda_{\min}(B^T B) \leq \lambda_{\max}(B^T B) \leq R, \tag{21}$$

where

$$r := \eta_{\min}^2 - \frac{2}{\pi} \|\eta\|_{L^\infty(\Omega)} \sum_{k=1}^{d} \|(\nabla \eta)_k\|_{L^\infty(\Omega)}, \tag{22}$$

$$R := \left(\|\eta\|_{L^\infty(\Omega)} + \frac{1}{\pi} \sum_{k=1}^{d} \|(\nabla \eta)_k\|_{L^\infty(\Omega)} \right)^2. \tag{23}$$

Moreover, given $1 - r/R < \delta < 1$ *and provided*

$$m \geq \widetilde{C}\, s \ln^3(s) \ln(N),$$

and $s \geq \widetilde{C} \ln(N)$, *where*

$$\widetilde{C} = C \cdot 2^d \left(\delta - \left(1 - \frac{r}{R}\right)\right)^{-2},$$

the matrix $A/\sqrt{R}$*, where* A *is defined as in* (16) *satisfies the restricted isometry property of order* s *and constant* δ *with probability at least* $1 - N^{-\ln^3(s)}$.

Proof Let us consider the matrices $S_n \in \mathbb{R}^{n\times n}$ defined as in (12) and $C_n \in \mathbb{R}^{n\times n}$ as

$$(C_n)_{ij} := \sqrt{\frac{2}{n+1}} \cos\left(\frac{\pi i j}{n+1}\right), \quad \forall i, j \in [N].$$

Using basic trigonometric formulas and Lagrange's trigonometric inequality, it is not difficult to show that

$$S_n^T S_n = I, \tag{24}$$

$$C_n^T C_n = I - \frac{2}{n+1} Q_n, \tag{25}$$

where $I \in \mathbb{R}^{n\times n}$ is the identity matrix and $Q_n \in \mathbb{R}^{n\times n}$ is a checkerboard-structured matrix whose entries are 1 on the diagonals of even order and 0 on the diagonals of odd order, namely

$$(Q_n)_{ij} = \frac{1-(-1)^{i+j+1}}{2}, \quad \forall i, j \in [n].$$

As already observed, S_n is orthogonal. On the other hand, C_n is "almost orthogonal", up to the corrective term $-\frac{2}{n+1}Q_n$. Now, using the chain rule

$$-\nabla\cdot(\eta\nabla\psi_j)(t_q) = -\eta(t_q)\Delta\psi_j(t_q) - \nabla\eta(t_q)\cdot\nabla\psi_j(t_q), \quad \forall q, j \in [n]^d, \tag{26}$$

we see that the full spectral collocation matrix B defined in (10) has the form

$$B = B_1 + B_2,$$

where

$$B_1 = -D_0 \bigotimes_{k=1}^{d} S_n,$$

$$B_2 = -\sum_{k=1}^{d} D_k \left(\bigotimes_{\ell=1}^{k-1} S_n \otimes (C_n E) \otimes \bigotimes_{\ell=k+1}^{d} S_n\right) F.$$

and

$$D_0 = \operatorname{diag}\left((\eta(t_q))_{q\in[n]^d}\right) \in \mathbb{R}^{N\times N},$$

$$D_k = \operatorname{diag}\left(((\nabla\eta)_k(t_q))_{q\in[n]^d}\right) \in \mathbb{R}^{N\times N}, \quad \forall k \in [d],$$

$$E = \operatorname{diag}\big((\pi j)_{j\in[n]}\big) \in \mathbb{R}^{n\times n},$$

$$F = \operatorname{diag}\left(\left(\frac{1}{\pi^2\|j\|_2^2}\right)_{j\in[n]^d}\right) \in \mathbb{R}^{N\times N}.$$

In particular, we have

$$\|Bv\|_2^2 = \|B_1 v\|_2^2 + \|B_2 v\|_2^2 + 2v^T B_1^T B_2 v, \quad \forall v \in \mathbb{R}^N.$$

Now, we estimate the three terms in the right-hand side separately. Recalling (24)-(25), the first term can be estimated as

$$\eta_{\min}\|v\|_2 \leq \|B_2 v\|_2 \leq \|\eta\|_{L^\infty(\Omega)}\|v\|_2, \quad \forall v \in \mathbb{R}^N. \tag{27}$$

As for the second term, we have

$$\|B_1 v\|_2 \leq \sum_{k=1}^d \|(\nabla\eta)_k\|_{L^\infty(\Omega)} \left\| \left(\bigotimes_{\ell=1}^{k-1} S_n \otimes (C_n E) \otimes \bigotimes_{\ell=k+1}^{d} S_n \right) F v \right\|_2 .$$

Recalling properties (24) and (25), noticing that Q_n is positive semidefinite, using standard properties of the Kronecker product, and defining $w := Fv$, we see that

$$\begin{aligned}
&\left\| \left(\bigotimes_{\ell=1}^{k-1} S_n \otimes (C_n E) \otimes \bigotimes_{\ell=k+1}^{d} S_n \right) F v \right\|_2^2 \\
&= w^T \left(\bigotimes_{\ell=1}^{k-1} S_n^T S_n \otimes (E C_n^T C_n E) \otimes \bigotimes_{\ell=k+1}^{d} S_n^T S_n \right) w \\
&= w^T \left(\bigotimes_{\ell=1}^{k-1} I_n \otimes (E C_n^T C_n E) \otimes \bigotimes_{\ell=k+1}^{d} I_n \right) w \\
&= w^T \left(\bigotimes_{\ell=1}^{k-1} I_n \otimes E^2 \otimes \bigotimes_{\ell=k+1}^{d} I_n \right) w - \frac{2}{n+1} w^T \left(\bigotimes_{\ell=1}^{k-1} I_n \otimes (E Q_n E) \otimes \bigotimes_{\ell=k+1}^{d} I_n \right) w \\
&\leq \left\| \operatorname{diag}\left(\left(\frac{\pi j_k}{\pi^2 \|j\|_2^2} \right)_{j\in[n]^d} \right) v \right\|_2^2 \leq \frac{1}{\pi^2}\|v\|_2^2.
\end{aligned}$$

As a result, we obtain

$$0 \leq \|B_2 v\|_2 \leq \frac{1}{\pi}\left(\sum_{k=1}^d \|(\nabla\eta)_k\|_{L^\infty(\Omega)} \right) \|v\|_2, \quad \forall v \in \mathbb{R}^N. \tag{28}$$

Finally, using the Chauchy-Schwartz inequality and combining the inequalities (27) and (28), the third term can be estimated as

$$|v^T B_1^T B_2 v| \leq \frac{1}{\pi}\|\eta\|_{L^\infty(\Omega)} \sum_{k=1}^{d} \|(\nabla\eta)_k\|_{L^\infty(\Omega)} \|v\|_2^2. \tag{29}$$

Finally, combining (27), (28), and (29) yields the spectral bound (21) under sufficient conditions (19)-(20) on the diffusion coefficient η.

The last step is the local coherence upper bound. Recalling (26) and the definition (23) of R, we see that

$$\begin{aligned}\max_{j\in[n]^d} (B_{qj})^2 &= \max_{j\in[n]^d} \left(-\eta(t_q)\Delta\psi_j(t_q) + \nabla\eta(t_q)\cdot\nabla\psi_j(t_q)\right)^2 \\ &\leq \left(\frac{2}{n+1}\right)^d \max_{j\in[n]^d} \left(|\eta(t_q)| + \sum_{k=1}^{d} |(\nabla\eta)_k(t_q)| \frac{\pi j_k}{\pi^2\|j\|_2^2}\right)^2 \leq \frac{2^d R}{N} =: \nu_q.\end{aligned}$$

This choice of the local coherence upper bound ν yields $\|\nu\|_1 \leq 2^d R$ and $p_q = \nu_q/\|\nu\|_1 = 1/N$, for every $q \in [N]$. Finally, a direct application of Theorem 1 completes the proof. □

Let us take a closer look to the sufficient condition (20) on the diffusion coefficient η. First of all, it is homogeneous in η, as it is natural to be expected. Moreover, this condition becomes more and more restrictive as the dimension d increases, which is another tangible effect of the curse of dimensionality. Nevertheless, the following example shows that (20) can be satisfied in practice.

Example 1 Let us consider an affine diffusion coefficient of the form

$$\eta(z) = 1 + w^T z, \quad \forall z \in \overline{\Omega},$$

where $w \in \mathbb{R}^d$ with $w \geq 0$. In this case, (20) is equivalent to

$$\|w\|_1 < \frac{1}{2}\left(\sqrt{1+2\pi} - 1\right) \approx 0.85.$$

As d gets larger, the above condition becomes more and more restrictive. One possible way to mitigate the effect of d on this condition is by requiring the gradient $w = \nabla\eta$ to be sparse. ■

We conjecture that condition (20) is suboptimal and that it could be improved. How to make it less restrictive is an object of future investigation. Equipped with a restricted isometry property result for the compressive spectral collocation matrix A, we can now discuss the recovery guarantees of the proposed approach.

4.2 Recovery Guarantees (Discussion)

In view of Theorem 2, the restricted isometry property is a sufficient condition for OMP to recover the best s-term approximation error to a given signal up to a universal constant in $K = 24s$ iterations. In this section, we discuss the implications of this result for the compressive spectral collocation approach. A fully rigorous analysis of the recovery guarantees is beyond the objectives of this paper and is left to future work.

In order to combine the restricted isometry result (Theorem 4) with the OMP recovery result (Theorem 2), one has to take into account the effect of the ℓ^2-normalization of the columns of A onto the restricted isometry constant. Let $\widetilde{A}$ be the normalized version of A, as defined in Algorithm 1. Then, it is not difficult to show that

$$\delta_s(\widetilde{A}) \leq \frac{2\delta_s(A)}{1 - \delta_s(A)}, \qquad \forall s \in \mathbb{N},\ s \leq N.$$

In particular, the condition $\delta_{26s}(\widetilde{A}) < 1/6$ required to apply Theorem 2 and ensure the recovery via OMP, is implied by $\delta_{26s}(A) < 1/13$.

Due to the additional constraint $\delta > 1 - r/R$ required by Theorem 4, we see that in order to be able to choose $\delta < 1/13$, we need

$$1 - \frac{r}{R} < \frac{1}{13} \implies \frac{r}{R} > \frac{12}{13} \approx 0.92, \tag{30}$$

where r and R are defined as in (22) and (23).

Now, let us notice that a solution x^{full} to the full system (9) is also a solution to the compressive system (15). Using Theorem 2 and the fact that $\{\Delta\xi_j\}_{j\in\mathbb{N}^d}$ is orthonormal in $L^2(\Omega)$, we can estimate the error between the full spectral approximation u^{full} and the compressive spectral approximation $\hat{u}$ computed by choosing m and K as in (17) as

$$\|\Delta(u^{\text{full}} - \hat{u})\|_{L^2(\Omega)} = \frac{\|x^{\text{full}} - \hat{x}\|_2}{(n+1)^{d/2}} \leq C \cdot \frac{\sigma_s(x^{\text{full}})_1}{(n+1)^{d/2}\sqrt{s}}, \tag{31}$$

where $C > 0$ is a universal constant. (C depends on the universal constant of Theorem 2 and on $\sqrt{1+\delta}$, due to the ℓ^2-normalization of the columns of A. Moreover, notice that we can fix, e.g., $\delta = 1/14 < 1/13$). It is worth observing that when x^{full} is s-sparse, the compressive spectral collocation recovers the coefficients of u^{full} exactly.

Remark 1 Combining (31) with the triangle and the Poincaré inequalities yields

$$
\begin{aligned}
|u - \hat{u}|_{H^1(\Omega)} &\leq |u - u^{\text{full}}|_{H^1(\Omega)} + |u^{\text{full}} - \hat{u}|_{H^1(\Omega)} \\
&\leq |u - u^{\text{full}}|_{H^1(\Omega)} + C_\Omega \cdot \|\Delta(u^{\text{full}} - \hat{u})\|_{L^2(\Omega)} \\
&\leq |u - u^{\text{full}}|_{H^1(\Omega)} + C_\Omega \cdot C \cdot \frac{\sigma_s(x^{\text{full}})_1}{(n+1)^{d/2}\sqrt{s}},
\end{aligned}
$$

where C_Ω is the Poincaré constant of Ω. In this way, an estimate of $|u - u^{\text{full}}|_{H^1(\Omega)}$ can be converted to an estimate of $|u - \hat{u}|_{H^1(\Omega)}$. The error term $|u - u^{\text{full}}|_{H^1(\Omega)}$ can be studied using the tools in [10, Section 6.4.2]. These tools allow to compare $|u-u^{\text{full}}|_{H^1(\Omega)}$ to the best linear approximation error of the solution u with respect to the basis $\{\psi_j\}_{j\in[n]^d}$ in the $H^1(\Omega)$-seminorm. In turn, the best linear approximation error can be estimated by assuming enough regularity of u with respect to standard or mixed Sobolev norms when fulfilling suitable boundary conditions (see [2, Lemma 3.4 and Lemma 3.5]). ■

5 Numerical Experiments

We conclude by illustrating some numerical experiments that show the robustness of the spectral collocation method described in Algorithm 1 for the numerical solution of the diffusion equation (6) when the solution is sparse or compressible. The experiments demonstrate that the compressive approach is able to outperform the full one both from the accuracy and the efficiency viewpoints when the solution is sparse. When the solution is compressible, the compressive method can reduce the computational cost of the full method while preserving good accuracy.

We underline that the comparison is made without using fast transforms, which could considerably accelerate performance of both methods.

Given the order n of the ambient multi-index set $[n]^d$ and a target sparsity $s \in \mathbb{N}$, in all the numerical experiments we define the number of collocation points and of OMP iterations as

$$
m = \lceil 2s \ln(N) \rceil \quad \text{and} \quad K = s, \tag{32}
$$

numerically showing that the sufficient condition (17) is rather pessimistic in practice. Moreover, we focus on a two-dimensional diffusion equation with nonconstant coefficient

$$
\eta(z) = 1 + \frac{1}{4}(z_1 + z_2), \quad \forall z \in \overline{\Omega}, \tag{33}
$$

satisfying condition (20).

All the numerical experiments have been performed in MATLAB® R2017b version 9.3 64-bit on a MacBook Pro equipped with a 3 GHz Intel Core i7 processor and with 8 GB DDR3 RAM. We have employed the OMP implementation provided by the MATLAB® package OMP-Box v10 [29].

5.1 *Recovery of Sparse Solutions*

We start by comparing the full and the compressive spectral collocation approaches for the recovery of sparse solutions.

Given $s, n \in \mathbb{N}$ with $s \leq n^2 =: N$, we consider s-sparse solutions x randomly generated as follows. First, we draw s indices from $[N]$ uniformly at random. Then, we fill the corresponding entries with s independent realizations of a standard Gaussian variable $N(0, 1)$. This is implemented in MATLAB® using the commands `randperm` and `randn`, respectively. For each randomly-generated vector x, we run the full and the compressive methods 5 times. The recovery error of the full and of the compressive solution is measured using the relative discrete ℓ^2-error of the coefficients, namely,

$$\frac{\|x^{\text{full}} - x\|_2}{\|x\|_2} \quad \text{and} \quad \frac{\|\hat{x} - x\|_2}{\|x\|_2}.$$

The results are shown in Fig. 1, where we plot the relative error as a function of the computational cost for $n = 32$ (corresponding to $N = 1024$) and $s = 2, 4, 8, 16, 32$. Recalling (32), these values correspond to $m = 28, 56, 111, 222, 444$.

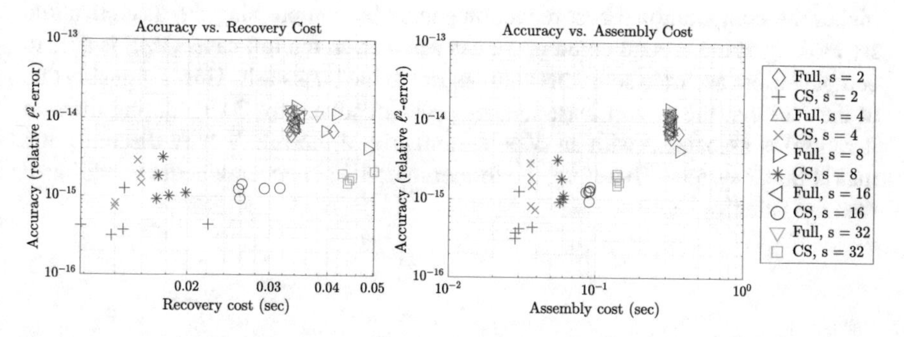

Fig. 1 Accuracy vs. cost plots for the full and the compressive spectral collocation approaches for the recovery of randomly generated s-sparse solutions to the diffusion equation with nonconstant coefficient η defined by (33). Different colors refer to different sparsity levels: $s = 2$ (red), $s = 4$ (green), $s = 8$ (blue), $s = 16$ (magenta), and $s = 32$ (orange). The markers diamond, triangle, right triangle, left triangle and down triangle correspond to the full approach with $s = 2, 4, 8, 16, 32$, respectively. The markers $+, \times, *, \circ$, and $\square$ refer to the compressive approach with $s = 2, 4, 8, 16, 32$, respectively

The computational cost is distinguished in assembly and recovery cost:

- For the full approach, the assembly cost is the time employed to build B and c as in (10) and the recovery cost is the time employed by the backslash MATLAB® command to solve the linear system (9).
- For the compressive approach, the assembly cost is the time employed to randomly generate the multi-indices $\tau_1, \ldots, \tau_m$ and to build A and b as in (16) and the recovery cost is the time needed to recover the solution to (15) via OMP, including the time to normalize the columns of A with respect to the ℓ^2 norm and the time to rescale the entries of the OMP solution accordingly, as prescribed by Algorithm 1.

Both approaches have a high level of accuracy, around 10^{-15} and 10^{-14}. The markers corresponding to the compressive approach are closer to the lower left corner of the plot. This shows that when dealing with exact sparsity, the compressive approach is more advantageous both in terms of accuracy and of computational cost.

Let us now assess cost and accuracy of both approaches in a more systematic way. Figure 2 shows the box plots generated after repeating the same random experiment as before 100 times and for $s = 2, 4, 8, 16, 32, 64$. Recalling (32), the corresponding numbers of collocation points are $m = 28, 56, 111, 222, 444, 888$. For the full approach we also compare backslash with OMP. In practice, the backslash approach simply computes a solution to (9) as $B\backslash c$, whereas the OMP-based approach computes an s-sparse approximate solution to (9) (up to normalization of the columns of B) via OMP.

The very good level of accuracy of both approaches is confirmed by this second experiment. It is remarkable that the recovery error of the compressive approach is slightly better than that of the full approach, especially for smaller sparsities. We observe that, in general, the compressive approach outperforms the full one both in terms of accuracy *and* computational cost. In general, the smaller the sparsity s, the higher the computational cost reduction gained by compressing the discretization. By looking at the second column, we can see that, in the full case, OMP is able to compute more accurate solutions with respect to the backslash. This is arguably due to the fact that the largest least-squares problem solved by OMP (during the s-th iteration) is associated with an $N \times s$ submatrix of the full $N \times N$ discretization matrix linear system. Therefore, the former matrix is, in general, better conditioned than the latter.[1]

[1]The substantial independence of the OMP recovery cost with respect to s for the full approach depends on two factors: the particular implementation of OMP in the package OMP-Box and the normalization step $\widetilde{A} = AM^{-1}$ in Algorithm 1. In fact, in order to speed up the OMP iteration, the function omp of OMP-Box used to produce these results takes $\widetilde{A}^T\widetilde{A}$ as input. When A is $N \times N$, the cost of computing the matrices $\widetilde{A}$ and $\widetilde{A}^T\widetilde{A}$ is independent of s and it turns out to be consistently larger than the cost of OMP itself. As a result, the effect of s on the overall computational cost is negligible. The same remark holds for Fig. 4.

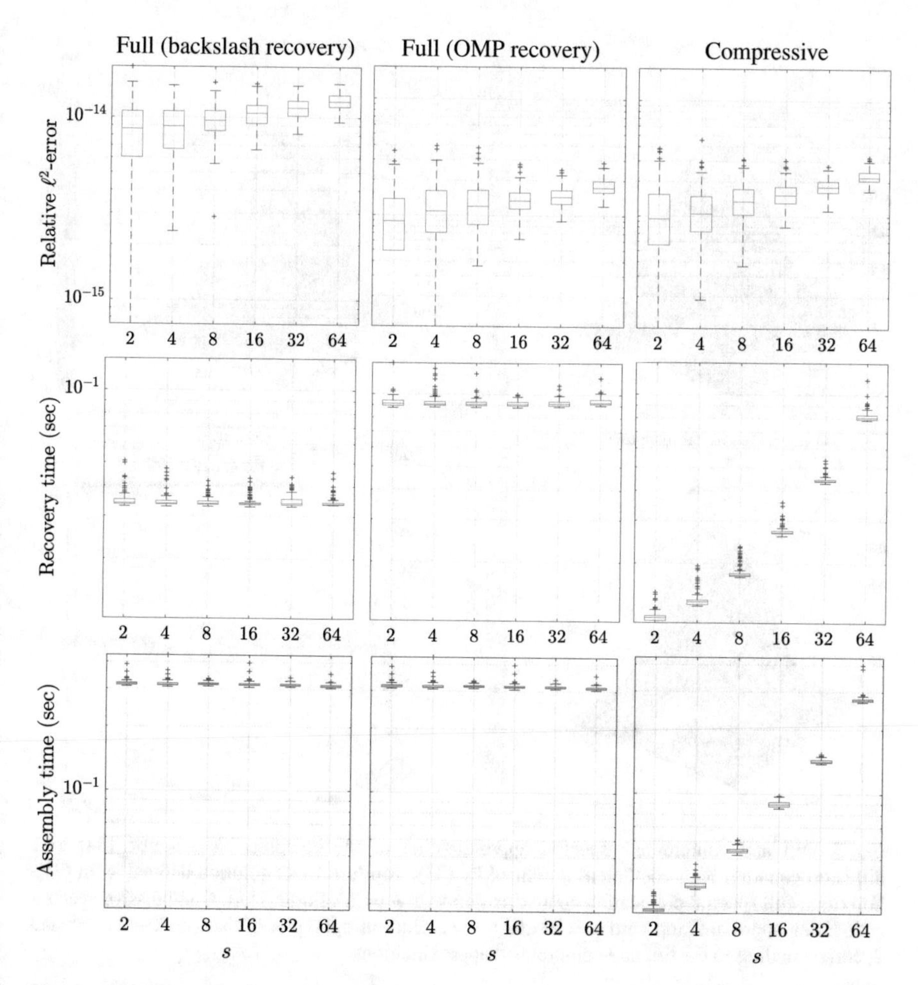

Fig. 2 Performance analysis of full and compressive spectral collocation from the accuracy and computational cost viewpoints for the recovery of randomly generated s-sparse solutions to the diffusion equation with nonconstant coefficient η defined by (33). The box plots are referred to 100 random runs

5.2 *Recovery of Compressible Solutions*

We compare the full and the compressive approaches for the recovery of compressible solutions. We will test the methods for the recovery of the exact solution

$$u(z) = (16\, z_1\, z_2\, (1 - z_1)(1 - z_2))^2, \quad \forall z \in \overline{\Omega}, \tag{34}$$

whose plot is shown in Fig. 3 (top left). The forcing term F in (6) is defined in order to have (34) as exact solution.

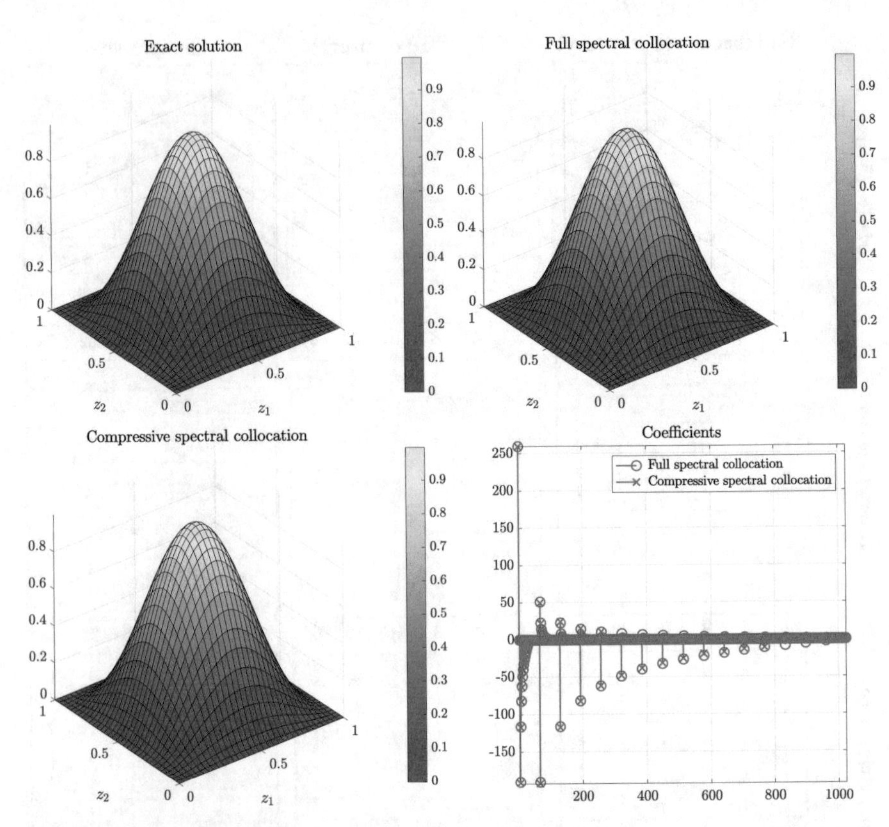

Fig. 3 Full and compressive spectral approximation of the compressible solution (34) to a diffusion equation with coefficient η defined by (33). Top left: exact solution defined as in (34). Top right: full spectral collocation approximation with $n = 32$. Bottom left: Compressive spectral collocation approximation with $n = 32$ and $s = 32$. Bottom right: plot of the coefficients x^{full} and $\hat{x}$, corresponding to the full and compressive approximations

Let us fix $n = 32$, corresponding to $N = 1024$, and $s = 32$. With this choice, and recalling (32), we have $m = 444$. Figure 3 shows the results of the full and the compressive spectral collocation approaches. Both methods produce a very good approximation to the exact solution. We can appreciate the ability of OMP to recover the largest absolute coefficients of the vector x^{full} in Fig. 3 (bottom right). Comparing Fig. 3 (top right) and Fig. 3 (bottom left), we see that computing a 32-sparse approximation to the 1024-dimensional vector x^{full} is sufficient to recover a compressive approximation that is visually indistinguishable from the full approximation, thanks to the compressibility of the solution.

In the same setting as before, we consider sparsity levels $s = 2, 4, 8, 16, 32, 64$ and carry out a more extensive numerical assessment, in the same spirit as Fig. 2. We repeat the previous experiment 100 times and show the corresponding box plots in Fig. 4. The recovery and assembly times are analogous to those of Fig. 2. In terms of accuracy, we are of course not able to obtain exact recovery, as in the sparse

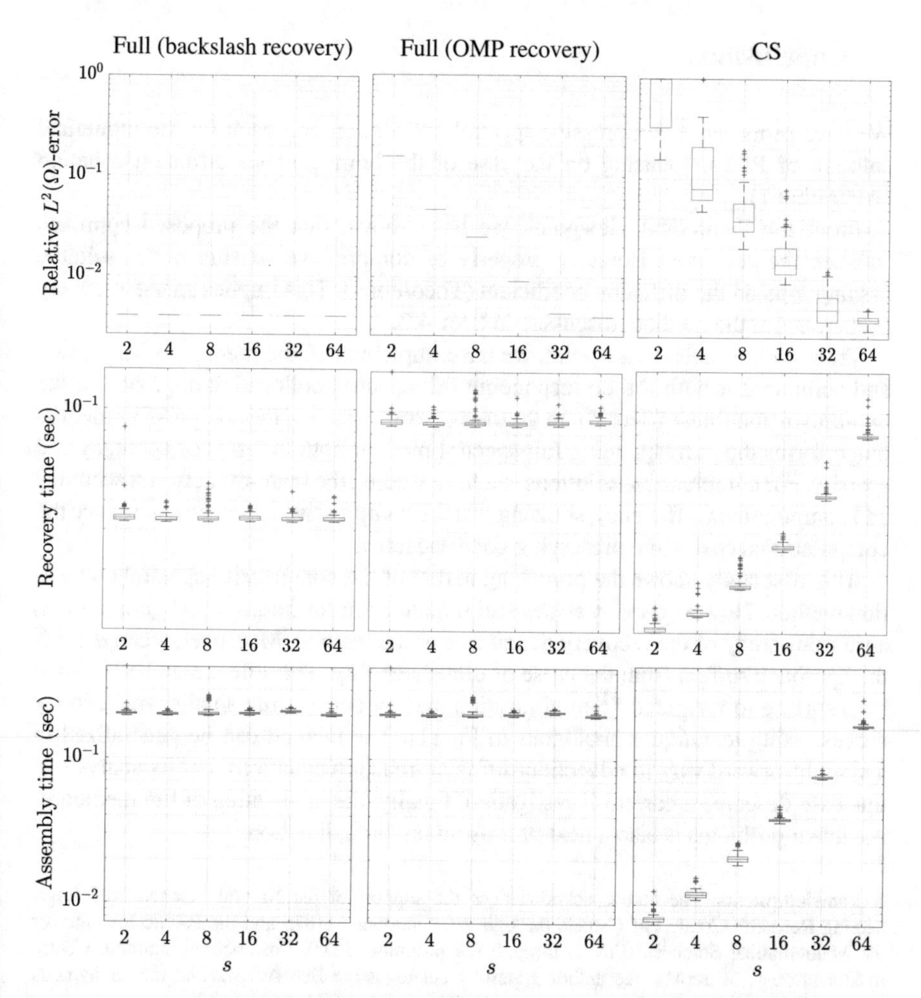

Fig. 4 Performance analysis of full and compressive spectral collocation from the accuracy and computational cost viewpoints for the recovery of the compressible solution (34) to the diffusion equation with nonconstant coefficient η defined by (33). The box plots are referred to 100 random runs

case. The relative $L^2(\Omega)$-error associated with the full spectral approximation is $4.0 \cdot 10^{-3}$. When performing s iterations of OMP on the full system (Fig. 4 top center), the error decays up to $s = 32$, when the accuracy saturates to the level of the full approximation. The situation is analogous for the compressive approach, and the decay of the recovery error shares the same trend as the full approach with OMP recovery, up to a distortion due to randomization and to subsampling. Of course, the assembly cost is always lower for the compressive approach. The recovery cost is lower for $s \leq 16$. The values $s = 8, 16, 32$ seem to be realize a good trade-off between accuracy and computational efficiency.

6 Conclusions

We have proposed a compressive spectral collocation approach for the numerical solution of PDEs, focusing on the case of the homogeneous diffusion equation (Algorithm 1).

From the theoretical viewpoint, we have shown that the proposed approach satisfies the restricted isometry property of compressive sensing under suitable assumptions on the diffusion coefficient (Theorem 4). This implies sparse recovery properties for the method, discussed in Sect. 4.2.

From the numerical viewpoint, we have implemented the method in MATLAB® and compared it with the corresponding full spectral collocation approach in the two-dimensional case (Sect. 5). In the case of exact sparsity, the compressive method outperforms the corresponding full spectral method both in terms of accuracy and sparsity. For compressible solutions, we have studied the trade-off between accuracy and computational efficiency, showing that the compressive approach can reduce the computational cost while preserving good accuracy.

This first study shows the promising nature of the compressive spectral collocation method. However, many issues still remain open for future investigation. First, a rigorous study of the recovery guarantees of the method. Moreover, when $d \gg 1$, the approach suffers from the curse of dimensionality. This effect may be lessened by resorting to weighted ℓ^1-minimization and by considering smaller multi-index spaces, using techniques analogous to [3, 11]. The method can be generalized in a straightforward way to advection-diffusion-reaction equations, but its analysis in this case deserves a careful investigation. Finally, the application of the method to nonlinear problems is also a next promising research direction.

Acknowledgments The author acknowledges the support of the Natural Sciences and Engineering Research Council of Canada through grant number 611675 and the Pacific Institute for the Mathematical Sciences (PIMS) through the program "PIMS Postdoctoral Training Centre in Stochastics". Moreover, the author gratefully acknowledge Ben Adcock and the anonymous reviewer for providing helpful comments on the first version of this manuscript.

References

1. Adcock, B.: Univariate modified fourier methods for second order boundary value problems. BIT Numer. Math. **49**(2), 49–280 (2009)
2. Adcock, B.: Multivariate modified Fourier series and application to boundary value problems. Numer. Math. **115**(4), 511–552 (2010)
3. Adcock, B., Brugiapaglia, S., Webster, C.G.: Compressed sensing approaches for polynomial approximation of high-dimensional functions. In: Compressed Sensing and its Applications, pp. 93–124. Springer, Cham (2017)
4. Alberti, G.S., Santacesaria, M.: Infinite dimensional compressed sensing from anisotropic measurements (2017). Preprint. arXiv:1710.11093
5. Bouchot, J.-L., Rauhut, H., Schwab, C.: Multi-level compressed sensing Petrov-Galerkin discretization of high-dimensional parametric PDEs (2017). Preprint. arXiv:1701.01671

6. Brugiapaglia, S.: COmpRessed SolvING: sparse approximation of PDEs based on compressed sensing. PhD thesis, Politecnico di Milano, Italy (2016)
7. Brugiapaglia, S., Micheletti, S., Perotto, S.: Compressed solving: a numerical approximation technique for elliptic PDEs based on compressed sensing. Comput. Math. Appl. **70**(6), 1306–1335 (2015)
8. Brugiapaglia, S., Nobile, F., Micheletti, S., Perotto, S.: A theoretical study of COmpRessed SolvING for advection-diffusion-reaction problems. Math. Comput. **87**(309), 1–38 (2018)
9. Candès, E.J., Romberg, J., Tao, T.: Robust uncertainty principles: exact signal reconstruction from highly incomplete frequency information. IEEE Trans. Inf. Theory **52**(2), 489–509 (2006)
10. Canuto, C., Hussaini, M.Y., Quarteroni, A.M., Zhang, T.A.: Spectral Methods in Fluid Dynamics. Springer Science & Business Media, New York (2012)
11. Chkifa, A., Dexter, N., Tran, H., Webster, C.G.: Polynomial approximation via compressed sensing of high-dimensional functions on lower sets. Math. Comput. **87**(311), 1415 (2018)
12. Daubechies, I., Runborg, O., Zou, J.: A sparse spectral method for homogenization multiscale problems. Multiscale Model. Simul. **6**(3), 711–740 (2007)
13. Donoho, D.L.: Compressed sensing. IEEE Trans. Inf. Theory **52**(4), 1289–1306 (2006)
14. Doostan, A., Owhadi, H.: A non-adapted sparse approximation of PDEs with stochastic inputs. J. Comput. Phys. **230**(8), 3015–3034 (2011)
15. Foucart, S., Rauhut, H.: A Mathematical Introduction to Compressive Sensing. Birkhäuser, Basel (2013)
16. Gottlieb, D., Orszag, S.A.: Numerical Analysis of Spectral Methods: Theory and Applications, vol. 26. SIAM, Philadelphia (1977)
17. Guermond, J.-L.: A finite element technique for solving first-order PDEs in L^p. SIAM J. Numer. Anal. **42**(2), 714–737 (2004)
18. Guermond, J.-L., Popov, B.: An optimal L^1-minimization algorithm for stationary Hamilton-Jacobi equations. Commun. Math. Sci. **7**(1), 211–238 (2009)
19. Iserles, A., Nørsett, S.P.: From high oscillation to rapid approximation I: modified Fourier expansions. IMA J. Numer. Anal. **28**(4), 862–887 (2008)
20. Iserles, A., Nørsett, S.P.: From high oscillation to rapid approximation III: multivariate expansions. IMA J. Numer. Anal. **29**(4), 882–916 (2009)
21. Jokar, S., Mehrmann, V., Pfetsch, M.E., Yserentant, H.: Sparse approximate solution of partial differential equations. Appl. Numer. Math. **60**(4), 452–472 (2010). Special Issue: NUMAN 2008
22. Krahmer, F., Ward, R.: Stable and robust sampling strategies for compressive imaging. IEEE Trans. Image Process. **23**(2), 612–622 (2014)
23. Lavery, J.E.: Nonoscillatory solution of the steady-state inviscid Burgers' equation by mathematical programming. J. Comput. Phys. **79**(2), 436–448 (1988)
24. Lavery, J.E.: Solution of steady-state one-dimensional conservation laws by mathematical programming. SIAM J. Numer. Anal. **26**(5), 1081–1089 (1989)
25. Mackey, A., Schaeffer, H., Osher, S.: On the compressive spectral method. Multiscale Model. Simul. **12**(4), 1800–1827 (2014)
26. Mathelin, L., Gallivan, K.A.: A compressed sensing approach for partial differential equations with random input data. Commun. Comput. Phys. **12**(4), 919–954 (2012)
27. Peng, J., Hampton, J., Doostan, A.: A weighted ℓ^1-minimization approach for sparse polynomial chaos expansions. J. Comput. Phys. **267**, 92–111 (2014)
28. Rauhut, H., Schwab, C.: Compressive sensing Petrov-Galerkin approximation of high-dimensional parametric operator equations. Math. Comput. **86**(304), 661–700 (2017)
29. Rubinstein, R., Zibulevsky, M., Elad, M.: Efficient implementation of the K-SVD algorithm using batch orthogonal matching pursuit. Technical report, Computer Science Department, Technion (2008)
30. Schaeffer, H., Caflisch, R., Hauck, C.D., Osher, S.: Sparse dynamics for partial differential equations. Proc. Natl. Acad. Sci. **110**(17), 6634–6639 (2013)
31. Temlyakov, V.N.: Nonlinear methods of approximation. Found. Comput. Math. **3**(1), 33 (2003)

32. Tran, G., Ward, R.: Exact recovery of chaotic systems from highly corrupted data. Multiscale Model. Simul. **15**(3), 1108–1129 (2017)
33. Yang, X., Karniadakis, G.E.: Reweighted ℓ^1 minimization method for stochastic elliptic differential equations. J. Comput. Phys. **248**, 87–108 (2013)
34. Zhang, T.: Sparse recovery with orthogonal matching pursuit under RIP. IEEE Trans. Inf. Theory **57**(9), 6215–6221 (2011)

Surrogate-Based Ensemble Grouping Strategies for Embedded Sampling-Based Uncertainty Quantification

M. D'Elia, E. Phipps, A. Rushdi, and M. S. Ebeida

Abstract The embedded ensemble propagation approach introduced in Phipps et al. (SIAM J. Sci. Comput. 39(2):C162, 2017) has been demonstrated to be a powerful means of reducing the computational cost of sampling-based uncertainty quantification methods, particularly on emerging computational architectures. A substantial challenge with this method however is ensemble-divergence, whereby different samples within an ensemble choose different code paths. This can reduce the effectiveness of the method and increase computational cost. Therefore grouping samples together to minimize this divergence is paramount in making the method effective for challenging computational simulations. In this work, a new grouping approach based on a surrogate for computational cost built up during the uncertainty propagation is developed and applied to model advection-diffusion problems where computational cost is driven by the number of (preconditioned) linear solver iterations. The approach is developed within the context of locally adaptive stochastic collocation methods, where a surrogate for the number of linear solver iterations, generated from previous levels of the adaptive grid generation, is used to predict iterations for subsequent samples, and group them based on similar numbers of iterations. The effectiveness of the method is demonstrated by applying it to highly anisotropic advection-dominated diffusion problems with a wide variation in solver iterations from sample to sample. It extends the parameter-based grouping approach developed in D'Elia et al. (SIAM/ASA J. Uncertain. Quantif. 6:87, 2017) to

M. D'Elia (✉)
Computational Science and Analysis, Sandia National Laboratories, Livermore, CA, USA
e-mail: mdelia@sandia.gov

E. Phipps · M. S. Ebeida
Center for Computing Research, Sandia National Laboratories, Albuquerque, NM, USA
e-mail: etphipp@sandia.gov; msebeid@sandia.gov

A. Rushdi
Northrop Grumman Corporation, Albuquerque, NM, USA
e-mail: ahmad.rushdi@ngc.com

M. D'Elia et al. (eds.), *Quantification of Uncertainty: Improving Efficiency and Technology*, Lecture Notes in Computational Science and Engineering 137,
https://doi.org/10.1007/978-3-030-48721-8_3

more general problems without requiring detailed knowledge of how the uncertain parameters affect the simulation's cost, and is also less intrusive to the simulation code.

1 Introduction

During the last decade the quantification of the uncertainty in predictive simulations has acquired great importance and is a topic of very active research in large scale scientific computing; we mention, e.g., random sampling methods [23, 35, 43–45], stochastic collocation [2, 46, 47, 59] and stochastic Galerkin methods [28, 29, 60]. It is very often the case that the source of uncertainty resides in the parameters of the mathematical models that describe phenomena of interest; when these parameters belong to a high-dimensional space or when the solution exhibits a non-smooth or localized behavior with respect to those parameters, sampling methods for uncertainty quantification (such as those mentioned above) may require a huge number of samples making the problem computationally intractable. For this reason, many methods, with the goal of reducing the number of samples, have been developed; among the others, we have locally adaptive sampling methods [26, 32, 58], multilevel methods [7–9, 14, 30], compressed sensing [19, 42], and tensor methods [1–3, 15, 25, 27, 46, 47, 55, 59].

Nevertheless, the problem remains that for large scale scientific computing most of the computational cost is in the sample evaluation which, in most cases, corresponds to the numerical solution of a partial differential equation (PDE). Previous work [51] demonstrated that solving for groups (ensembles) of samples at the same time through forward simulations can dramatically reduce the cost of sampling-based uncertainty quantification (UQ) methods. However, fundamental to the success of this approach is the grouping of samples into ensembles to further reduce the computational work; in fact, the total number of iterations of the ensemble system is strongly affected by which samples are grouped together.

In a previous work [17] we investigated sample grouping strategies for local adaptive stochastic collocation methods applied to highly anisotropic diffusion problems where the uncertain diffusion coefficient is modeled by a truncated Karhunen-Loéve (KL) expansion. There, we investigated PDE-dependent and location-dependent grouping techniques and demonstrated that a measure of the total anisotropy of the diffusion coefficient provides an effective metric for grouping samples as it is a good proxy for the number of iterations associated with each sample. We referred to this approach as parameter-based as it depends on the parameters, or coefficients (the diffusion tensor in this case), of the PDE. However, accessing problem-related information can be non-trivial or time/memory consuming.

The main contribution of this follow-up work is the design of new grouping strategies that are cheaper and independent of the PDE. In the context of adaptive selection of the samples in the parameter space we propose a grouping strategy based on the construction of a (polynomial) surrogate for the number of solver iterations; the key idea of this approach is to group together samples with a similar predicted number of iterations. At each level of the adaptive grid generation algorithm we utilize data at previous levels to build surrogates of increasing accuracy so to maximize the computational saving. For the construction of the surrogate we consider standard polynomial sparse grid surrogates (SGS). This technique proves to be as successful as the parameter-based strategy for diffusion problems, whereas, as expected, it outperforms the parameter-based one in case of advection-dominated problems. It is also less intrusive and requires less computational work.

The use of surrogates is certainly not new in UQ methods for the solution of a variety of high-dimensional problems; we mention computational mechanics [12, 34, 54], computational fluid dynamics [24, 53], microwave circuit design [5, 6, 16], as well as system reliability and failure analysis [11, 20, 38, 39]. Surrogate models have received more research attention as cheap approaches to deal with model uncertainties in scientific computing applications. More specifically, surrogate models (also known as emulators, metamodels, or response surface models) would fit a set of expensive evaluations, introducing an appealing alternative to running costly/lengthy numerical simulations and they mimic the behavior of the high-fidelity simulation model as closely as possible while being computationally cheap to evaluate.

The paper is structured as follows. In Sect. 2 we introduce PDEs with random input parameters and stochastic collocation methods. We also describe SGSs and recall the main aspects of the numerical solution via embedded ensemble propagation. In Sect. 3 we describe the construction of the surrogates for the number of iterations using sparse grid approximations and show how to use them for sample grouping. We also report the results of analytic test cases that illustrate the proposed approach. In Sect. 4 we present the results of numerical tests for anisotropic advection-diffusion problems in three-dimensional spatial domains and multi-dimensional parameter spaces. Here we demonstrate the efficacy of the surrogate-based grouping and its overall better performance with respect to the parameter-based grouping approach. Finally, in Sect. 5, we draw conclusions and present future research plans.

2 Preliminaries

In this section we briefly introduce stochastic PDEs (SPDEs) and specify the mathematical model and its uncertain parameters. Following [33] we introduce stochastic collocation methods and sparse grid approximations. Also, based on [51], we recall the principal aspects of embedded ensemble propagation for the solution of groups of parameter dependent deterministic PDEs.

2.1 PDEs with Random Input Parameters

Let $D \subset \mathbb{R}^d$ ($d = 1, 2, 3$) be a bounded domain with boundary ∂D and let $(\Omega, \mathscr{F}, \mathbb{P})$ be a complete probability space.[1] We consider the following stochastic elliptic boundary value problem. Find $u : \overline{D} \times \Omega$ such that almost surely we have that

$$\begin{cases} \mathscr{L}(a, v)u = f & \mathbf{x} \in D \\ u = 0 & \mathbf{x} \in \partial D, \end{cases} \tag{1}$$

where $\mathscr{L}$ is an elliptic operator defined on D parametrized by the uncertain parameters $a(\mathbf{x}, \omega)$ and $v(\mathbf{x}, \omega)$, and $f(\mathbf{x})$ is a forcing term with $\mathbf{x} \in D$ and $\omega \in \Omega$. We make the following assumptions.[2]

1. $a(\mathbf{x}, \omega)$ and $v(\mathbf{x}, \omega)$ are bounded from above and below with probability 1.
2. $a(\mathbf{x}, \omega)$ can be written as $a(\mathbf{x}, \omega) = a(\mathbf{x}, \mathbf{y}(\omega))$ in $\overline{D} \times \Omega$, where $\mathbf{y}(\omega) = (y_1(\omega) \ldots y_N(\omega)) \in \mathbb{R}^N$ is a random vector with uncorrelated components. The same representation holds for $v(\mathbf{x}, \omega) = v(\mathbf{x}, \mathbf{y}(\omega))$.

A classical example of random parameter that satisfies (1) and (2) is given by a truncated KL expansion [40, 41], i.e.

$$a(\mathbf{x}, \omega) = \mathbb{E}_a + \sum_{n=1}^{N} \sqrt{\lambda_n}\, b_n(\mathbf{x}) y_n(\omega). \tag{2}$$

The latter corresponds to the approximation of a second order correlated random field with expected value $\mathbb{E}_a$ and covariance $cov(\mathbf{x}, \mathbf{x}')$ with eigenvalues (in decreasing order) λ_n and eigenfunctions $b_n(\mathbf{x})$. Note that the random variables $\{y_n(\omega)\}_{n=1}^N$ map the sample space Ω into $\mathbb{R}^N$; for $\Gamma_n = y_n(\Omega) \subset \mathbb{R}$, we define the parameter space as $\Gamma = \prod_{n=1}^N \Gamma_n$.

A Stochastic Linear Elliptic PDE In this work we consider the following SPDE in $(D \times \Gamma) \subset (\mathbb{R}^d \times \mathbb{R}^N)$

$$\begin{cases} \mathscr{L}(a(\mathbf{x}, \mathbf{y}))u = -\nabla \cdot (A(\mathbf{x}, \mathbf{y})\nabla u) + \mathbf{v}(\mathbf{x}, \mathbf{y}) \cdot \nabla u = f & \mathbf{x} \in D, \mathbf{y} \in \Gamma \\ u = 0 & \mathbf{x} \in \partial D \end{cases} \tag{3}$$

where $A(\cdot, \mathbf{y}) : \mathbb{R}^N \to \mathbb{R}^{d\times d}$ is a diffusivity tensor and $\mathbf{v}(\cdot, \mathbf{y}) : \mathbb{R}^N \to \mathbb{R}^d$ is an advection coefficient. In the first part of the paper we consider $\mathbf{v} = \mathbf{0}$; we introduce and test the parameterized advection case in Sect. 4. As an example, for $d = 3$, A

[1]Here, Ω is a set of realizations, $\mathscr{F}$ is a σ-algebra of events and $\mathbb{P} : \mathscr{F} \to [0, 1]$ is a probability measure.

[2]For details regarding the functional spaces and the well-posedness of problem (1) we refer to [33].

may be defined as $A(\mathbf{x}, \mathbf{y}) = diag(a(\mathbf{x}, \mathbf{y}), a_y, a_z)$ with $a_y, a_z \in \mathbb{R}_+$ and

$$a(\mathbf{x}, \mathbf{y}) = a_{\min} + \widehat{a} \exp\left\{ \sum_{n=1}^{N} \sqrt{\lambda_n} b_n(\mathbf{x}) y_n \right\}. \tag{4}$$

Here, instead of using the classical KL expansion, to preserve the positive-definiteness of the diffusion tensor required for the well-posedness of problem (3), we consider the expansion of the logarithm of the random field.

Quantity of Interest In the SPDE context the goal of uncertainty quantification is to determine statistical information about an output of interest that depends on the solution. In most of the cases the output of interest is not the solution itself but a functional $G_u(\mathbf{y})$, e.g., the spatial average of $u(\cdot, \mathbf{y})$: $G_u(\mathbf{y}) = \frac{1}{|D|} \int_D u(\mathbf{x}, \mathbf{y})\, d\mathbf{x}$. Then, the statistical information may come in the form of moments of $G_u(\mathbf{y})$; as an example, the quantity of interest (QoI) could be the expected value of $G_u(\mathbf{y})$ with respect to the probability density function of $\mathbf{y}$. Note that UQ methods aim to find accurate approximations of G_u, say $\widehat{G}_u$, that are used to cheaply evaluate the QoI.

2.2 *Numerical Solution via Stochastic Collocation Methods*

For the finite-dimensional approximation of problem (3) we focus on stochastic collocation (SC) methods; these are non-intrusive stochastic sampling methods based on decoupled deterministic solves. For a detailed description of stochastic collocation methods we refer to [33] and for a summary, with results relevant to this paper, to [17].

Given a Galerkin method for spatial discretizations of (3), we denote by $u_h(\cdot, \mathbf{y})$ the semi-discrete approximation of $u(\mathbf{x}, \mathbf{y})$ for all random vectors $\mathbf{y} \in \Gamma$. The main idea of stochastic collocation methods is to collocate $u_h(\cdot, \mathbf{y})$ on a suitable set of samples $\{\mathbf{y}_m\}_{m=1}^M \subset \Gamma$ to determine M semi-discrete solutions and then use the latter to construct a global or piecewise polynomial to represent the fully SC discrete approximation $u_{hM}(\mathbf{x}, \mathbf{y})$, i.e.

$$u_{hM}(\mathbf{x}, \mathbf{y}) = \sum_{m=1}^{M} c_m(\mathbf{x}) \psi_m(\mathbf{y}),$$

where $\{\psi_m\}_{m=1}^M$ are polynomial basis functions and $c_m(\mathbf{x})$ are coefficients that depend on the semi-discrete solutions. Because we are mainly interested in problems where the solution has an irregular dependence on the random parameters, we consider only local stochastic collocation methods, which use locally supported piecewise polynomials to approximate the dependence of the solution on the random parameters.

We consider a generalized version of sparse grids, introduced by Smolyak in [56], used in [3, 46, 47] that relies on tensor products of one-dimensional approximations. We choose the basis $\{\psi_m\}_{m=1}^M$ to be a piecewise hierarchical polynomial basis [13, 31].

For the generation of the grid and of the SGS we proceed level by level using the adaptive procedure described in [33] and summarized in [17]. We indicate the total number of collocation (or sparse grid) points at level l by M_l. The adaptivity is based on *surpluses*; these are error indicators of the accuracy of the SGS at level l at any point of the sparse grid at level $(l+1)$. A surplus-based adaptive algorithm adds points at level $(l+1)$ only if the surplus-based error indicator is higher than a given tolerance.

Remark 1 As already mentioned, the output of interest is usually a functional of the solution; in such cases at each step of the adaptive grid generation we construct an approximation, or surrogate, of $G_u(\mathbf{y})$ and base the stopping criterion on the accuracy of the surrogate itself.

Also, for the purpose of this paper, it must be noted that at each step we can compute multiple surrogates for functionals of interest, i.e., approximations of $I_u(\mathbf{y})$, to be used for different tasks; in our case, a functional representing the number of linear solver iterations is computed to perform the grouping.

2.3 Numerical Solution via Ensembles

Sampling-based uncertainty quantification methods such as the stochastic collocation methods described above are attractive since they can be applied to any scientific simulation code with little to no modification of the code. Furthermore, these methods are trivially parallelizable since each sample can be evaluated independently and therefore in parallel. However, in many cases of interest to large-scale scientific computing, each sample evaluation consumes a large fraction of the available computational resources due to the extremely high fidelity and complexity of the simulations. Therefore it is often possible to only parallelize a small fraction of the required sample evaluations, with the remaining fraction evaluated sequentially. Moreover, in many cases a large amount of data and computation is the same in each sample evaluation and in principle could be reused across samples that are being evaluated sequentially, potentially reducing aggregate computational cost.

In this context, an intrusive sample propagation scheme called *embedded ensemble propagation* was introduced in [51] where small groups of samples (called ensembles) are propagated together through the simulation. Given a user-chosen ensemble size S (typically in the range of 4–32), this approach requires modifying the simulation code to replace each sample-dependent scalar with a length-S array and mapping arithmetic operations on those scalars to the corresponding operation on each component of the array. In [51] it was demonstrated that this approach can substantially reduce the cost of evaluating S samples compared to evaluating them sequentially for several reasons:

- Sample independent quantities (for example spatial meshes and sparse matrix graphs are often sample independent) are automatically reused. This reduces computation by only computing these quantities once per ensemble, reduces memory usage by only storing them once per ensemble, and reduces memory traffic by only loading/storing them once per ensemble.
- Random memory accesses of sample-dependent quantities are replaced by contiguous accesses of ensemble arrays. This amortizes the latency costs associated with these accesses over the ensemble, since consecutive memory locations can usually be accessed with no additional latency cost. It was demonstrated in [51] that this effect, combined with reuse of the sparse matrix graph can result in 50% reduction in cost of matrix-vector products associated with sparse iterative linear system solvers on emerging computational architectures, when applied to scalar diffusion problems such as those considered here.
- Arithmetic on ensemble arrays can be naturally mapped to fine-grained vector parallelism present in most computer architectures today, and this vector parallelism can be more easily extracted by compilers than can typically be extracted from the simulation itself.
- The number of distributed memory communication steps of sample-dependent information (e.g., within sparse iterative linear system solvers) is reduced by a factor of S, with the size of each communication message increased by a factor of S. This both reduces the latency cost associated with these messages by S as well as improves the throughput of each message since larger messages can often be communicated with higher bandwidth. It was demonstrated in [51] that this can substantially improve scalability to large processor counts when the costs associated with distributed memory communication become significant.

Furthermore, it was also shown in [51] that the translation from scalar to ensemble propagation within C++ simulation codes can be facilitated through the use of a template-based generic programming approach [48, 49] whereby the traditional floating point scalar type is replaced by a template parameter. This template code can then be instantiated on the original floating point type to recover the original simulation, as well as a new C++ *ensemble scalar type* that internally stores the length-S ensemble array to implement the ensemble propagation. Such a scalar type is provided by the Stokhos [50] package within Trilinos [36, 37] and has been integrated with the Kokkos [21, 22] package for portable shared-memory parallel programming as well as the Tpetra package [4] for distributed linear algebra.

In [51] it was shown that the ensemble propagation method was equivalent to solving commuted Kronecker product systems. To be precise, consider a finite element discretization of (3). For every sample y_m, $m = 1, \ldots M$, we write the resulting algebraic system as follows

$$L_m \mathbf{U}_m = \mathbf{F}_m, \quad L_m \in \mathbb{R}^{J \times J}, \ \mathbf{U}_m \in \mathbb{R}^J, \ \mathbf{F}_m \in \mathbb{R}^J, \tag{5}$$

where J is the number of spatial degrees of freedom.[3] Consider solving (5) for S samples $\boldsymbol{y}_{m_1}, \ldots, \boldsymbol{y}_{m_S}$:

$$L_{m_1}\mathbf{U}_{m_1} = \mathbf{F}_{m_1}, \\ \vdots \\ L_{m_S}\mathbf{U}_{m_S} = \mathbf{F}_{m_S}, \tag{6}$$

which can be written more compactly through Kronecker product notation:

$$\left(\sum_{i=1}^{S} e_i e_i^T \otimes L_{m_i}\right)\left(\sum_{i=1}^{S} e_i \otimes \mathbf{U}_{m_i}\right) = \sum_{i=1}^{S} e_i \otimes \mathbf{F}_{m_i}. \tag{7}$$

Here e_i is the i-th column of the $S \times S$ identity matrix. Furthermore, a symmetric permutation may be applied to (7) which results in commuting the order of the terms in each Kronecker product:

$$\left(\sum_{i=1}^{S} L_{m_i} \otimes e_i e_i^T\right)\left(\sum_{i=1}^{S} \mathbf{U}_{m_i} \otimes e_i\right) = \sum_{i=1}^{S} \mathbf{F}_{m_i} \otimes e_i. \tag{8}$$

Systems (7) and (8) are mathematically equivalent, but have different orderings of degrees of freedom. In (7), all spatial degrees of freedom for a given sample $\boldsymbol{y}_{m_i}$ are ordered consecutively, whereas in (8) degrees of freedom for all samples are ordered consecutively for a given spatial degree of freedom.

The embedded ensemble propagation method described in [51] produces linear systems equivalent to the commuted Kronecker product system (8) by storing each nonzero entry (i, j) in the ensemble matrix as a length-S array $\{(L_{m_1})_{ij}, \ldots, (L_{m_S})_{ij}\}$. Furthermore, to maintain consistency with the Kronecker-product formulation, norms and inner products of ensemble vectors produce scalar results by summing the components for the norm/inner-product across the ensemble. In terms of sparse iterative solvers such as the conjugate gradient (CG), this has the effect of coupling the systems in (6) together, causing them all to converge at the same rate.

Note that this makes it impossible to determine when each system would have converged when solved independently, which is required for the surrogate-based grouping strategy described below. To remedy this, we changed the implementation of norms and inner products to not sum contributions across the ensemble, and instead compute an ensemble norm/inner-product. This breaks the equivalency to a Kronecker-product formulation, but is instead equivalent to solving the systems independently as in (5) and allows each component system to converge at its own

[3]Note that here we allow the forcing term f to be sample dependent.

rate. However since the component systems are stored through the ensemble arrays, the iterative solver must continue until all systems have converged. Through a custom implementation of the iterative solver convergence test, we are able to determine when each system would have converged when solved independently. How this is implemented in the software is described in the numerical tests section.

The amount of performance improvement enabled by the embedded ensemble propagation approach is highly problem, problem size, and computer architecture dependent. For an in-depth examination of performance, see [51]. However as motivation for the usefulness of the ensemble propagation approach, as well as to provide a quantitative means of evaluating the impact of the grouping approaches described below, Fig. 1 displays the speed-up observed when solving (5) using the ensemble technique for several choices of ensemble size S relative to solving S systems sequentially. In these calculations an isotropic diffusion parameter is modeled by the truncated KL expansion $a(\mathbf{x}, \mathbf{y}) = a_{\min} + \widehat{a} \sum_{n=1}^{N} \sqrt{\lambda_n} b_n(\mathbf{x}) y_n$, where λ_n and b_n are the eigenvalues and eigenfunctions of an exponential covariance, see Sect. 4, and $y_n \in [-1, 1]$. A spatial mesh of 32^3 mesh cells was used for the spatial discretization and the resulting linear equations are solved by CG preconditioned with algebraic multigrid (AMG). The calculations were implemented on a single node of the Titan CPU architecture (16 core AMD Opteron processors using 2 MPI ranks and 8 OpenMP threads per MPI rank). In this case, due to the isotropy and the fact that we are using a uniform grid, the number of CG iterations is independent of the sample

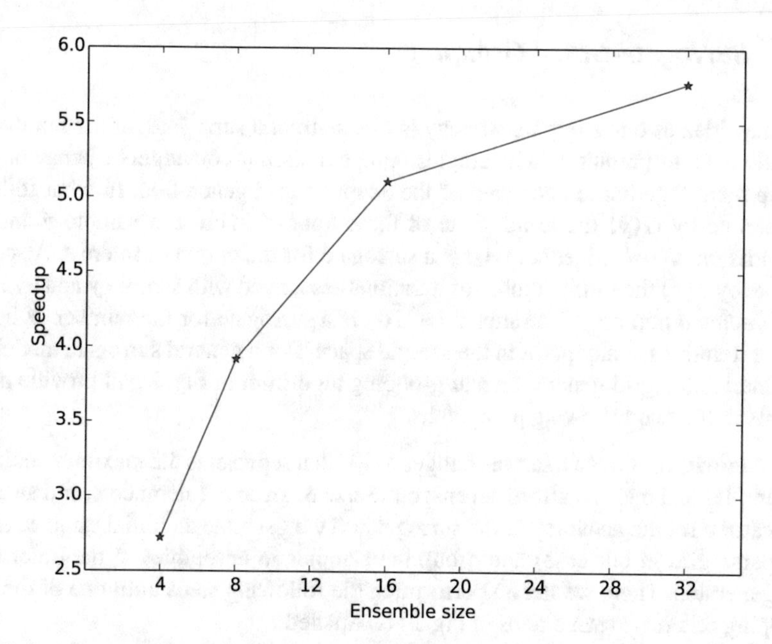

Fig. 1 Speed-up for the embedded ensemble propagation approach for various ensemble sizes S when applied to a simple isotropic diffusion problem where the number of solver iterations is sample independent, implemented on a single node of the Titan architecture

value and therefore the number of CG iterations for each ensemble is independent of the choice of which samples are grouped together in each ensemble. Essentially, this curve indicates the maximum speed-up possible for the ensemble propagation approach (for the given problem, problem size, and computer architecture) with perfect grouping, obtained when all samples within every ensemble require exactly the same number of preconditioned CG iterations. Variation in the number of iterations will reduce this speed-up due to increased computational work, which the grouping approaches discussed next attempt to mitigate.

3 Grouping Strategies

In this section we introduce an ensemble grouping approach with the goal of maximizing the performance of the embedded ensemble propagation algorithm described in the previous section when the number of linear solver iterations varies dramatically from sample to sample; we describe the grouping algorithm and apply it to analytic test cases for illustration. Also, for the sake of comparison, we recall the parameter-based grouping strategy, a successful technique introduced in [17] where the grouping depends on the diffusion parameter characterizing the PDE.

3.1 *Surrogate-Based Grouping*

The key idea of our grouping strategy is to construct a surrogate for the number of iterations so to predict which samples induce a similar convergence behavior and group them together at each step of the adaptive grid generation. In what follows we denote by $G(\widetilde{\mathbf{y}})$ the *exact* value of the output of interest at sample $\widetilde{\mathbf{y}}$ and by $\widehat{G}(\widetilde{\mathbf{y}})$ its *predicted* value, i.e. $\widehat{G}(\cdot)$ is a surrogate for the output of interest. Also, we denote by $I(\widetilde{\mathbf{y}})$ the *exact* number of iterations associated with sample $\widetilde{\mathbf{y}}$ and by $\widehat{I}(\widetilde{\mathbf{y}})$ the *predicted* number of iterations, i.e. $\widehat{I}(\cdot)$ is a surrogate for the number of linear solver iterations at any point in the sample space. For a general surrogate model we summarize the grid generation and grouping algorithm in Fig. 2 and provide more details in the two following paragraphs.

The Algorithm Given a sample budget $N_{\max}$, that represents the maximum number of samples that one can afford, an ensemble size S, an error tolerance τ, and an error indicator e for the accuracy of the surrogate $\widehat{G}$, we generate an initial sample set $\mathscr{Y}_0$ (a sparse grid in our case) and group the samples in ensembles in the order they are generated. Then, we iterate performing the following steps until one of the two stopping criteria (green circles in Fig. 2) is satisfied.

1. Solve the PDEs in ensembles and evaluate $G(\mathbf{y}_i)$ and $I(\mathbf{y}_i)$ for all $\mathbf{y}_i$ in $\mathscr{Y}_l$.
2. Build the surrogates $\widehat{G}$ and $\widehat{I}$ based on the values of G and I at $\mathbf{y}_i \in \cup_{l=0}^{L}\mathscr{Y}_l$.
   If the current surrogate does not satisfy the accuracy requirement, i.e. $e \geq \tau$

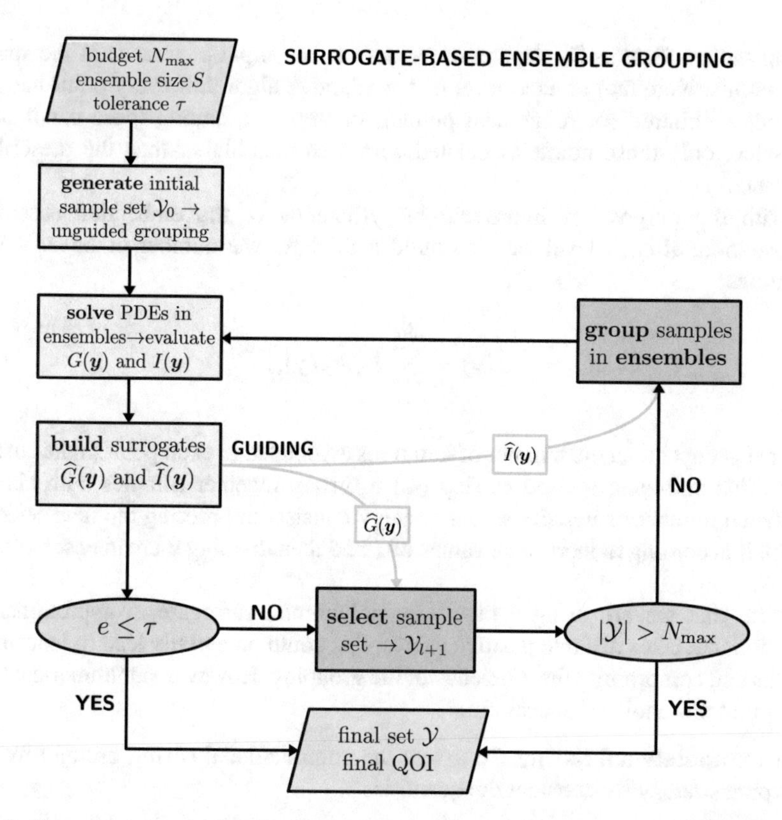

Fig. 2 Flow chart summarizing the general grid generation and grouping algorithm; here $\mathscr{Y} = \cup_{l=0}^{L} \mathscr{Y}_l$, $N_{\max}$ is the maximum number of samples that one can afford and τ is a user-defined error tolerance

3. Use $\widehat{G}$ to select the new sample set $\mathscr{Y}_{l+1}$.
   If the total number of samples is below the sample budget, i.e. $|\mathscr{Y}| < N_{\max}$
4. Update $\mathscr{Y}$, use $\widehat{I}$ to group the samples in ensembles and go back to 1.

Note that the number of ensembles to be solved in (1.) is not known a priori but it depends on the sparse grid generator. In fact, the number of samples at every level is determined by the adaptive grid generation algorithm and it depends on the accuracy of the surrogate $\widehat{G}$. In the next paragraph we describe how to perform (3.) and (4.) using SGS.

SGS-Based Grouping As described in Sect. 2.2, at every level of the adaptive grid generation we can construct a sparse grid approximation of the solution of (3) or we can compute an approximation (a surrogate) of an output of interest. We express such surrogate in terms of sparse grid basis functions as follows:

$$\widehat{G}(y) = \sum_{m=1}^{M_l} \widetilde{c}_m \psi_m(y), \tag{9}$$

where the coefficients $\widetilde{c}_m$ depend on values of the output of interest at the sparse grid points. Note that at each level of the adaptive algorithm every point has $2N$ candidate children (or refinement points); in step (3.), among those candidates, we select only those points associated with a surplus higher than the prescribed tolerance.

With the purpose of improving the efficiency of the embedded ensemble propagation, at each level we also build a SGS for the number of linear solver iterations:

$$\widehat{I}(\boldsymbol{y}) = \sum_{m=1}^{M_l} \overline{c}_m \psi_m(\boldsymbol{y}), \tag{10}$$

where $\overline{c}_m$ depend on the number of iterations associated to each point in the current grid. This surrogate is used in step (4.) to group together samples with similar predicted number of iterations. Our strategy consists in ordering the new selected samples according to increasing values of $\widehat{I}$ and then dividing them in ensembles of size S.

Note that we are using a piecewise polynomial surrogate to approximate a function that takes discrete positive values; this could potentially lead to inaccurate results and compromise the efficiency of the grouping. However, our numerical tests show that this choice is successful.

The Computational Saving To assess the computational saving brought by our grouping strategy we consider the quantities

$$R_l = \frac{S \sum_{k=1}^{K_l} \mathbf{I}_k}{\sum_{k=1}^{K_l} \sum_{i=1}^{S} I(\boldsymbol{y}_{k,i})}, \quad \text{or, equivalently} \quad R_l = \frac{S \sum_{k=1}^{K_l} \max_{i=1,\dots S} I(\boldsymbol{y}_{k,i})}{\sum_{k=1}^{K_l} \sum_{i=1}^{S} I(\boldsymbol{y}_{k,i})} \tag{11}$$

$$R = \frac{S \sum_{k=1}^{K} \mathbf{I}_k}{\sum_{k=1}^{K} \sum_{i=1}^{S} I(\boldsymbol{y}_{k,i})}, \quad \text{or, equivalently} \quad R = \frac{S \sum_{k=1}^{K} \max_{i=1,\dots S} I(\boldsymbol{y}_{k,i})}{\sum_{k=1}^{K} \sum_{i=1}^{S} I(\boldsymbol{y}_{k,i})} \tag{12}$$

where $\mathbf{I}_k$ is the number of iterations required by the kth ensemble, $I(\boldsymbol{y}_{k,i})$ is the number of iterations required by the ith sample in the kth ensemble, K_l is the number of ensembles at level l and K is the total number of ensembles. R_l represents the increase in computational work (as indicated by the number of solver iterations) induced by the ensemble propagation at level l, whereas R represents the same quantity over all levels. This increase in work is mitigated by the computational savings induced by the ensemble propagation technique, referred to as speed-up. The achieved speed-up in practice is then reduced by a factor of R.

Note that the equivalence in (11) and (12) follows from the implementation of the embedded ensemble propagation described in Sect. 2.3.

Illustrative Tests We perform some illustrative tests using quantities of interest represented by analytic functions. More specifically, using continuous and discontinuous functions we test the efficacy of the surrogate-based grouping. For $N = 2$, we consider the following quantities

$$G_1(\mathbf{y}) = -e^{-(y_1-1)^2} + e^{-0.8(y_1+1)^2} e^{-(y_2-1)^2} + e^{-0.8(y_2+1)}, \quad (y_1, y_2) \in [-2, 2]^2$$

$$G_2(\mathbf{y}) = \begin{cases} 1 & y_1^2 + y_2^2 < r_1 \\ 0 & r_1 \le y_1^2 + y_2^2 \le r_2 \\ 1 & y_1^2 + y_2^2 > r_2, \end{cases} \quad (y_1, y_2) \in [0, 1]^2,$$

$$I(\mathbf{y}) = e^{-a_1^2(y_1-u_1)^2 - a_2^2(y_2-u_2)^2} + 1. \tag{13}$$

The functions G_1 and G_2 (only used in this section for testing purposes) play the role of outputs of interest and are used to perform adaptivity, whereas I plays the role of the number of iterations and it is used to test the effectiveness of the surrogate-based grouping, i.e. we build a surrogate for I, we use it to order the samples for increasing predicted values of I, and we group the samples in ensembles of size S.

For the generation of the sparse grid and the adaptive refinement based on $\widehat{G}$ we use TASMANIAN [57] (toolkit for adaptive stochastic modeling and non-intrusive approximation), a set of libraries for high-dimensional integration and interpolation, and parameter calibration, sponsored by the Oak Ridge National Laboratory. TASMANIAN implements a wide class of one-dimensional rules (and extends them to the multi-dimensional case by tensor products) based on global and local basis functions. In this work the sparse grid is obtained using piecewise linear local basis functions and classic refinement.

It is common practice to apply the adaptive refinement to a grid of level $l > 1$; in these experiments we set the initial level to $l = 2$. Also, we set $\tau = 5 \cdot 10^{-4}$ and $N_{\max} = 1000$ or 2000 and we use the second definition of R in (12). Results are reported in Table 1, for G_1 on the left and G_2 on the right. Values of R are very close to 1, the optimal value that corresponds to perfect grouping; this is expected due to the ability of SGS to well approximate smooth functions. However, in our application I is not only discontinuous, but it takes values in $\mathbb{N}_+$; nonetheless, results in the next section show the performance improvement enabled by the SGS-based grouping for SPDEs.

Table 1 SGS: For G_1 (left) and G_2 (right), values of R for different ensemble sizes and maximum number of points

S	N_{max}	R	S	N_{max}	R
8	2000	1.428	8	2000	1.039
16	2000	1.501	16	2000	1.059
20	2000	1.470	20	2000	1.067
8	1000	1.064	8	1000	1.072
16	1000	1.362	16	1000	1.112
20	1000	1.075	20	1000	1.114

3.2 Parameter-Based Grouping

We recall that we are interested in the solution of anisotropic diffusion problems; common choices of solvers include PGC with AMG preconditioners. However, AMG methods exhibit poor performance when applied to diffusion problems featuring pronounced anisotropy; this suggests that a measure of the anisotropy is a good indicator of the solver convergence behavior. In [17] we proposed as an indicator of slow convergence (high number of iterations) for $\mathbf{y}_i$ the quantity

$$H(\widetilde{\mathbf{y}}) = \|r(\mathbf{x}, \widetilde{\mathbf{y}})\|_{L^\infty(D)} \quad \text{where} \quad r(\mathbf{x}, \widetilde{\mathbf{y}}) = \frac{\lambda_{\max}(A(\mathbf{x}, \widetilde{\mathbf{y}}))}{\lambda_{\min}(A(\mathbf{x}, \widetilde{\mathbf{y}}))}. \tag{14}$$

As in the surrogate-based grouping, this strategy consists in ordering the samples according to increasing values of H and then dividing them in ensembles of size S.

The idea of this approach is to identify the intensity of the anisotropy at each point in the spatial domain with the ratio between the maximum and minimum eigenvalues of the diffusion tensor; the maximum value of this quantity over D then provides a measure of the anisotropy associated with the sample $\widetilde{\mathbf{y}}$. Note that the computation of this indicator comes at a cost. In fact, prior to the assembling of the ensemble matrix we need to compute for each sample the diffusion tensor and its eigenvalues.

The Algorithm Given $N_{\max}$, S, τ, and e, we generate an initial sample set $\mathscr{Y}_0$. Then, we iterate performing the following steps until one of the two stopping criteria is satisfied.

1. Evaluate $H(\mathbf{y}_i)$ for all $\mathbf{y}_i$ in $\mathscr{Y}_l$ and group the samples.
2. Solve the PDEs in ensembles.
   If the current surrogate does not satisfy the accuracy requirement, i.e. $e \geq \tau$
3. Use $\widehat{G}$ to select the new sample set $\mathscr{Y}_{l+1}$.
   If the total number of samples is below the sample budget, i.e. $|\mathscr{Y}| < N_{\max}$
4. Update $\mathscr{Y}$ and go back to 1.

4 Numerical Tests

In this section we present the results of numerical tests performed on a spatial domain of dimension $d = 3$ and a sample space of dimension $N = 4$, first for the anisotropic diffusion problem (3) in Sect. 4.1, followed by an anisotropic advection-diffusion problem in Sect. 4.2.

4.1 *Anisotropic Diffusion*

For the solution of the anisotropic diffusion problem (3) we let $D = [0, 1]^3$ and $\Gamma = [-1, 1]^4$. We consider the following exponential covariance function

$$cov(\mathbf{x}, \mathbf{x}') = \sigma_0 \exp\left\{-\frac{\|\mathbf{x} - \mathbf{x}'\|_1}{\delta}\right\}, \tag{15}$$

where δ is the characteristic distance of the spatial domain, i.e. the distance for which points in the spatial domain are significantly correlated. In all our simulations we set $\delta = 1/4$ and $\sigma_0 = \sqrt{300}$. We discretize (3) using trilinear finite elements and 32^3 mesh cells. The Kokkos [21, 22] and Tpetra [4] packages within Trilinos [36, 37] are used to assemble and solve the linear systems for each sample value using hybrid shared-distributed memory parallelism via OpenMP and MPI. The equations are solved via CG implemented by the Belos package [10] with a linear solver tolerance of 10^{-7}. CG is preconditioned via smoothed-aggregation AMG as provided by the MueLu package [52]. A second-order Chebyshev smoother is used at each level of the AMG hierarchy and a sparse-direct solve for the coarsest grid. The linear system assembly, CG solve, and AMG preconditioner are templated on the scalar type for the template-based generic programming approach to implement the embedded ensemble propagation as described in Sect. 2.3, allowing the code to be instantiated on `double` for single sample evaluation and the ensemble scalar type provided by Stokhos [50] for ensembles. As before, the calculations were implemented on a single node of the Titan CPU architecture (16 core AMD Opteron processors using 2 MPI ranks and 8 OpenMP threads per MPI rank).

For the adaptive grid generation we use TASMANIAN and we consider a classic refinement; we set the initial sparse grid level to $l = 1$, $\tau = 10^{-3}$ and $N_{\max} = 2000$. We choose the ℓ^2-norm of the vector of the values of the discrete solution at the degrees of freedom as the output of interest, i.e. $G(\mathbf{y}) = \|\mathbf{u}(\mathbf{y})\|_{\ell^2}^2$, where $\mathbf{u}(\mathbf{y})$ is the discrete solution in correspondence of the sample $\mathbf{y}$.

The surrogate-based grouping strategy described above requires access to the number of CG iterations for each sample within an ensemble in order to build up the iterations surrogate $\widehat{I}$. As described in Sect. 2.3 we modified the ensemble propagation implementation to compute inner products and norms of ensemble vectors as ensembles (instead of scalars). Thus the CG residual norm computed during the CG iteration becomes an ensemble value. The CG iteration must continue until each ensemble residual norm satisfies the supplied tolerance, and thus all samples within the ensemble must take the same number of iterations. However we modified the convergence decision implementation in Belos (through partial specialization of the Belos convergence test abstraction on the ensemble scalar type) to keep track of when each sample within the ensemble would have converged, based on its component of the residual norm. These values are then reported to TASMANIAN to build the iterations surrogate.

Note that because of the high variation in CG iterations from sample to sample present in the following tests, and the requirement that the CG iteration continue until each component of the ensemble satisfies the convergence criteria, numerical underflow in the A-conjugate norm calculation may occur for some samples within an ensemble. When this occurs, the corresponding components of the norm appear numerically as zero, resulting in invalid floating point values (i.e., `NaN`) in the next search direction. While the grouping strategies generally mitigate this, it may occur with a poor grouping of samples. To alleviate this, we also modified the CG iteration logic (again via partial specialization of the Belos iteration abstraction) to replace the update with zero for each ensemble component when the A-conjugate norm is zero. For these samples, the CG algorithm continues until the remaining samples have converged, but the approximate solution no longer changes.

Test 1 For $S = 4, 8, 16, 32$ we report the results of our tests in Table 2. The adaptive algorithm generates a sparse grid of size $|\mathscr{Y}| = 1372$ after achieving the prescribed error tolerance τ with eight levels of refinement. The strategies "sur", "par" and "nat" correspond to the SGS-based grouping, the parameter-based grouping and the one based on the order in which the samples are generated by TASMANIAN. We also include the best hypothetical grouping based on the actual iterations for each sample in the rows labeled "its". For each method and ensemble size, Table 2 displays the calculated R_l for each level l of the adaptive grid generation (see (11)), the final R for the entire sample propagation (see (12)), and the total measured speedup for the ensemble linear system solves defined to be the time for all linear solves computed sequentially divided by the time for all ensemble solves (for the "its" method, a speedup is not computed since it is a hypothetical grouping constructed after all solves have been completed). To validate the measured speed-ups, we also computed the predicted speed-up determined by the iteration independent speed-up given by Fig. 1 divided by R.

Because of the large variation in number of CG iterations from sample to sample, we observe large values of R for the natural ordering based on the order in which samples are generated, particularly for larger ensemble sizes; this reduces the performance of the ensemble propagation method.[4] However the

[4]Note that at each level, the number of samples is not usually evenly divisible by the ensemble size. To use a uniform ensemble size for all ensembles, samples are added by replicating the last sample in the last ensemble. This results in larger R values when the number of samples is small and the ensemble size is large, as can be seen in the results for R_1.

Table 2 Computational results for Test 1, displaying R_l for each level l, the final R, measured ensemble linear solver speed-up and predicted speed-up based on R and Fig. 1, for the iterations-based ("its"), surrogate-based ("sur"), parameter-based ("par"), and natural grouping ("nat") methods

I	S	R_1	R_2	R_3	R_4	R_5	R_6	R_7	R_8	R	Speed-up	Pred. speed-up
its	4	1.68	1.43	1.23	1.05	1.01	1.04	1.06	1.10	1.06	–	2.56
sur	4	2.04	1.44	1.27	1.32	1.13	1.09	1.10	1.14	1.15	2.35	2.37
par	4	2.04	1.71	1.52	1.30	1.34	1.20	1.12	1.12	1.24	1.81	2.20
nat	4	2.04	1.66	1.44	1.23	1.34	1.28	1.25	1.24	1.29	2.15	2.11
its	8	2.83	1.60	1.27	1.08	1.10	1.08	1.10	1.44	1.14	–	3.44
sur	8	2.83	1.67	1.33	1.14	1.13	1.11	1.12	1.44	1.17	3.35	3.35
par	8	2.83	2.15	1.71	1.39	1.49	1.27	1.18	1.47	1.35	2.84	2.89
nat	8	2.83	2.29	1.90	1.48	1.49	1.51	1.41	1.64	1.52	2.61	2.58
its	16	3.11	1.94	1.60	1.12	1.28	1.25	1.25	1.56	1.30	–	3.94
sur	16	3.11	1.94	1.69	1.19	1.30	1.29	1.28	1.56	1.33	3.70	3.84
par	16	3.11	2.59	1.83	1.46	1.65	1.41	1.30	1.57	1.49	3.33	3.43
nat	16	3.11	2.59	2.12	1.87	1.80	1.83	1.67	2.16	1.84	2.73	2.78
its	32	6.22	3.07	2.62	1.25	1.67	1.32	1.61	2.63	1.66	–	3.46
sur	32	6.22	3.07	2.75	1.30	1.70	1.36	1.64	2.64	1.70	3.47	3.39
par	32	6.22	3.07	2.76	1.59	2.11	1.57	1.65	2.64	1.87	3.05	3.08
nat	32	6.22	3.07	2.99	2.37	2.24	2.16	2.07	2.74	2.28	2.06	2.53

parameter and surrogate-based orderings reduce R and therefore lead to larger speed-ups. The surrogate and iterations-based orderings lead to similar R values, demonstrating that the surrogate approach is predicting the number of solver iterations well. This is particularly evident for higher refinement levels where a more accurate surrogate for the number of iterations has been constructed. In Fig. 3 the

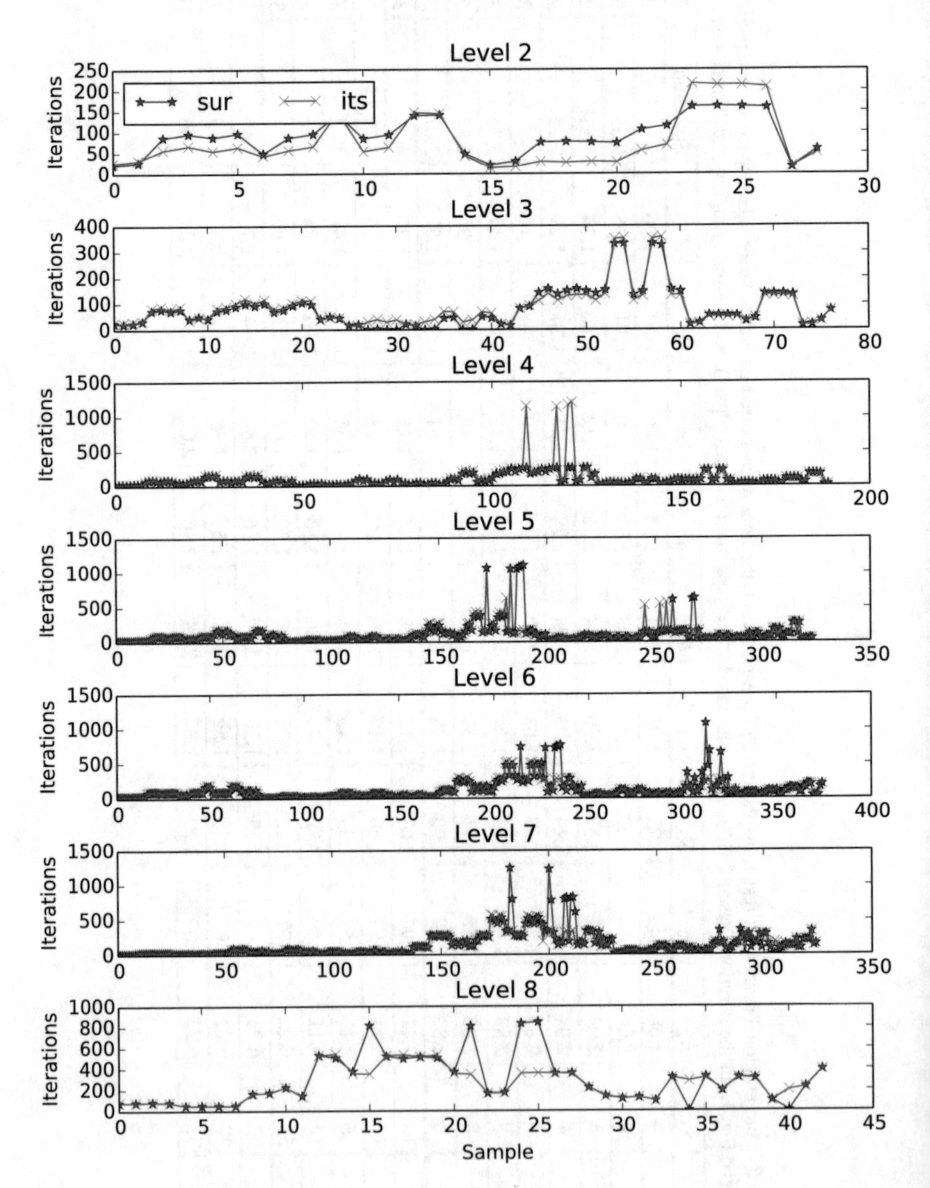

Fig. 3 Actual ("its") and surrogate-predicted ("sur") linear solver iterations for each sample at each adaptive level for Test 1

number of solver iterations for each sample at each level as well as the iterations predicted by the surrogate are displayed, demonstrating that the surrogate generally predicts the number of iterations well, and its accuracy generally improves as the stochastic grid is refined. Spikes in the graph correspond to samples generating parameter combinations that result in highly anisotropic random fields, where the preconditioner performs very poorly.

Test 2 Next we consider a diffusion parameter with a discontinuous behavior with respect to the uncertain variable $\mathbf{y}$. Specifically, we define $\widehat{a}$ as follows

$$\widehat{a}(\mathbf{y}) = \begin{cases} 1 & r(\mathbf{y}) < \frac{d}{4} \\ 100 & \frac{d}{4} \leq r(\mathbf{y}) < \frac{d}{2} \\ 10 & r(\mathbf{y}) \geq \frac{d}{2}, \end{cases} \tag{16}$$

with $d = \sqrt{3}$ and $r(\mathbf{y}) = \sqrt{\sum_i y_i^2}$. Due to the discontinuous nature of the problem, the adaptive algorithm is unable to achieve the error tolerance τ within the maximum number of points, and stops after reaching a size of $|\mathscr{Y}| = 1009$ and five levels of refinement. We report the results in Table 3; generally, results similar to the continuous case above are observed. However we do see larger differences

Table 3 Computational results for Test 2, displaying R_l for each level l, the final R, measured ensemble linear solver speed-up and predicted speed-up based on R and Fig. 1, for the iterations-based ("its"), surrogate-based ("sur"), parameter-based ("par"), and natural grouping ("nat") methods

I	S	R_1	R_2	R_3	R_4	R_5	R	Speed-up	Pred. speed-up
its	4	1.77	1.06	1.20	1.06	1.06	1.08	–	2.51
sur	4	2.08	1.13	1.24	1.14	1.25	1.22	2.09	2.23
par	4	2.08	1.44	1.62	1.36	1.37	1.41	1.93	1.94
nat	4	2.08	1.51	1.32	1.34	1.35	1.36	2.00	2.01
its	8	2.91	1.23	1.56	1.16	1.14	1.22	–	3.22
sur	8	2.91	1.29	1.64	1.27	1.21	1.30	2.80	3.02
par	8	2.91	1.74	2.01	1.49	1.79	1.74	2.29	2.25
nat	8	2.91	1.56	1.80	1.55	1.55	1.59	2.41	2.46
its	16	3.33	1.79	1.64	1.22	1.17	1.29	–	3.97
sur	16	3.33	1.79	1.69	1.33	1.24	1.36	3.22	3.74
par	16	3.33	2.38	2.37	1.60	1.66	1.77	2.87	2.88
nat	16	3.33	2.38	2.10	1.99	1.81	1.93	2.60	2.65
its	32	6.65	2.88	2.28	1.38	1.28	1.54	–	3.74
sur	32	6.65	2.88	2.34	1.46	1.37	1.62	3.04	3.55
par	32	6.65	2.88	2.53	1.75	1.77	1.94	2.87	2.96
nat	32	6.65	2.88	2.87	2.56	2.16	2.43	2.39	2.38

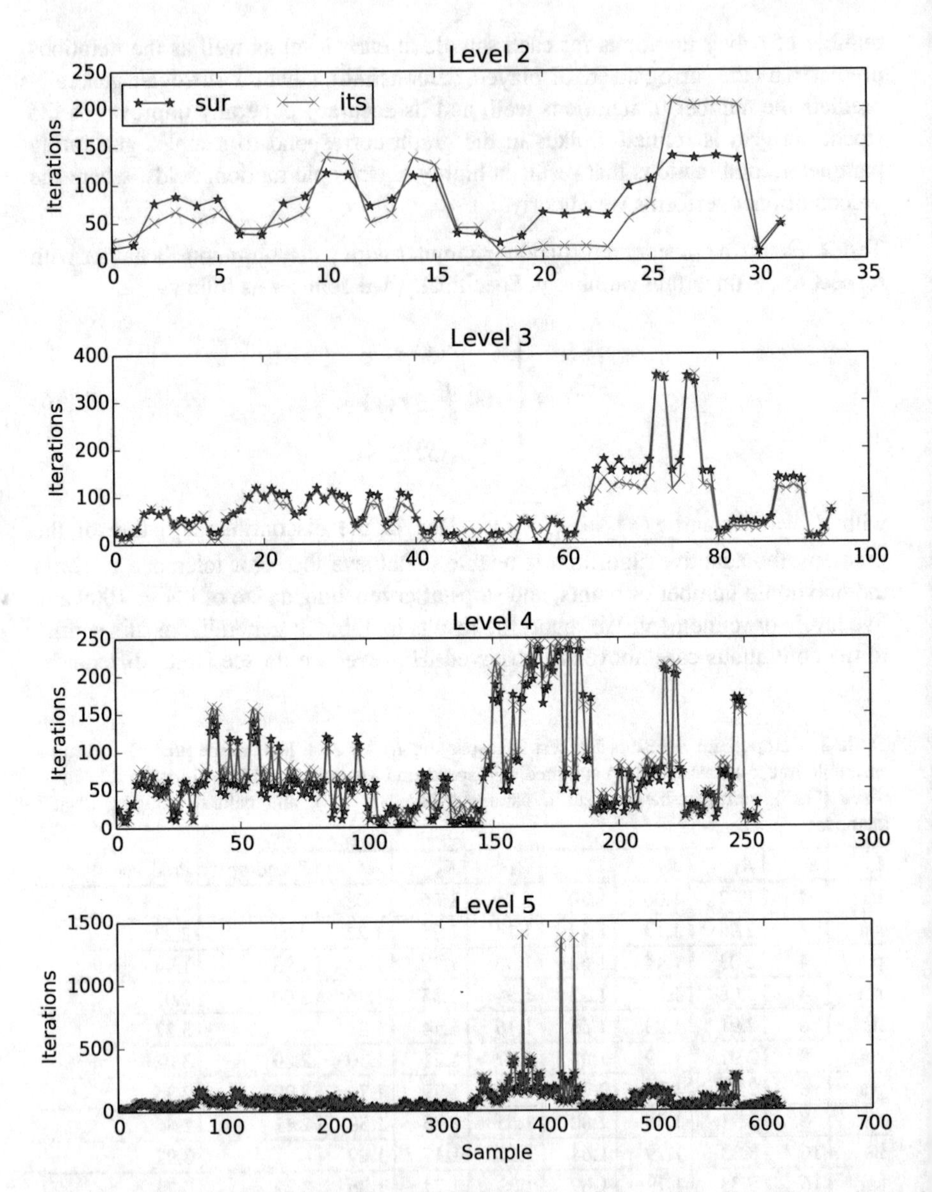

Fig. 4 Actual ("its") and surrogate-predicted ("sur") linear solver iterations for each sample at each adaptive level for Test 2

between the R values at higher levels between the iteration and surrogate-based groupings. As before, the solver iterations at each level as well as the iterations predicted by the surrogate are displayed in Fig. 4. Again, the surrogate predicts the number of iterations for most samples reasonably well, even for this more difficult discontinuous case.

4.2 Anisotropic Advection-Diffusion

As a final numerical test we consider applying the above grouping strategies to the advection-diffusion problem (3), where $A(\mathbf{x}, \mathbf{y})$ is represented by the same anisotropic diffusivity tensor (4):

$$A(\mathbf{x}, \mathbf{y}) = diag(a(\mathbf{x}, \mathbf{y}), a_y, a_z), \quad a(\mathbf{x}, \mathbf{y}) = a_{\min} + \widehat{a} \exp\left\{ \sum_{n=1}^{N_a} \sqrt{\lambda_n} b_n(\mathbf{x}) y_n \right\},$$

with the same exponential covariance (15) as above. The advection coefficient $\mathbf{v}(\mathbf{x}, \mathbf{y}) = (v(\mathbf{x}, \mathbf{y}), v(\mathbf{x}, \mathbf{y}), v(\mathbf{x}, \mathbf{y}))$ has a fixed direction $(1, 1, 1)$ but a varying magnitude also modeled by a truncated KL expansion

$$v(\mathbf{x}, \mathbf{y}) = v_{\min} + \sum_{n=N_a+1}^{N_a+N_v} \sigma_0 \sqrt{\lambda_n} b_n(\mathbf{x}) y_n \tag{17}$$

derived from the exponential covariance (15) with $\delta = 1/4$, $\sigma_0 = 10^3$, and $v_{\min} = 0$. For this test, we set $N_a = N_v = 2$ so that $N = N_a + N_v = 4$. As above, we discretize (4) using trilinear finite elements and 32^3 mesh cells, however now including SUPG stabilization [18] because of the large advection coefficient. The solver and sparse grid parameters are the same as in Test 1 above, however this time we use GMRES as the linear solver since the problem is no longer symmetric.

In Table 4 we display R_l for each level $l = 1, \ldots, 5$ and the final R for each of the four grouping methods and $S = 4, 8, 16$, and 32.[5] The convergence behavior of the preconditioned linear solver is now sensitive to both the level of anisotropy and the magnitude of the advection, which vary independently according to their respective random field representations. Thus we see the surrogate approach significantly outperforms the parameter-based grouping which is based on the level of anisotropy of the diffusion tensor alone. And because the diffusion and advection coefficients vary independently, deriving an effective parameter-based grouping for this problem would be extremely difficult. This example shows that using a surrogate-based technique is more efficient and more accurate at the same time.

[5]Note that we do not include timing results in Table 4 since the ensemble implementation is currently not optimized for GMRES. A significant cost within GMRES is the orthogonalization of each new Krylov vector against the previous set of vectors, which in Trilinos is implemented through GEMV dense matrix-vector product BLAS routine. An optimized implementation of this routine for ensembles is currently being developed.

Table 4 Computational results for the advection-diffusion problem, displaying R_l for each level l and the final R for the iterations-based ("its"), surrogate-based ("sur"), parameter-based ("par"), and natural grouping ("nat") methods

I	S	R_1	R_2	R_3	R_4	R_5	R
its	4	1.94	1.26	1.05	1.13	1.01	1.06
sur	4	1.97	1.36	1.27	1.33	1.14	1.21
par	4	1.97	1.47	1.54	1.49	1.42	1.45
nat	4	1.97	1.62	1.50	1.55	1.40	1.46
its	8	3.16	1.39	1.34	1.31	1.06	1.17
sur	8	3.16	1.66	1.54	1.58	1.24	1.37
par	8	3.16	1.65	2.04	1.77	1.63	1.71
nat	8	3.16	2.08	2.06	1.91	1.78	1.85
its	16	3.41	1.75	1.47	1.68	1.09	1.30
sur	16	3.41	1.76	1.58	1.96	1.33	1.53
par	16	3.41	1.75	2.37	2.03	1.82	1.93
nat	16	3.41	2.93	2.45	2.51	2.24	2.35
its	32	6.81	3.12	1.91	1.84	1.17	1.49
sur	32	6.81	3.12	2.01	2.15	1.45	1.74
par	32	6.81	3.12	2.90	2.50	1.99	2.25
nat	32	6.81	3.12	4.02	3.24	2.70	2.99

5 Conclusion

The embedded ensemble propagation approach introduced in [51] has been demonstrated to be a powerful means of reducing the computational cost of sampling-based uncertainty quantification methods, particularly on emerging computational architectures. A substantial challenge with this method however is ensemble-divergence, whereby different samples within an ensemble choose different code paths. This can reduce the effectiveness of the method and increase computational cost. Therefore grouping samples together to minimize this divergence is paramount in making the method effective for challenging computational simulations.

In this work, a new grouping approach based on a surrogate for computational cost built up during the uncertainty propagation was developed and applied to model advection-diffusion problems where computational cost is driven by the number of (preconditioned) linear solver iterations. The approach was developed within the context of locally adaptive stochastic collocation methods, where an iterations surrogate generated from previous levels of the adaptive grid generation is used to predict iterations for subsequent samples, and group them based on similar numbers of iterations. While the approach was developed within the context of stochastic collocation methods, we believe the idea is general and could be easily applied to any adaptive uncertainty quantification algorithm. In principle it could even be applied to non-adaptive algorithms by pre-selecting a set of samples, evaluating those samples, and generating an appropriate iterations surrogate from those results. The method was applied to two highly anisotropic diffusion problems with a wide variation in solver iterations from sample to sample, one continuous with respect to the uncertain parameters, and one discontinuous, and the method was demonstrated to significantly improve grouping and increase performance of the ensemble propagation method. Numerical tests performed on advection-dominated problems show that this technique extends the (less effective) parameter-based grouping approach developed in [17] to more general problems without requiring detailed knowledge of how the uncertain parameters affect the simulation's cost, and is also less intrusive to the simulation code.

The idea developed here could be further improved by allowing for variation in the ensemble size within each ensemble step. Given a prediction of how each ensemble size affects performance (e.g., from Fig. 1) and a surrogate for computational cost as developed here, ensembles of varying sizes could be selected to maximize performance through a constrained combinatorial optimization. Furthermore, the adaptive uncertainty quantification method could be modified to select new points not only based on the PDE quantity-of-interest, but also choose points that minimize divergence of computational cost/iterations. These ideas will be pursued in future works.

Acknowledgments Sandia National Laboratories is a multimission laboratory managed and operated by National Technology & Engineering Solutions of Sandia, LLC, a wholly owned subsidiary of Honeywell International Inc., for the U.S. Department of Energy's National Nuclear Security Administration under contract DE-NA0003525. This paper describes objective technical results and analysis. Any subjective views or opinions that might be expressed in the paper do not necessarily represent the views of the U.S. Department of Energy or the United States Government.

This material is based upon work supported by the U.S. Department of Energy, Office of Science, and Office of Advanced Scientific Computing Research (ASCR), as well as the National Nuclear Security Administration, Advanced Technology Development and Mitigation program. This research used resources of the Oak Ridge Leadership Computing Facility, which is a DOE Office of Science User Facility.

The authors would like to thank Dr. Miro Stoyanov for useful conversations and for providing great support with TASMANIAN.

References

1. Babuška, I., Tempone, R., Zouraris, G.E.: Galerkin finite element approximations of stochastic elliptic partial differential equations. SIAM J. Numer. Anal. **42**(2), 800–825 (2004)
2. Babuška, I., Nobile, F., Tempone, R.: A stochastic collocation method for elliptic partial differential equations with random input data. SIAM J. Numer. Anal. **45**(3), 1005–1034 (2007)
3. Bäck, J., Nobile, F., Tamellini, L., Tempone, R.: Stochastic spectral Galerkin and collocation methods for PDEs with random coefficients: a numerical comparison. In: Spectral and High Order Methods for Partial Differential Equations, pp. 43–62. Springer, Berlin (2011)
4. Baker, C.G., Heroux, M.A.: Tpetra, and the use of generic programming in scientific computing. Sci. Program. **20**(2), 115–128 (2012)
5. Bakr, M.H., Bandler, J.W., Madsen, K., Rayas-Sanchez, J.E., Søndergaard, J.: Space-mapping optimization of microwave circuits exploiting surrogate models. IEEE Trans. Microwave Theory Tech. **48**(12), 2297–2306 (2000)
6. Bandler, J.W., Cheng, Q., Gebre-Mariam, D.H., Madsen, K., Pedersen, F., Sondergaard, J.: Em-based surrogate modeling and design exploiting implicit, frequency and output space mappings. In: Microwave Symposium Digest, 2003 IEEE MTT-S International, vol. 2, pp. 1003–1006. IEEE, Piscataway (2003)

7. Barth, A., Lang, A.: Multilevel Monte Carlo method with applications to stochastic partial differential equations. Int. J. Comput. Math. **89**(18), 2479–2498 (2012)
8. Barth, A., Schwab, C., Zollinger, N.: Multi-level Monte Carlo finite element method for elliptic PDEs with stochastic coefficients. Numer. Math. **119**(1), 123–161 (2011)
9. Barth, A., Lang, A., Schwab, C.: Multilevel Monte Carlo method for parabolic stochastic partial differential equations. BIT Numer. Math. **53**(1), 3–27 (2013)
10. Bavier, E., Hoemmen, M., Rajamanickam, S., Thornquist, H.: Amesos2 and Belos: direct and iterative solvers for large sparse linear systems. Sci. Program. **20**(3), 241–255 (2012)
11. Bichon, B.J., McFarland, J.M., Mahadevan, S.: Efficient surrogate models for reliability analysis of systems with multiple failure modes. Reliab. Eng. Syst. Saf. **96**(10), 1386–1395 (2011)
12. Breitkopf, P., Coelho, R.F.: Multidisciplinary Design Optimization in Computational Mechanics. Wiley, Hoboken (2013)
13. Bungartz, H.J., Griebel, M.: Sparse grids. Acta Numer. **13**(1), 147–269 (2004)
14. Cliffe, K.A., Giles, M.B., Scheichl, R., Teckentrup, A.L.: Multilevel Monte Carlo methods and applications to elliptic PDEs with random coefficients. Comput. Vis. Sci. **14**(1), 3–15 (2011)
15. Cohen, A., DeVore, R., Schwab, C.: Analytic regularity and polynomial approximation of parametric and stochastic elliptic PDE's. Anal. Appl. **9**(01), 11–47 (2011)
16. Couckuyt, I., Declercq, F., Dhaene, T., Rogier, H., Knockaert, L.: Surrogate-based infill optimization applied to electromagnetic problems. Int. J. RF Microwave Comput. Aided Eng. **20**(5), 492–501 (2010)
17. D'Elia, M., Edwards, H.C., Hu, J., Phipps, E., Rajamanickam, S.: Ensemble grouping strategies for embedded stochastic collocation methods applied to anisotropic diffusion problems. SIAM/ASA J. Uncertain. Quantif. **6**, 87 (2017)
18. Donea, J., Huera, A.: Finite Element Methods for Flow Problems. Wiley, New York (2003)
19. Doostan, A., Owhadi. H.: A non-adapted sparse approximation of PDEs with stochastic inputs. J. Comput. Phys. **230**(8), 3015–3034 (2011)
20. Ebeida, M.S., Mitchell, S.A., Swiler, L.P., Romero, V.J., Rushdi, A.A.: POF-darts: geometric adaptive sampling for probability of failure. Reliab. Eng. Syst. Saf. **155**, 64–77 (2016)
21. Edwards, H.C., Sunderland, D., Porter, V., Amsler, C., Mish, S.: Manycore performance-portability: Kokkos multidimensional array library. Sci. Program. **20**(2), 89–114 (2012)
22. Edwards, H.C., Trott, C.R., Sunderland, D.: Kokkos: enabling manycore performance portability through polymorphic memory access patterns. J. Parallel Distrib. Comput. **74**, 3202–3216 (2014)
23. Fishman, G.S.: Monte Carlo: Concepts, Algorithms, and Applications. Springer Series in Operations Research. Springer, New York (1996)
24. Forrester, A.I.J., Bressloff, N.W., Keane, A.J.: Optimization using surrogate models and partially converged computational fluid dynamics simulations. Proc. R. Soc. Lond. A Math. Phys. Eng. Sci. **462**, 2177–2204 (2006). The Royal Society
25. Frauenfelder, P., Schwab, C., Todor, R.A.: Finite elements for elliptic problems with stochastic coefficients. Comput. Methods Appl. Mech. Eng. **194**(2), 205–228 (2005)
26. Galindo, D., Jantsch, P., Webster, C.G., Zhang, G.: Accelerating stochastic collocation methods for PDEs with random input data. Technical Report TM–2015/219, Oak Ridge National Laboratory (2015)
27. Ganapathysubramanian, B., Zabaras, N.: Sparse grid collocation schemes for stochastic natural convection problems. J. Comput. Phys. **225**(1), 652–685 (2007)
28. Ghanem, R.G., Spanos, P.D.: Polynomial chaos in stochastic finite elements. J. Appl. Mech. **57**, 197 (1990)
29. Ghanem, R.G., Spanos, P.D.: Stochastic Finite Elements: A Spectral Approach. Springer, New York (1991)
30. Giles, M.B.: Multilevel Monte Carlo path simulation. Oper. Res. **56**(3), 607–617 (2008)
31. Griebel, M.: Adaptive sparse grid multilevel methods for elliptic PDEs based on finite differences. Computing **61**(2), 151–179 (1998)

32. Gunzburger, M., Webster, C.G., Zhang, G.: An adaptive wavelet stochastic collocation method for irregular solutions of partial differential equations with random input data. In: Sparse Grids and Applications – Munich 2012, pp. 137–170. Springer International Publishing, Cham (2014)
33. Gunzburger, M.D., Webster, C.G., Zhang, G.: Stochastic finite element methods for partial differential equations with random input data. Acta Numer. **23**, 521–650 (2014)
34. Hao, P., Wang, B., Li, G.: Surrogate-based optimum design for stiffened shells with adaptive sampling. AIAA J. **50**(11), 2389–2407 (2012)
35. Helton, J.C., Davis, F.J.: Latin hypercube sampling and the propagation of uncertainty in analyses of complex systems. Reliab. Eng. Syst. Saf. **81**, 23–69 (2003)
36. Heroux, M.A., Willenbring, J.M.: A new overview of the Trilinos project. Sci. Program. **20**(2), 83–88 (2012)
37. Heroux, M.A., Bartlett, R.A., Howle, V.E., Hoekstra, R.J., Hu, J.J., Kolda, T.G., Lehoucq, R.B., Long, K.R., Pawlowski, R.P., Phipps, E.T., Salinger, A.G., Thornquist, H.K., Tuminaro, R.S., Willenbring, J.M., Williams, A.B., Stanley, K.S.: An overview of the Trilinos package. ACM Trans. Math. Softw. **31**(3), 397 (2005)
38. Li, J., Xiu, D.: Evaluation of failure probability via surrogate models. J. Comput. Phys. **229**(23), 8966–8980 (2010)
39. Li, J., Li, J., Xiu, D.: An efficient surrogate-based method for computing rare failure probability. J. Comput. Phys. **230**(24), 8683–8697 (2011)
40. Loève, M.: Probability Theory I. Graduate Texts in Mathematics, 4th edn., vol. 45. Springer, New York (1977)
41. Loève, M.: Probability Theory II. Graduate Texts in Mathematics, 4th edn., vol. 46. Springer, New York (1978)
42. Mathelin, L., Gallivan, K.A.: A compressed sensing approach for partial differential equations with random input data. Commun. Comput. Phys. **12**(04), 919–954 (2012)
43. McKay, M.D., Beckman, R.J., Conover, W.J.: A comparison of three methods for selecting values of input variables in the analysis of output from a computer code. Technometrics **21**(2), 239–245 (1979)
44. Metropolis, N., Ulam, S.: The Monte Carlo method. J. Am. Stat. Assoc. **44**(247), 335–341 (1949)
45. Niederreiter, H.: Quasi-Monte Carlo methods and pseudo-random numbers. Bull. Amer. Math. Soc **84**(6), 957–1041 (1978)
46. Nobile, F., Tempone, R., Webster, C.G.: A sparse grid stochastic collocation method for partial differential equations with random input data. SIAM J. Numer. Anal. **46**(5), 2309–2345 (2008)
47. Nobile, F., Tempone, R., Webster, C.G.: An anisotropic sparse grid stochastic collocation method for partial differential equations with random input data. SIAM J. Numer. Anal. **46**(5), 2411–2442 (2008)
48. Pawlowski, R.P., Phipps, E.T., Salinger, A.G.: Automating embedded analysis capabilities and managing software complexity in multiphysics simulation, Part I: Template-based generic programming. Sci. Program. **20**, 197–219 (2012)
49. Pawlowski, R.P., Phipps, E.T., Salinger, A.G., Owen, S.J., Siefert, C.M., Staten, M.L.: Automating embedded analysis capabilities and managing software complexity in multiphysics simulation part II: application to partial differential equations. Sci. Program. **20**, 327–345 (2012)
50. Phipps, E.T.: Stokhos Stochastic Galerkin Uncertainty Quantification Methods (2015). Available online: http://trilinos.org/packages/stokhos/
51. Phipps, E., D'Elia, M., Edwards, H.C., Hoemmen, M., Hu, J., Rajamanickam, S.: Embedded ensemble propagation for improving performance, portability and scalability of uncertainty quantification on emerging computational architectures. SIAM J. Sci. Comput. **39**(2), C162 (2017)
52. Prokopenko, A., Hu, J.J., Wiesner, T.A., Siefert, C.M., Tuminaro, R.S.: MueLu user's guide 1.0. Technical Report SAND2014-18874, Sandia National Laboratories (2014)

53. Razavi, S., Tolson, B.A., Burn, D.H.: Review of surrogate modeling in water resources. Water Resour. Res. **48**(7), 7401 (2012)
54. Rikards, R., Abramovich, H., Auzins, J., Korjakins, A., Ozolinsh, O., Kalnins, K., Green, T.: Surrogate models for optimum design of stiffened composite shells. Compos. Struct. **63**(2), 243–251 (2004)
55. Roman, L.J., Sarkis, M.: Stochastic Galerkin method for elliptic SPDEs: a white noise approach. Discrete Contin. Dynam. Syst. B **6**(4), 941 (2006)
56. Smolyak, S.A.: Quadrature and interpolation formulas for tensor products of certain classes of functions. Dokl. Akad. Nauk SSSR **4**, 240–243 (1963)
57. Stoyanov, M.: Hierarchy-direction selective approach for locally adaptive sparse grids. Technical Report TM–2013/384, Oak Ridge National Laboratory (2013)
58. Stoyanov, M., Webster, C.G.: A dynamically adaptive sparse grid method for quasi-optimal interpolation of multidimensional analytic functions. Technical Report TM–2015/341, Oak Ridge National Laboratory (2015)
59. Xiu, D.B., Hesthaven, J.S.: High-order collocation methods for differential equations with random inputs. SIAM J. Sci. Comput. **27**(3), 1118–1139 (2005)
60. Xiu, D.B., Karniadakis, G.E.: The Wiener-Askey polynomial chaos for stochastic differential equations. SIAM J. Sci. Comput. **24**(2), 619–644 (2002)

Conservative Model Order Reduction for Fluid Flow

Babak Maboudi Afkham, Nicolò Ripamonti, Qian Wang, and Jan S. Hesthaven

Abstract In the past decade, model order reduction (MOR) has been successful in reducing the computational complexity of elliptic and parabolic systems of partial differential equations (PDEs). However, MOR of hyperbolic equations remains a challenge. Symmetries and conservation laws, which are a distinctive feature of such systems, are often destroyed by conventional MOR techniques which result in a perturbed, and often unstable reduced system. The importance of conservation of energy is well-known for a correct numerical integration of fluid flow. In this paper, we discuss model reduction, that exploits skew-symmetry of conservative and centered discretization schemes, to recover conservation of energy at the level of the reduced system. Moreover, we argue that the reduced system, constructed with the new method, can be identified by a reduced energy that mimics the energy of the high-fidelity system. Therefore, the loss in energy, associated with the model reduction, remains constant in time. This results in an, overall, correct evolution of the fluid that ensures robustness of the reduced system. We evaluate the performance of the proposed method through numerical simulation of various fluid flows, and through a numerical simulation of a continuous variable resonance combustor model.

1 Introduction

Model order reduction (MOR), and in particular reduced basis (RB) methods, has emerged as a powerful approach to cope with the complex and computationally intensive models in engineering and science. Such techniques construct a reduced ordered representation for the state of a model which accurately approximates the

B. Maboudi Afkham (✉) · N. Ripamonti · Q. Wang · J. S. Hesthaven
EPFL, Lausanne, Switzerland
e-mail: babak.maboudi@epfl.ch

M. D'Elia et al. (eds.), *Quantification of Uncertainty: Improving Efficiency and Technology*, Lecture Notes in Computational Science and Engineering 137,
https://doi.org/10.1007/978-3-030-48721-8_4

configuration of the system. The evaluation of this representation is then possible with considerable acceleration.

Although RB methods are successful in reducing the computational complexity of models with elliptic and parabolic partial differential equations (PDEs), MOR of systems of hyperbolic equations, or models with strong advective terms, remains a challenge. Often, such models arise from a set of invariants and conservation laws, some of which are violated by MOR which result in a qualitatively wrong, and sometimes unstable, solution.

Constructing MOR techniques and RB methods that preserve intrinsic structures has recently attracted attention [3, 4, 14, 26, 28, 33]. Structure preservation can recover a physically meaningful reduced model, rather than a pure algebraic coupling of equations. This enforces robustness and can help with the stability of the reduced model. Preserving time-symmetries of Lagrangian, Hamiltonian, and port-Hamiltonian systems can be found in the works of [3, 8, 10, 19, 28, 33]. Conserving inf-sup stability, in the context of finite element methods, can be found in [1, 14]. Furthermore, a flux preserving model reduction for finite volume methods is presented in [7].

Large scale simulations of fluid flows arise in a wide range of disciplines and industries. Therefore, MOR of fluid flows, specially when advective terms are dominant, is important. It is well known that conservation of the energy, specially kinetic energy, is essential for a qualitatively correct numerical integration of fluid flows. Conventional model reduction techniques often violates conservation of mass, momentum [7], or energy in fluid flows which result in an unstable reduced system, in particular for long time-integration. In [2] an entropy stable model reduction method for linear compressible flows is presented by considering an entropy-stable formulation of linearized compressible flows. Furthermore, a conservative model reduction for finite-volume models is presented in [7] that conserves any quantity conserved by the finite-volume scheme. This method finds a reduced linear subspace that ensures conservation of quantities by solving an optimization problem with, generally nonlinear, equality constraints. The constrained optimization problem is solved online, and is only slightly more expensive than solving the unconstrained optimization problem associated with a typical Galerkin MOR.

Skew-symmetric formulation of fluid flows constructs a skew-symmetric differential operator, acting on the momentum vector field, that ensures conservation of quadratic invariants, such as energy. Combined with centered time and space discretization schemes, typically a finite differences discretization method, they recover time-symmetries of a fluid at the discrete level. Such discretization schemes are studies comprehensively over the past few decades and can be found in the works of [11, 30, 31, 35, 38] and the references therein.

In this paper we discuss how to preserve skew-symmetry of the differential operators at the level of the reduced system. This results in conservation of quadratic invariants. The conservation of quantities in the proposed method is guaranteed through the mathematical formulation of the reduced system, for any orthonormal reduced basis. Therefore, the offline and online computational costs of this method is comparable with conventional MOR techniques. However, other conservative model reduction methods, e.g. [7], often require solving multiple nonlinear optimization problems to ensure conservation which can increase the computational costs. Furthermore, we show that the reduced system, as a system of coupled differential equations, contains quadratic invariants and an associated energy which approximates the energy of the high-fidelity system. Therefore, a proper time stepping scheme preserves the reduced representation of the energy, and therefore, the loss in energy due to model reduction remains constant in time. Furthermore, we demonstrate, through numerical experiments, that a quasi-skew-symmetric form of fluid flow, i.e. a formulation where only spacial differential operators are in a skew-symmetric form, offer remarkable stability properties in terms of MOR. This allows an explicit time-integration to be utilized while recovering robustness of skew-symmetric forms at the reduced level.

The rest of this paper is organized as follows. In Sect. 2 we summarize the theory on MOR and introduce the proper orthogonal decomposition (POD) as a conventional RB method. We discuss skew-symmetric and conservatives methods for compressible and incompressible fluid flows in Sect. 3. Conservative and energy-preserving model reduction of fluid flows is discussed in Sect. 4. We evaluate the performance of the method through numerical simulations of incompressible and compressible fluid flow in Sect. 5. We also apply the method to construct a reduced system for the continuous variable resonance combustor, a one dimensional reaction-diffusion model for a rocket engine. Finally, we present conclusive remarks in Sect. 6.

2 Model Order Reduction for Time Dependent Problems

Consider a dynamical system of the form

$$\begin{cases} \frac{d}{dt}u(t) = f(t, u(t)), \\ u(0) = u_0. \end{cases} \tag{1}$$

Here, $u(t), u_0 \in \mathbb{R}^n$ and $f : [0, T] \times \mathbb{R}^n \to \mathbb{R}^n$, for some $T < \infty$, is a Lipschitz function. We may apply the *method of lines* [13] to a system of partial differential equations to obtain a dynamical system of the form (1). The *solution manifold* for (1) is defined as

$$\mathcal{M}_u := \{u(t) | t \in [0, T]\}. \tag{2}$$

When $\mathcal{M}_u$ has a low-dimensional representation, it is referred to as *reducible*. Assume that $\mathcal{M}_u$ can be well approximated by a k-dimensional linear subspace $\mathcal{V}_k$, with $k \ll n$ and let $E_k = \{v_1, \dots, v_k\}$ be the basis vectors for $\mathcal{V}_k$ and V_k the basis matrix that contains these vectors in its columns. A reduced basis (RB) method assumes that $u \approx \tilde{u} = V_k v$, where $v \in \mathbb{R}^k$ is the expansion coefficients of $\tilde{u}$ in the basis V_k. Substituting this into (1) yields

$$V_k \frac{d}{dt} v(t) = f(t, V_k v) + r(t, u). \tag{3}$$

Here, r is the error vector in this approximation. The Petrov–Galerkin projection of (1) onto $\mathscr{V}_k$ requires r to be orthogonal to a k-dimensional subspace $\mathscr{W}_k$. One can construct a projection operator $P_{\mathscr{V}_k,\mathscr{W}_k}$ that projects elements of $\mathbb{R}^n$ onto $\mathscr{V}_k$, orthogonal to $\mathscr{W}_k$ as $P_{\mathscr{V}_k,\mathscr{W}_k} = V_k(W_k^T V_k)^{-1} W_k^T$, where W_k is the basis matrix that contains the basis vectors of $\mathscr{W}_k$ in its columns and $W_k^T V_k$ is assumed to be invertible. With this projection, (1) reduces to the k-dimensional problem

$$\begin{cases} \frac{d}{dt} v(t) = (W_k^T V_k)^{-1} f(t, V_k v), \\ v(0) = (W_k^T V_k)^{-1} u_0. \end{cases} \tag{4}$$

When we require $W_k = V_k$, then (4) is referred to as the *Galerkin* projection of (1) onto $\mathscr{V}_k$. Since (4) has a smaller size, as compared to (1), one can expect an accelerated evaluation. To numerically identify the best possible subspace $\mathscr{V}_k$ we first discretize the solution manifold to obtain

$$\mathscr{M}_u^\Delta = \{u(t_i) | i \in \{1, \dots, N_t\}\}. \tag{5}$$

Members of $\mathscr{M}_u^\Delta$ are referred to as *snapshots* of (1). One can obtain these snapshots by applying a time-integration scheme, e.g., the Runge–Kutta methods, to (1) to obtain $\tilde{\mathscr{M}}_u^\Delta$ as an approximation to $\mathscr{M}_u^\Delta$. Throughout this paper, we assume that we can choose $\tilde{\mathscr{M}}_u^\Delta$ arbitrary close to $\mathscr{M}_u^\Delta$. By an abuse of notation, we drop the overscript "$\sim$". For a Galerkin projection, the best possible basis V_k is the one that minimizes the collective projection error [22], i.e., the solution to the minimization problem

$$\begin{aligned} \underset{V_k \in \mathbb{R}^{n\times k}}{\text{minimize}} \quad & \| S - V_k V_k^T S \|_F, \\ \text{subject to} \quad & V_k^T V_k = I_k. \end{aligned} \tag{6}$$

Here S collects vectors in $\mathscr{M}_u^\Delta$ in its columns, referred to as the *snapshot matrix*, $\| \cdot \|_F$ is the Frobenius norm [40], and I_k is the identity matrix of size k. Note that the constraint in (6) requires V_k to be orthonormal. The basis matrix V_k that solves the minimization problem (6) is referred to as the *proper orthogonal decomposition* (POD) of S of size k [22] and, according to the Schmidt–Mirsky theorem, can be constructed using the left singular vectors of S as

$$V_k = [u_i]_{i=1}^k. \tag{7}$$

Here u_i, for $i = 1, \dots k$, are the first k singular vectors of S.

Model order reduction is often studied in the parametric setting, where the vector u, u_0, and the right hand side function f of (1) are of the form $u(t; \mu)$, $u_0(\mu)$, and $f(t, u; \mu)$, respectively. Here μ belongs to $\mathbb{P}$, a closed subset of $\mathbb{R}^d$. In this case, the reduced system can recover quantities of interest at low cost. Since the nature of time, as a parameter, is different from other spacial and physical parameters, in this paper we solely focus on t as the parameter. Nevertheless, it is straight forward to extend the results of this paper to the parameter setting by using POD in time and parametric space, or by using the POD-Greedy [20, 22, 34] method to generate a basis V_k.

Since the approximate solution $\tilde{u}$ is a linear combination of the POD basis vectors, $\tilde{u}$ inherits linear properties of these basis vectors. However, when the solution u to (1) satisfies some nonlinear invariants, there is no guarantee that, in general, $\tilde{u}$ also satisfy such invariants [8, 28, 29, 33]. This results in a qualitatively wrong and often unstable solution to (4). In the later sections, we discuss how the skew-symmetric formulation of the fluid flow allows conservation of quadratic invariants, e.g. the kinetic energy, at the level of the reduced system.

3 Skew Symmetric and Centered Schemes for Fluid Flows

In this section we summarize the conservation properties of skew-symmetric forms and discretization schemes, following, closely, the works of [30, 31, 35, 38].

3.1 Conservation Laws

In the context of fluid flows, transport of conserved quantities, can be expressed as

$$\frac{\partial}{\partial t}\rho\varphi + \nabla \cdot (\rho u \varphi) = \nabla \cdot F_\varphi \quad \text{defined in} \quad \Omega \subset \mathbb{R}^d. \tag{8}$$

Here, $d = 1, 2$ or 3, $\rho : \Omega \to \mathbb{R}$ is the density, $u \in \Omega \to \mathbb{R}^d$ is the velocity vector field, φ is a measured scalar quantity of the flow, and F_φ is the flux function associated to φ. Integration of (8) over Ω yields

$$\frac{d}{dt}\int_\Omega \rho\varphi \, dx = \int_{\partial\Omega} (F_\varphi - \rho u \varphi) \cdot \hat{n} \, ds, \tag{9}$$

where $\partial\Omega$ is the boundary of Ω, and $\hat{n}$ is the unit outward normal vector to $\partial\Omega$. This means that the quantity $(\rho\varphi)$ is explicitly conserved over control volumes. Therefore, (8) is referred to as the *conservative form* and the convective term in (8) is referred to as the *divergence form*. However, using the *continuity equation*

$$\frac{\partial}{\partial t}\rho + \nabla \cdot (\rho u) = 0, \tag{10}$$

we can rewrite (8) as

$$\rho \frac{\partial}{\partial t}\varphi + (\rho u) \cdot \nabla\varphi = \nabla \cdot F_{\varphi}. \tag{11}$$

The convective term in this formulation is referred to as the *advective form*. The *skew-symmetric* form of the convective term is obtained by the arithmetic average of the divergent and the advective form:

$$\frac{1}{2}\left(\rho \frac{\partial}{\partial t}\varphi + \frac{\partial}{\partial t}(\rho\varphi)\right) + \frac{1}{2}\left((\rho u) \cdot \nabla\varphi + \nabla \cdot (\rho u \varphi)\right) = \nabla \cdot F_{\varphi}. \tag{12}$$

Multiplying (12) with φ yields

$$\frac{\partial}{\partial t}\rho\varphi^2 + \nabla \cdot (\rho u \varphi^2) = \varphi \nabla \cdot F_{\varphi}. \tag{13}$$

Therefore, $\rho\varphi^2$ is a conserved quantity for a flux-free φ. Since the divergence, the advective and the skew-symmetric forms are identical at the continuous level, φ^2 is a conserved quantity for all forms. However, the equivalence of these forms is not preserved through a general discretization scheme and we can not expect φ^2 to be a conserved quantity at the discrete level. To motivate numerical advantages of the skew-symmetric form consider the operator

$$S_{\rho u}(\cdot) = \frac{1}{2}([\nabla \cdot \rho u] + (\rho u) \cdot \nabla)(\cdot). \tag{14}$$

With a proper set of boundary condition, this operator is a skew-adjoint operator on L^2. Here, $[\cdot]$ indicates that the inside of the brackets act as a differential operator. This skew-adjoint property is used later to show the conservation of some quadratic quantities in (8). Similarly, we can define a skew-adjoint operator with respect to the time variable as

$$S_{\rho,\partial_t} = \frac{1}{2}\left(\rho \frac{\partial}{\partial t} + [\frac{\partial}{\partial t}\rho]\right). \tag{15}$$

Here, the subscript ∂t is to emphasize that S_{ρ,∂_t} is a differential operator with respect to t. A proper time and space discretization of $S_{\rho u}$ and S_{ρ,∂_t} can preserve the skewness property.

Numerical time integration of (12) can be challenging since the time differentiation of different variables is present. Following [30], we rewrite (12) as

$$\sqrt{\rho}\frac{\partial}{\partial t}(\sqrt{\rho}\varphi) + S_{\rho u}(\varphi) = \nabla \cdot F_{\varphi}. \tag{16}$$

Time integration of this form is presented in [30, 35]. Note that one can also generate a quasi-skew-symmetric form [6, 32] of (8) as

$$\frac{\partial}{\partial t}(\rho\varphi) + \frac{1}{2}\left(\nabla \cdot (\rho u \varphi) + \rho u \cdot \nabla\varphi + \varphi\nabla \cdot (\rho u)\right) = \nabla \cdot F_{\varphi}. \tag{17}$$

Even though this is not a fully skew-symmetric form (skew-symmetric only in space), the numerical stability of this form is significantly better than the divergence and advective form [6, 30, 32]. Note that this quasi-skew-symmetric form is identical to the skew-symmetric form in the incompressible limit.

3.2 *Incompressible Fluid*

Consider the governing equations of an incompressible fluid with skew-symmetric convective term:

$$\begin{cases} \nabla \cdot u = 0, \\ \dfrac{\partial}{\partial t}u + S_u(u) + \nabla p = \nabla \cdot \tau, \end{cases} \tag{18}$$

defined on Ω. Here, $p : \Omega \to \mathbb{R}^+$ is the pressure, $\tau : \Omega \to \mathbb{R}^{d\times d}$ is the viscous stress tensor, and $S_u = \frac{1}{2}([\nabla \cdot u] + u \cdot \nabla)$. It is straight forward to check

$$\frac{d}{dt}K + \nabla \cdot (Ku) + \nabla \cdot (pu) = \nabla \cdot (\tau u) - (\tau\nabla) \cdot u, \tag{19}$$

where $K = \frac{1}{2}\sum_{i=1}^{d} u_i^2$ is the kinetic energy and we used

$$u \cdot S_u(u) = \nabla \cdot (Ku). \tag{20}$$

The only non-conservative term in (19) is $-(\tau\nabla) \cdot u$, which corresponds to dissipation of kinetic energy. Therefore, in the absence of the viscous terms, K is a conserved quantity of the system, and $\frac{d}{dt}\int_{\Omega} K \, dx < 0$ when $\tau \neq 0$. Note that as long as $\nabla \cdot u = 0$, as discussed in Sect. 3.1, the divergence, the convective, and the skew-symmetric forms are identical for the incompressible fluid equation. Thus, kinetic energy is conserved for all forms. However, for a general discretization scheme, these forms are not identical and often the conservation of kinetic energy (in the discrete sense) is be violated.

A skew-symmetric discretization of (18) is a centered scheme that exploits the skew-adjoint property of S_u, and ensures conservation of kinetic energy at the discrete level. We uniformly discretize Ω into N points and denote by $\mathbf{u} \in \mathbb{R}^{N\times d}$, $\mathbf{p} \in \mathbb{R}^N$, and $T \in R^{N\times d\times d}$ the discrete representation of u, p, and τ, respectively. Let D_j be the centered finite difference scheme for $\partial/\partial x_j$, and for $j = 1, \dots, d$. The momentum equation in (18) is discretized as

$$\frac{d}{dt}\mathbf{u}_i + S_\mathbf{u}\mathbf{u}_i + D_i\mathbf{p} = \sum_{j=1}^d D_j T_{ij}, \quad i = 1, \dots, d, \tag{21}$$

where $S_\mathbf{u}$ is the discretization of S_u given by

$$S_\mathbf{u} = \sum_{j=1}^d D_j U_j + U_j D_j, \tag{22}$$

and U_i contains components of $\mathbf{u}_i$ on its diagonal. We require D_i to satisfy

1. $D_i = -D_i^T$
2. $D_i\mathbf{1} = \mathbf{0}$, where $\mathbf{1}$ and $\mathbf{0}$ are vectors of ones and zeros, respectively.

Conditions 1 and 2 yield

$$S_\mathbf{u} = -S_\mathbf{u}^T, \quad \mathbf{1}^T S_\mathbf{u}\mathbf{u}_i = 0, \quad i = 1, \dots, d. \tag{23}$$

Conservation of momentum in the discrete sense is expressed as

$$\frac{d}{dt}\sum_{i=1}^d \mathbf{1}^T\mathbf{u}_i = \sum_{i=1}^d \left(-\mathbf{1}^T S_\mathbf{u}\mathbf{u}_i - \mathbf{1}^T D_i\mathbf{p} + \sum_{j=1}^d \mathbf{1}^T D_j T_{ij} \right) = 0. \tag{24}$$

Similarly, it is verified that

$$\frac{d}{dt}\sum_{i=1}^d \left(\frac{1}{2}\mathbf{u}_i^T\mathbf{u}_i \right) = -\sum_{i,j=1}^d T_{ij} D_j \mathbf{u}_i \leq 0. \tag{25}$$

Conditions 1 and 2 for D_i are easily checked for a centered finite differences scheme on a periodic domain. For other types of boundaries, e.g., wall boundary and inflow/outflow, we refer the reader to [11, 31] for the construction of the proper discrete centered differentiation operator. We note that the finite differences schemes are chosen here for illustration purposes. It is easily checked that any discrete differentiation operator that satisfies discrete integration by parts, e.g. summation by part (SBP) methods and discontinuous Galerkin (DG) methods, also satisfies conditions 1 and 2 and can be used to construct a skew-symmetric discretization.

3.3 Compressible Fluid

Consider the equations governing the evolution of a compressible fluid in a skew-symmetric form in one spacial dimension

$$\begin{cases} \frac{\partial}{\partial t}\rho + \frac{\partial}{\partial x}(\rho u) = 0, \\ S_{\rho,\partial_t}(u) + S_{\rho u}(u) + \frac{\partial}{\partial x}p = \frac{\partial}{\partial x}\tau, \\ \frac{\partial}{\partial t}\rho E + \frac{\partial}{\partial x}(uE + up) = \frac{\partial}{\partial x}(u\tau - \phi). \end{cases} \tag{26}$$

Here $E = e + u^2/2$ is the total energy per unit mass, with $e = p/\rho(\gamma - 1)$ being the internal energy, γ the adiabatic gas index, and $\phi = -\lambda\frac{\partial T}{\partial x}$ is the heat flux, with λ the heat conductivity. The remaining variables are the same as those discussed in Sect. 3.2. Following [35], the evolution of the momentum equation is

$$\begin{aligned} \frac{\partial}{\partial t}(\frac{\rho u^2}{2}) + \frac{\partial}{\partial x}(\rho u\frac{u^2}{2}) &= \frac{1}{2}u(\frac{d}{dt}\rho u + \rho\frac{d}{dt}u) + \frac{1}{2}u([\frac{\partial}{\partial x}\rho u]u + \rho u\frac{\partial}{\partial x}u) \\ &= -u\frac{\partial}{\partial x}p + u\frac{\partial}{\partial x}\tau. \end{aligned} \tag{27}$$

Substituting this into the energy equation in (26), while assuming a constant adiabatic index, yields

$$\frac{1}{\gamma - 1}\frac{d}{dt}p + \frac{\gamma}{\gamma - 1}\frac{\partial}{\partial x}up - u\frac{\partial}{\partial x}(p) = -u\frac{\partial}{\partial x}\tau + \frac{\partial}{\partial x}(u\tau - \phi). \tag{28}$$

We discretize the real line, uniformly, into N grid points and denote by $\mathbf{r}, \mathbf{u}, \mathbf{p} \in \mathbb{R}^N$, the discrete representations of ρ, u, and p, respectively. Using the matrix differentiation operator $D \in \mathbb{R}^{N\times N}$ (we omit the subscript "i" for the one dimensional case), introduced in Sect. 3.2, we define the skew-symmetric matrix operator $S_{\mathbf{ru}} = \frac{1}{2}(DUR + RUD)$, where R is the matrix that contains r in its diagonal. Semi-discrete expression of (26) and (28) takes the form

$$\begin{cases} \frac{d}{dt}\mathbf{r} + DU\mathbf{r} = 0, \\ S_{\mathbf{r},\partial_t}(\mathbf{u}) + S_{\mathbf{ru}}\mathbf{u} + D\mathbf{p} = DT, \\ \frac{1}{\gamma - 1}\frac{d}{dt}\mathbf{p} + \frac{\gamma}{\gamma - 1}DU\mathbf{p} - UD\mathbf{p} = -UDT + D(UT - \boldsymbol{\phi}). \end{cases} \tag{29}$$

Recalling conditions 1 and 2 for D, discussed in Sect. 3.2, it is easily verified that

$$S_{\mathbf{ru}}^T = -S_{\mathbf{ru}}, \quad \mathbf{1}^T S_{\mathbf{ru}}\mathbf{u} = -\mathbf{u}^T DU\mathbf{r}. \tag{30}$$

Conservation of mass is expressed as

$$\frac{d}{dt}(\mathbf{1}^T \mathbf{r}) = -\mathbf{1}^T DR\mathbf{u} = 0. \tag{31}$$

Furthermore, we recover conservation of momentum in the discrete sense as

$$\begin{aligned}
\frac{d}{dt}(\mathbf{r}^T\mathbf{u}) &= \frac{1}{2}\frac{d}{dt}(\mathbf{r}^T\mathbf{u}) + \frac{1}{2}\left(\mathbf{r}^T\frac{d}{dt}\mathbf{u} + \mathbf{u}^T\frac{d}{dt}\mathbf{r}\right)\\
&= \frac{1}{2}u^T\frac{d}{dt}\mathbf{r} + \mathbf{1}^T S_{\mathbf{r},\partial_t}(\mathbf{u})\\
&= -\frac{1}{2}\mathbf{u}^T DU\mathbf{r} - \frac{1}{2}\mathbf{1}^T S_{\mathbf{ru}}\mathbf{u} - \mathbf{1}^T D\mathbf{p} + \mathbf{1}^T DT = 0.
\end{aligned} \tag{32}$$

Here we used (30) and the mass and the momentum equation in (29). Similarly, for conservation of the total energy, we have

$$\frac{d}{dt}\left(\frac{1}{\gamma-1}\mathbf{1}^T\mathbf{p} + \frac{1}{2}(R\mathbf{u})^T\mathbf{u}\right) = \frac{d}{dt}\left(\frac{1}{\gamma-1}\mathbf{1}^T\mathbf{p}\right) + \frac{1}{2}\mathbf{u}^T S_{\mathbf{r},\partial_t}(\mathbf{u}) = 0. \tag{33}$$

In addition to the conservation of the total energy, the skew-symmetric form of (29) also conserves the evolutions of the kinetic energy:

$$\begin{aligned}
\frac{d}{dt}(\frac{1}{2}\mathbf{u}^T R\mathbf{u}) = \frac{1}{2}\mathbf{u}^T S_{\mathbf{r},\partial_t}(\mathbf{u}) &= -\mathbf{u}^T S_{\mathbf{ru}}\mathbf{u} + \mathbf{u}^T Dp + \mathbf{u}^T DT\\
&= \mathbf{u}^T Dp + \mathbf{u}^T DT.
\end{aligned} \tag{34}$$

Here, we used the skew-symmetry of $S_{\mathbf{ru}}$. Therefore, only the pressure and the viscous terms contribute to a change in the kinetic energy.

We point out that there are other methods to obtain a skew-symmetric form for (26), that result in the conservation of other quantities. An entropy preserving skew-symmetric form can be found in [36]. Furthermore, a fully quasi-skew-symmetric form for (26), where all quadratic fluxes are in a skew-symmetric form, is shown to minimize aliasing errors [23, 24]

3.4 Time Integration

Following [30, 35] we can construct a fully discrete second order accurate scheme for (3.3) as

$$\begin{cases} \frac{1}{2}\sqrt{\mathbf{r}}^{n+1/2}\frac{\sqrt{\mathbf{r}}^{n+1}-\sqrt{\mathbf{r}}^{n}}{\Delta t}+DU^{n+1/2}\mathbf{r}^{n}=0, \\ \sqrt{\mathbf{r}}^{n+1/2}\frac{\sqrt{\mathbf{R}}^{n+1}\mathbf{u}^{n+1}-\sqrt{\mathbf{R}}^{n}\mathbf{u}^{n}}{\Delta t}+S_{\mathbf{r}^{n}\mathbf{u}^{n}}\mathbf{u}_{\alpha}^{n+1/2}+D\mathbf{p}^{n}=DT^{n}, \\ \frac{1}{\gamma-1}\frac{\mathbf{p}^{n+1}-\mathbf{p}^{n}}{\Delta t}+\frac{\gamma}{\gamma-1}DU^{n}\mathbf{p}^{n}-U^{n}D\mathbf{p}^{n}=-U^{n}DT^{n}+D(U^{n}T^{n}-\phi^{n}), \end{cases} \tag{35}$$

and all the variables are stored at the same location with the same resolution (colocated). Here, $\mathbf{R}$ contains elements of $\mathbf{r}$ on its diagonal, Δt is the time step, superscript n denotes evaluating at $t = n\Delta t$, superscript $n + 1/2$ denotes the arithmetic average of a variable evaluated at $t = n\Delta t$ and $t = (n + 1)\Delta t$, the square root sign denotes element-wise application of square root, and

$$\mathbf{u}_{\alpha}^{n+1/2}=\frac{\sqrt{\mathbf{R}}^{n+1}\mathbf{u}^{n+1}+\sqrt{\mathbf{R}}^{n}\mathbf{u}^{n}}{2\sqrt{\mathbf{r}}^{n+1/2}}. \tag{36}$$

As discussed in [35], this time discretization scheme preserves the symmetries expressed in (25), (32), (33), and (34). In the incompressible case, the method reduces to the implicit mid-point scheme [21]. For further information see [30, 35].

4 Model Reduction of Fluid Flow

A straight-forward model reduction of (18) and (26) does not, in general, preserve symmetries and conservation laws, presented in Sect. 3. In this section we discuss how to exploit the discrete skew-symmetric structure of (21) and (29) to recover conservation of mass, momentum, and energy at the level of the reduced system.

Let $V_{\mathbf{r}}$, $V_{\mathbf{ru}}$, and $V_{\mathbf{u}_i}$ be the reduced bases for the snapshots of $\mathbf{r}$, $R\mathbf{u}$, and $\mathbf{u}_i$, respectively. For the one dimensional case, the subscript "i" is omitted and for an incompressible fluid, $V_{\mathbf{r}}$ and $V_{\mathbf{ru}}$ are not computed. For the purpose of simplicity, we assume that all bases have the size k. We seek to project $S_{\mathbf{u}}$ and $S_{\mathbf{ru}}$ onto the reduced space, such that the projection preserves the skew-symmetric property. The projected operators, using a Galerkin projection, read

$$S_{\mathbf{u}}^{r}=V_{\mathbf{u}_i}^{T}S_{\mathbf{u}}V_{\mathbf{u}_i}, \quad i=1,\ldots,d, \tag{37}$$

and

$$S^r_{\mathbf{r},\partial_t} = V^T_{\mathbf{ru}} S_{\mathbf{r},\partial_t} V_{\mathbf{u}}, \quad S^r_{\mathbf{ru}} = V^T_{\mathbf{ru}} S_{\mathbf{ru}} V_{\mathbf{u}}. \tag{38}$$

Note that $S^r_{\mathbf{r},\partial_t}$ is not computed explicitly. It is clear that $S^r_{\mathbf{u}}$ is already in a skew-symmetric form. On the other hand, $S^r_{\mathbf{r},\partial_t}$ and $S^r_{\mathbf{ru}}$ are not, in general, skew-adjoint and skew-symmetric, respectively. This can be ensured, however, by requiring $V_{\mathbf{ru}} = V_{\mathbf{u}}$. We denote such a basis by $V_{\mathbf{ru},\mathbf{u}}$.Using (37) and (38), a Galerkin projection of the momentum equation in (21) and the governing equations for a compressible fluid in (29) take the form

$$\frac{d}{dt}\mathbf{u}^r{}_i + S^r_{\mathbf{u}}\mathbf{u}^r_i + V^T_{\mathbf{u}_i} D_i \mathbf{p} = \sum_{j=1}^{d} V^T_{k_3,\mathbf{u}_i} D_j T_{ij}(V_{\mathbf{u}_i}\mathbf{u}^r_i), \quad i = 1, \dots, d, \tag{39}$$

and

$$\begin{cases} \dfrac{d}{dt}\mathbf{r}^r + \displaystyle\sum_{i=1}^{k} V^T_{\mathbf{r}} D U_i V_{\mathbf{r}} \mathbf{r}^r = 0, \\ S^r_{\mathbf{r},\partial_t} + S^r_{\mathbf{ru}}\mathbf{u}^r + V^T_{\mathbf{ru},\mathbf{u}} D V_{\mathbf{p}} \mathbf{p}^r = V^T_{\mathbf{ru},\mathbf{u}} D T, \\ \dfrac{1}{\gamma-1}\dfrac{d}{dt}\mathbf{p}^r + \dfrac{\gamma}{\gamma-1} V^T_{\mathbf{p}} D U V_{\mathbf{p}} \mathbf{p}^r - V^T_{\mathbf{p}} U D V_{\mathbf{p}} \mathbf{p}^r = -V^T_{\mathbf{p}} U D T + V^T_{\mathbf{p}} D(UT - \boldsymbol{\phi}), \end{cases} \tag{40}$$

respectively. Note that in (40), dependency of T on $V_{\mathbf{ru},\mathbf{u}}$ is not shown for abbreviation. In (39) and (40), D_i is always multiplied from the left with a basis matrix or a diagonal matrix. Therefore, the telescoping sum, discussed in Condition 2 in Sect. 3.1, cannot be used to show conservation of mass and momentum. However, POD preserves linear properties of snapshots. To demonstrate this, let the overscript "$\sim$" denote the representation of a reduced variable in the high-fidelity space. An approximated variable, e.g. density, can be represented as a linear combination of some snapshots as $\mathbf{r} \approx \tilde{\mathbf{r}} = \sum_{i=1}^k c_i \mathbf{r}_i$, for some snapshots $\mathbf{r}_i$ and some coefficients $c_i \in \mathbb{R}$, for $i = 1, \dots, k$. Conservation of mass, evaluated by $\tilde{\mathbf{r}}$, reads

$$\frac{d}{dt}\mathbf{1}^T\tilde{\mathbf{r}} = \sum_{i=1}^{k} c_i \left(\mathbf{1}^T \frac{d}{dt}\mathbf{r}_i\right) = -\sum_{i=1}^{k} c_i \left(\mathbf{1}^T D R_i \mathbf{u}_i\right) = 0, \tag{41}$$

where we used the fact that $\mathbf{1}^T D = \mathbf{0}^T$. Similarly, we recover conservation of momentum

$$
\begin{aligned}
\frac{d}{dt}(\tilde{\mathbf{r}}^T\tilde{\mathbf{u}}) &= \frac{1}{2}\frac{d}{dt}(\tilde{\mathbf{r}}^T\tilde{\mathbf{u}}) + \frac{1}{2}\left(\tilde{\mathbf{r}}^T\frac{d}{dt}\tilde{\mathbf{u}} + \tilde{\mathbf{u}}^T\frac{d}{dt}\tilde{\mathbf{r}}\right) \\
&= \sum_{i,j=1}^{k} d_i c_j \left(\mathbf{u}_i^T\frac{d}{dt}\mathbf{r}_j + \left(\mathbf{r}_j^T\frac{d}{dt}\mathbf{u}_i + \mathbf{u}_i^T\frac{d}{dt}\mathbf{r}_j\right)\right) = 0.
\end{aligned}
\tag{42}
$$

Here, $\tilde{\mathbf{u}} = \sum_{i=1}^{k} d_i\mathbf{u}_i$, for some snapshot $\mathbf{u}_i$ and coefficients $d_i \in \mathbb{R}$. Denoting by $\{R\mathbf{u}\}^r$ the reduced representation of $R\mathbf{u}$ in basis $V_{\mathbf{ru},\mathbf{u}}$, the evolution of kinetic energy is expressed as

$$
\begin{aligned}
\frac{d}{dt}\left(\frac{1}{2}\tilde{\mathbf{u}}^T R\tilde{\mathbf{u}}\right) &= \frac{d}{dt}\left(\frac{1}{2}\mathbf{u}^{rT} V_{\mathbf{ru},\mathbf{u}}^T V_{\mathbf{ru},\mathbf{u}}\{R\mathbf{u}\}^r\right) = \frac{d}{dt}\left(\frac{1}{2}\mathbf{u}^{rT}\{R\mathbf{u}\}^r\right) \\
&= \frac{1}{2}\left(\mathbf{u}^{rT}\frac{d}{dt}\{R\mathbf{u}\}^r + \{R\mathbf{u}\}^r\frac{d}{dt}\mathbf{u}^r\right) \\
&= \frac{1}{2}\left(\mathbf{u}^{rT} V_{\mathbf{ru},\mathbf{u}}^T V_{\mathbf{ru},\mathbf{u}}\frac{d}{dt}\{R\mathbf{u}\}^r + \{R\mathbf{u}\}^r\frac{d}{dt}V_{\mathbf{ru},\mathbf{u}}^T V_{\mathbf{ru},\mathbf{u}}\mathbf{u}^r\right) \\
&= \mathbf{u}^{rT} S_{\mathbf{r},\partial_t}^r \mathbf{u}^r = \mathbf{u}^{rT} V_{\mathbf{ru},\mathbf{u}} D V_{\mathbf{p}}\mathbf{P}^r + \mathbf{u}^{rT} V_{\mathbf{ru},\mathbf{u}}^T DT.
\end{aligned}
\tag{43}
$$

In the missing steps in the last line, skew-symmetry of $S_{\mathbf{ru}}^r$ is used. Note, that only the reduced pressure and the viscous term contribute to the evolution of kinetic energy. Furthermore, the quantity $\frac{1}{2}\mathbf{u}^{rT}\{R\mathbf{u}\}^r$ is the kinetic energy associated with the reduced system (40), approximating the kinetic energy of the high-fidelity system (29), and is a quadratic form with respect to the reduced variables. Conservation of kinetic energy for (39) follows similarly. It is straight-forward to check that

$$
\frac{d}{dt}\left(\frac{1}{\gamma - 1}\mathbf{1}^T\tilde{\mathbf{p}} + \frac{1}{2}\tilde{\mathbf{u}}^T R\tilde{\mathbf{u}}\right) = 0,
\tag{44}
$$

i.e., the total energy is conserved. We immediately recognize that $\mathbf{p}^r/(\gamma - 1)$ is the internal energy of the reduced system. However, the total internal energy of (40) is a weighted sum, $b^T\mathbf{p}^r/(\gamma - 1)$, with $b = V_{\mathbf{p}}^T\mathbf{1}$ which is an approximation of the total internal energy in (29). From (41), (42), (43), and (44) we conclude the following proposition.

Proposition 1 *The loss in the mass, momentum and energy associated with the model reduction in* (40) *is constant in time, and therefore, bounded.*

4.1 Assembling Nonlinear Terms and Time Integration

Nonlinear terms that appear in (39) and (40) are of quadratic nature. These terms can be evaluated exactly using a set of precomputed matrices as proposed in [5]. As an example, consider

$$S_{\mathbf{u}}^{r} = V_{\mathbf{u}}^{T}(DU + UD)V_{\mathbf{u}}^{T}. \tag{45}$$

We write U as a linear combination of matrices as $U = \sum_{j=1}^{k} \mathbf{u}_j^r U_j$, where $\mathbf{u}_j^r$ is the jth component of $\mathbf{u}^r$, and U_j contains the jth column of $V_{\mathbf{u}}$ on its diagonal. It follows

$$S_{\mathbf{u}}^{r} = \sum_{j=1}^{k} \mathbf{u}_j^r \left(V_{\mathbf{u}}^{T}(DU_j + U_j D)V_{\mathbf{u}}^{T} \right). \tag{46}$$

The matrices $V_{\mathbf{u}}^{T}(DU_j + U_j D)V_{\mathbf{u}}^{T}$ can be computed prior to the time integration of the reduced system. However, the form of the fully discrete system in (35) introduces cubic and even quartic terms. In principle, the same method can be applied to assemble the nonlinear terms. However, the number of precomputed matrices grows proportional to the order of the nonlinear term.

To accelerate assembly of the nonlinear terms we may approximately evaluate them using the discrete empirical interpolation method (DEIM). This approximation can affect the accuracy of conserved quantities in (40). Therefore, the accuracy of the DEIM approximation must be chosen higher than the one of POD.

To integrate (40) in time, the fully discrete system (35) is modified prior to model reduction, by dividing the mass and momentum equation with $\sqrt{\mathbf{r}}^{n+1}$. Note that since the new form is identical to (35), it does not affect the conserved quantities. Subsequently, a basis for $\sqrt{\mathbf{r}}$, denoted by $V_{\sqrt{\mathbf{r}}}$, is constructed. Since the reduced system shows the same structure of (35), the same numerical integrator is used to compute the reduced solution, without decoupling the modes related to velocity, density and pressure. The nonlinear terms are evaluated exactly using the quadratic expansion or approximated using the DEIM.

5 Numerical Experiments

5.1 Vortex Merging

Consider the two-dimensional incompressible Euler equation (18) on a square domain $\Omega = [0, 2\pi]^2$, with periodic boundary conditions. Spatial derivatives are discretized using a Fourier spectral method. To capture the fine details characterizing the solution, 256 × 256 modes are used. We consider the evolution of three vortices, with the initial structure given by

$$\omega = \omega_0 + \sum_{i=1}^{3} \alpha_i e^{-\frac{(x-x_i)^2+(y-y_i)^2}{\beta^2}}. \tag{47}$$

Here, $\omega = \nabla \times u$ is the vorticity, (x, y) represents the spatial coordinates, (x_i, y_i) is the center of the ith vortex, α_i its maximum amplitude, and β controls the effective radius of the vortex. In this example, the center of three vortices are

$$(x_1, y_1) = (0.75\pi, \pi), (x_2, y_2) = (1.25\pi, \pi), (x_3, y_3) = (1.25\pi, 1.5\pi), \tag{48}$$

close to the center of the domain. Two of the vortices have a positive spin with $\alpha_1 = \alpha_2 = \pi$ and the third rotates in the opposite direction with $\alpha_3 = -0.5\pi$. The effective radius of all the vortices is set to $\beta = 1/\pi$. This arrangement of vortices is an interesting initial condition to study the process of vortex merging. This phenomenon is often a result of fast-moving dipoles of vortices with the same spin facing another vortex [12] of opposite spin. The merging process transfers the vorticity from the initial configuration into long, narrow, and spiral-shaped strips of intense vorticity [27]. The formation of such thin vorticity filaments in the fluid may pose numerical challenges, due to aliasing.

In the context of MOR, conservation of energy and stability is crucial to capturing fine structures. With the absence of natural dissipation, straight forward application of MOR techniques for the Euler equation is often unstable.

To define the initial conditions in terms of the velocity components u and the pressure p, we define a stream-function Ψ, the solution to the equation

$$-\Delta\Psi = \omega. \tag{49}$$

The initial velocity is then given by $\nabla \times \Psi$. To solve the stream-function problem (49), we require $\int_\Omega \omega \, dx = 0$. It is easily verified that this requirement implies $\omega_0 = 0.038$. The pressure is recovered by solving the related Poisson pressure equation

$$\Delta p = -\nabla \cdot S_u(u),$$

obtained by applying the divergence operator to (18) and using the incompressibility condition. The implicit midpoint scheme, to mimic the time integration scheme presented in (35), is used to integrate in time. The merging phenomenon is simulated for a total of 18 time units using a temporal step $\Delta t = 0.004$.

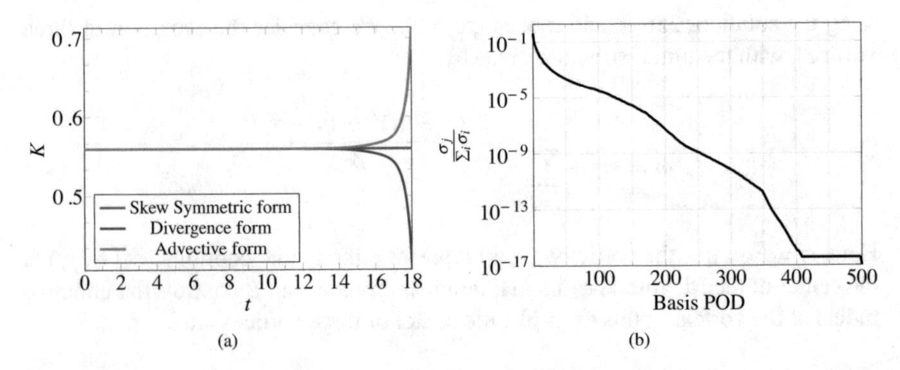

Fig. 1 (**a**) The kinetic energy K for the advective, divergence and the skew-symmetric formulations. (**b**) The decay of the singular values for the vortex merging

Figure 1a illustrates the evolution of the kinetic energy for the advective, divergence, and the skew-symmetric form of the high-fidelity system. It is observed that only the skew-symmetric form preserves the kinetic energy, confirming the discussion in Sect. 3.2.

A total of 5000 temporal snapshots is used to construct a reduced basis, following the process discussed in Sect. 2. The decay of the singular values, used as an indication of the reducibility of the problem, is presented in Fig. 1b. The first 35 POD modes corresponds to over 99% of the modes of the high-fidelity solution. This suggests that an accurate reduced system can be constructed using a small number of basis vectors. To illustrate the effectiveness of the method, smaller bases are also considered.

For a qualitative analysis, in Fig. 2, four solutions at different times are shown for the high fidelity system and the reduced system with $k = 17$ and $k = 35$ modes. The overall dynamics of the problem, and in particular the formation and development of vorticity filaments, are correctly represented, even with a moderate number of basis vectors. Although small details are not captured by the reduced system with a small number of basis vectors, the position and the spreading of the vortices are comparable.

Figure 3a shows the L^2 error between the high-fidelity solution and the reduced solution. The error decreases, consistently, as the number of basis vectors increases. Furthermore, the accuracy is maintained over the period of time integration.

The conservation of the kinetic energy is presented in Fig. 3b. Even for a small number of basis vectors, where the solution is not well approximated, the kinetic energy remains constant. Furthermore, the error in the kinetic energy, due to MOR, is constant in time. This is central for the robustness of the reduced system during long time-integration.

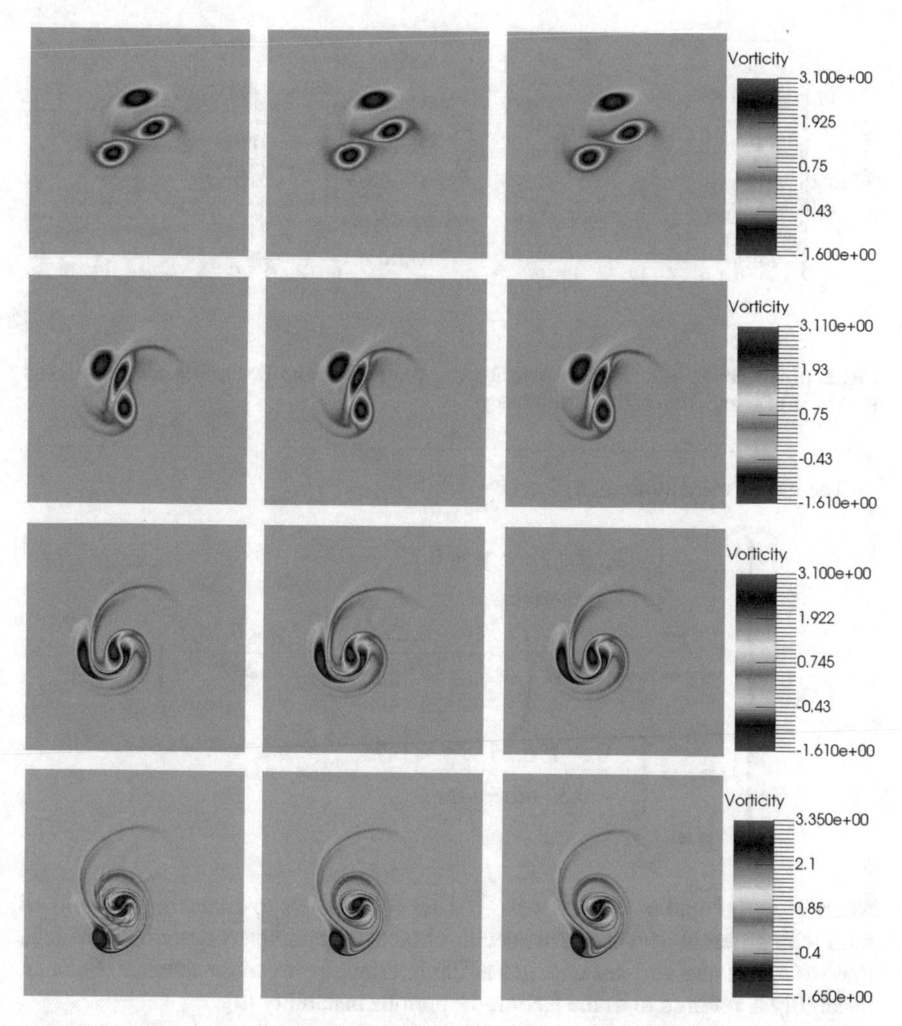

Fig. 2 Snapshots of the high-fidelity system and the reduced system at $t = \{4, 8, 12, 18\}$. From left to right: the solution of the reduced model with $k = 17$, $k = 35$ and the high fidelity solution

5.2 2D Kelvin–Helmholtz Instability

Consider the two-dimensional compressible Euler equation (26) in a periodic square box $[0, 1]^2$. Unlike the incompressible example in Sect. 5.1, a centered finite difference scheme of fourth order is used to discretize (26). The physical domain is discretized into a grid of 256×256 nodes.

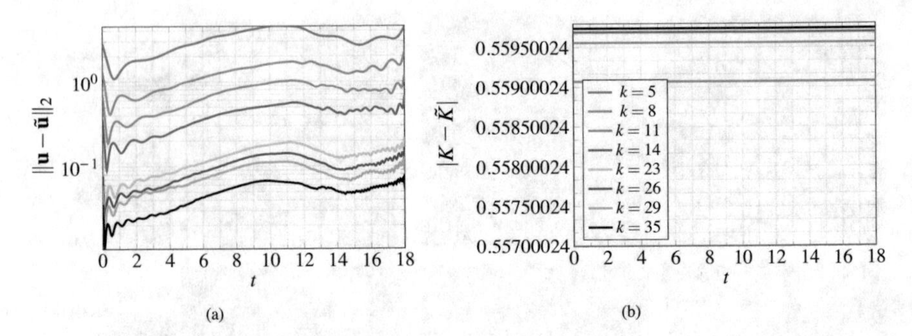

Fig. 3 (**a**) Evolution of L^2 error in velocity, between the high-fidelity system and the reduced system. (**b**) Conservation of the kinetic energy

The initial velocity field is given by

$$\begin{cases} \mathbf{r} = \begin{cases} 2, & \text{if } 0.25 < y < 0.75, \\ 1, & \text{otherwise}, \end{cases} \\ \mathbf{u}_x = a \sin(4\pi y) \left(e^{-\dfrac{(y-0.25)^2}{2\sigma^2}} + e^{-\dfrac{(y-0.75)^2}{2\sigma^2}} \right), \\ \mathbf{u}_y = \begin{cases} 0.5, & \text{if } 0.25 < y < 0.75, \\ -0.5, & \text{otherwise}, \end{cases}, \\ \mathbf{p} = 2.5, \end{cases}$$

where $a = 0.1$ and $\sigma = 5\sqrt{2} \cdot 10^{-3}$. This corresponds to contacting streams of fluid with different densities. For specific choices of parameters describing the jets, fine structures and vortices emerges at the interface between the streams. Such an instability is referred to as the Kelvin–Helmholtz instability [9].

As centered schemes are often dissipation free, resolving the discontinuous initial data requires some artificial viscosity. In the high-fidelity model, the method discussed in [42], based on a derivative-based model, is used as an artificial viscosity. However, at the level of the reduced system, this is replaced with a low pass filter on the expansion coefficients of POD basis vectors. The last 5% of the POD modes are put to zero every 20 time iterations. The reason for the different treatment is that we want to avoid the reconstruction of the derivative of the solution in the full space during the integration of the reduced system.

The fully discrete skew-symmetric form (35) is used as a time marching scheme with $\Delta t = 5 \cdot 10^{-4}$ over a period of 1 time unit. A total of 500 snapshots have been used for the computation of the basis in the offline stage.

Figure 4 illustrates that the accuracy of the method consistently improves as a higher number of POD basis modes are considered. Furthermore, the skew-symmetric form preserves the accuracy over the period of time integration (Fig. 5).

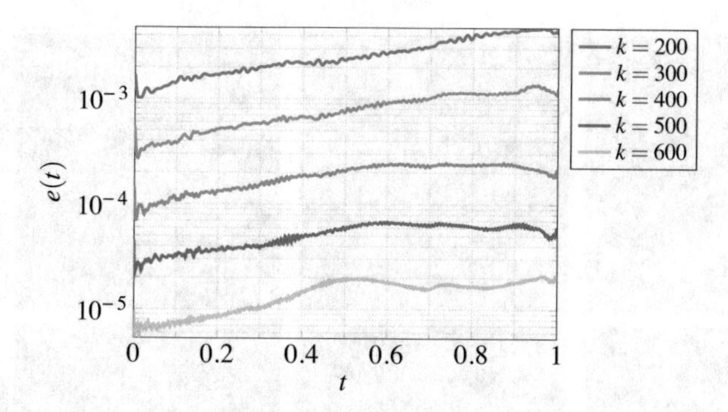

Fig. 4 Evolution in time of the error between the high-fidelity solution of the Kelvin–Helmoltz and the reduced solution for different number of basis k. As error measure we use $e(t) = \sqrt{\|\mathbf{r} - \mathbf{r}^r\|^2 + \|\mathbf{u_{xr}} - \mathbf{u}^r_{\mathbf{xr}}\|^2 + \|\mathbf{u_{yr}} - \mathbf{u}^r_{\mathbf{yr}}\|^2 + \|\mathbf{p} - \mathbf{p}^r\|^2}$

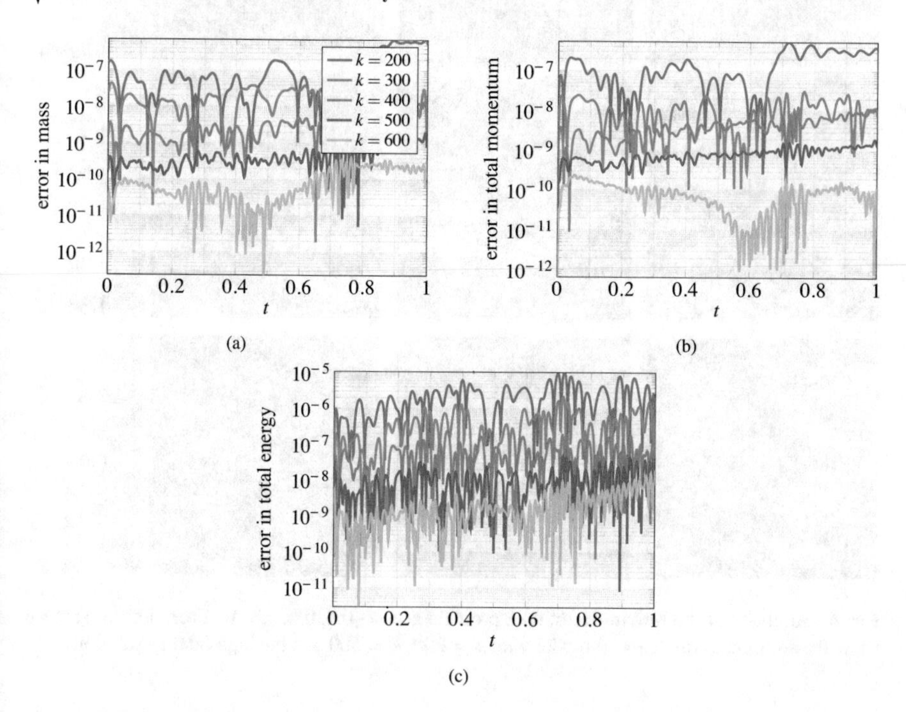

Fig. 5 Difference between the high-fidelity solution of the Kelvin–Helmholtz problem and the reduced solution of the mass (**a**), the momentum (**b**), and the total energy (**c**)

It is observed in Fig. 6 that all features of the flow are correctly represented in the reduced system, even with a low number of basis vectors.

Conservation of mass, momentum and energy is presented in Fig. 5. The accuracy of the method in approximating these invariants improves as the size of the basis is increased. Furthermore, Fig. 5c shows how the kinetic energy associated with the

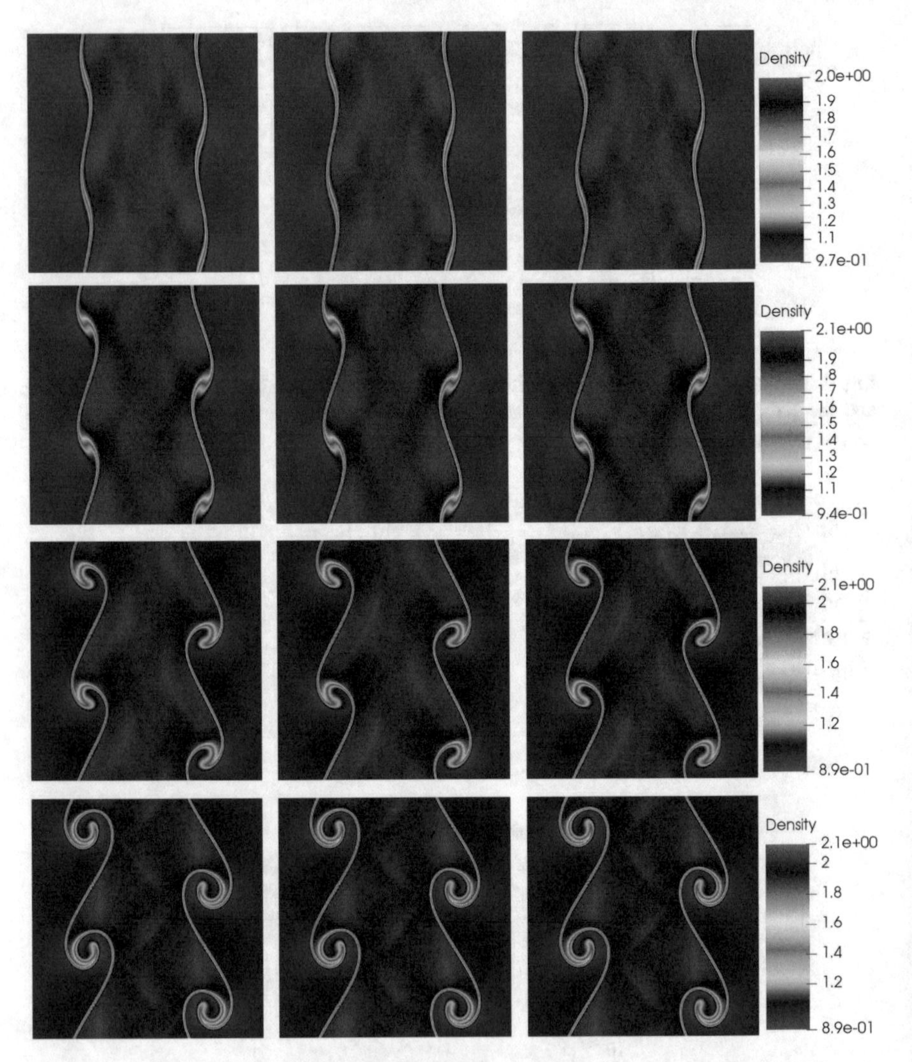

Fig. 6 Solutions of the Kelvin–Helmholtz problem at $t = \{0.4, 0.6, 0.8, 1\}$. From left to right we show the solution of the reduced model with $k = 200$, $k = 500$ and the high fidelity solution

reduced system mimics the kinetic energy of the high-fidelity system. This helps to ensure the correct evolution of kinetic energy, and thus, the internal energy.

5.3 1D Shock Problem

In this section we study the one-dimensional compressible Euler problem, (26) without viscous terms, with a steady state discontinuous solution. This is in preparation for Sect. 5.4, where development and propagation of shock waves is discussed. Here we asses how the skew-symmetric form of (26) can recover moving discontinuities at the level of the reduced system. Consider a periodic boundary conditions on $\Omega = [0, 1]$ with the initial condition

$$\begin{cases} \mathbf{r} = 0.5 + 0.2\cos(2\pi x), \\ \mathbf{u} = 1.5, \\ \mathbf{p} = 0.5 + 0.2\sin(2\pi x). \end{cases}$$

The domain is discretized into $N = 2000$ nodes and a centered finite differences scheme is used to assemble the discrete Euler equation in skew-symmetric form, as discussed in Sect. 3.3.

The fully discrete skew-symmetric form (35) is used for time integration over a time interval $[0, 0.3]$. To resolve the discontinuous solution we use an artificial viscosity with $\tau = \mu \partial u/\partial x$, where $\mu = 0.5 \cdot 10^{-4}$. Regarding the reduction procedure, 1000 snapshots of the numerical solution given by the high fidelity method have been collected.

Figure 7 shows the evolution of conserved quantities for the high-fidelity and reduced system. Here, the high-fidelity model is also considered in the divergence and advective form in addition to the skew-symmetric form. It is clear that when the reduced systems is not in skew-symmetric form, it violates conservation of mass, momentum, and energy. Even while the high-fidelity systems in divergence and advective forms are stable, the constructed reduced system is unstable, independent on the number of basis vectors. On the other hand, the skew-symmetric form yields a stable and conservative reduce system. Note that the loss in the energy associated with the skew-symmetric form, illustrated in Fig. 7b, d, f, is due to the application of an artificial viscosity.

Figure 8 shows the total error, when the reduced system captures a discontinuous solution at $t = 0.16$. It is observed that the formation of a discontinuity affects the accuracy of the method. This is expected as a sharp gradient is approximated by a relatively few POD modes. However, the method remains robust and stable during the period of time integration.

In Fig. 9 we compare the numerical artifacts of different formulations of the Euler equation. The advective formulation is not presented since it does not yield a stable reduced system. It is observed that the reduced system based on the skew-symmetric formulation accurately represent the overall behavior of the high-fidelity solution. On the other hand, a Gibbs-type error [39] appears near sharp gradients, for the reduced system based on the divergence form of the Euler equation. The well-representation of the skew-symmetric form is due the low aliasing error property.

As discussed in Sect. 4, the DEIM approximation needed for an efficient evaluation of the nonlinear components of (26), can affect the conservation properties of the skew-symmetric form. Figure 10 shows the decay of the singular values of the nonlinear snapshots. The decay of these snapshots is significantly slower than the temporal snapshots of (26). This indicates that to maintain the accuracy

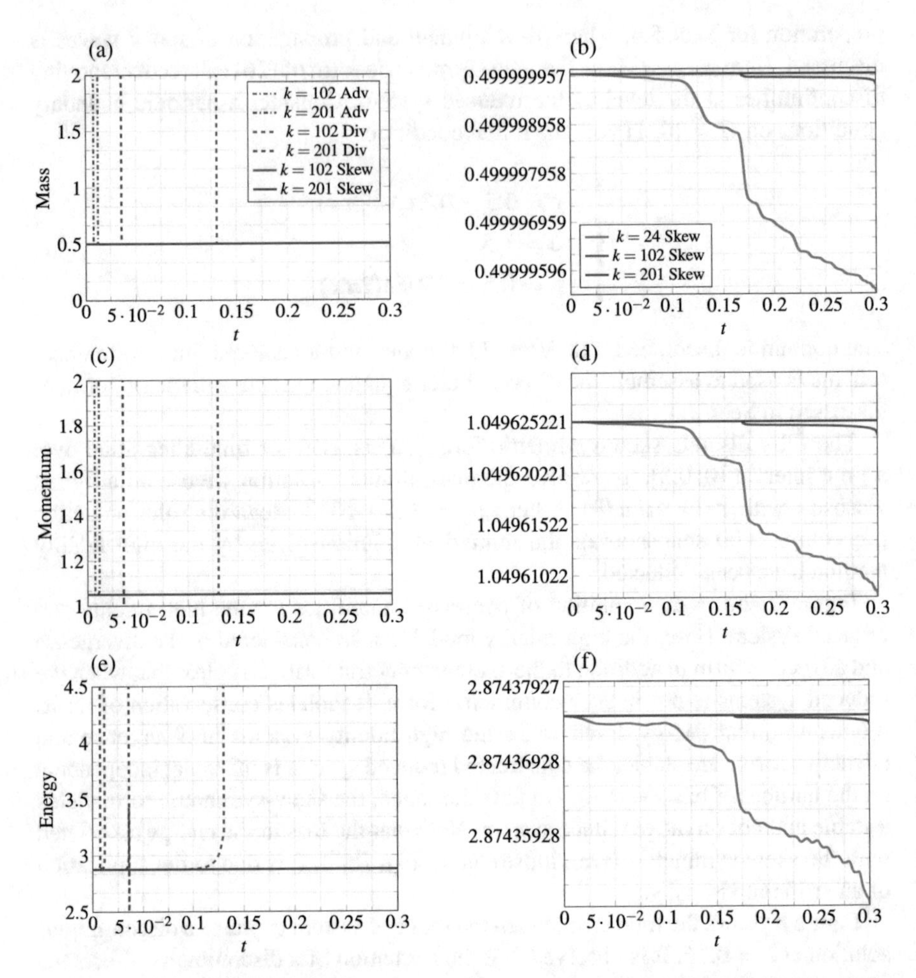

Fig. 7 **(a,c,e)** Evolution of the three conserved quantities for the reduced solution of the compressible Euler equation (mass, total momentum and total energy). The divergent, advective and skew-symmetric formulations have been considered and $k = 102, 204$ basis are used in the reduced model. **(b,d,f)** Evolution of the conserved quantity for a stable reduced model using the skew-symmetric formulation

of the reduced system, the DEIM basis should be chosen richer than the POD basis. Figure 11a, b present the error and the conservation of total energy when the DEIM is used to approximate the nonlinear term. The conservation of energy is recovered once DEIM approximates the nonlinear terms with enough accuracy. In this numerical experiment, evaluation of the nonlinear terms in (26) using the DEIM is ten times faster than the high-fidelity evaluation.

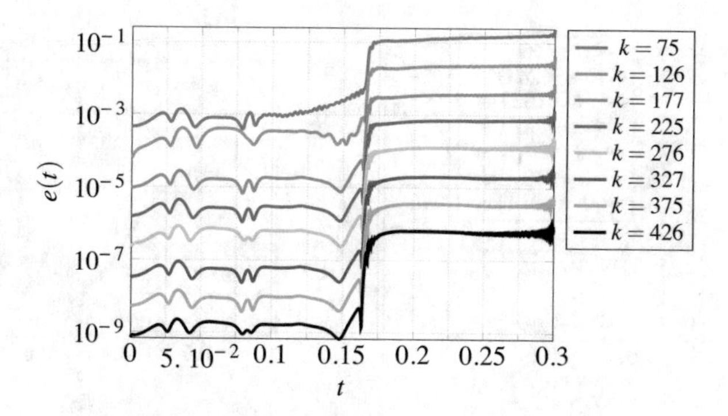

Fig. 8 Evolution in time of the error between the high-fidelity solution of the 1D compressible Euler and the reduced solution for different number of basis k. As error measure we consider $e(t) = \sqrt{\|\mathbf{r} - \mathbf{r}^r\|^2 + \|\mathbf{ur} - \mathbf{ur}^r\|^2 + \|\mathbf{p} - \mathbf{p}^r\|^2}$

5.4 *Continuous Variable Resonance Combustor*

CVRC is a model rocket combustor, designed and operated at Purdue University (Indiana, USA) to investigate combustion instabilities [43]. This setup is called the Continuously Variable Resonance Combustor (CVRC) because the length of the oxidizer injector can be varied continuously, allowing for a detailed investigation of the coupling between acoustics and combustion in the chamber [18]. The 2D/3D high-fidelity simulations of CVRC are expensive. Thus to get a fast analysis tool, a quasi-1D model has been proposed by Smith et al. [37] and further developed by Frezzotti et al. [15–17].

The CVRC consists of three parts: oxidizer post, combustion chamber and exit nozzle, as shown in Fig. 12. The oxidizer is injected from the left end of the oxidizer post and meets the fuel, injected through an annular ring around the oxidizer injector, at the back-step. The combustion happens in a region around the back-step. The combustion products flow through the chamber and exit the system from the nozzle. Both the injector and the nozzle are operated at choked condition during the experiment. The length of the oxidizer post L_{op} of the CVRC can be varied continuously, leading to different dynamics. Here, we focus on the case with $L_{op} = 14.0\,\text{cm}$, in which the combustion is unstable.

The geometry parameters of the quasi-1D CVRC with a oxidizer post length $L_{op} = 14.0\,\text{cm}$ are shown in Table 1. The back-step and the converging part of the nozzle are sinusoidally contoured to avoid a discontinuity of the radius that will invalidate the quasi-1D governing equations presented in the next subsection.

The fuel is pure gaseous methane. The oxidizer is a mixture of 42% oxygen and 58% water (per unit mass), and is injected in the oxidizer post at a temperature $T_{ox} = 1030\,\text{K}$ so that both water and oxygen are in the gaseous phase. The operating conditions are listed in Table 2.

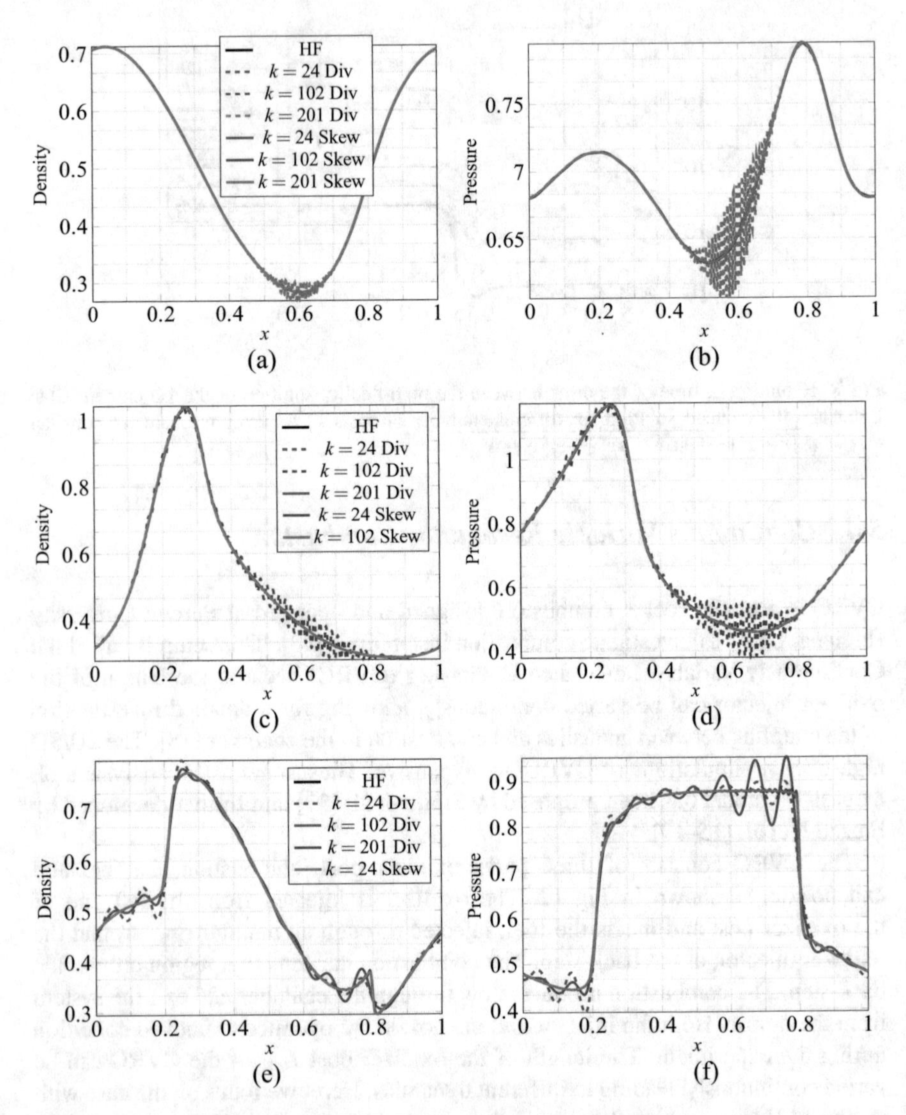

Fig. 9 Qualitative comparison between different formulations for the reduced model in terms of density and pressure at $t = 0.1$ (**a–b**), 0.3 (**c–d**) and 1s (**e–f**). Results for the advective formulation are not showed here because the related reduced solutions are unstable after a few time steps

For the combustion, we consider the one-step reaction model

$$CH_4 + 2O_2 \rightarrow CO_2 + 2H_2O.$$

We assume that the fuel reacts instantaneously to form products, allowing us to neglect intermediate species and finite reaction rates. As the equivalence ratio is

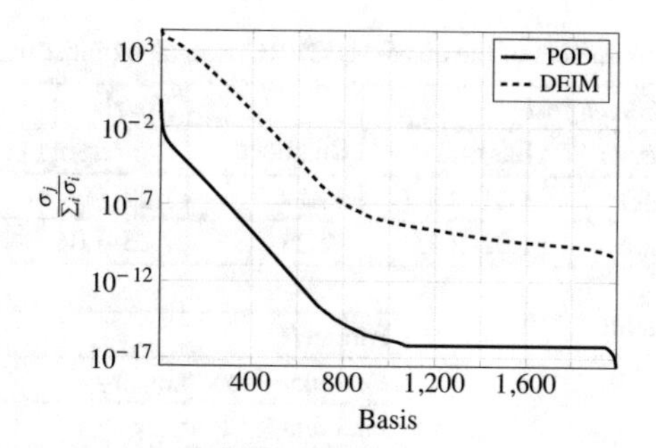

Fig. 10 Decay of the singular values of the snapshot matrix related to POD and DEIM algorithms for the 1D compressible Euler problem

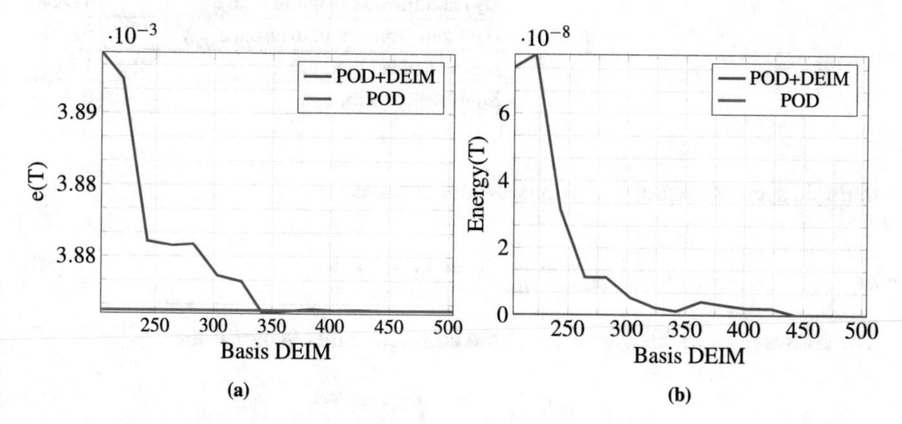

Fig. 11 Comparison between standard POD and POD with DEIM treatment of the nonlinear term in terms of the error (**a**) and the total energy (**b**)

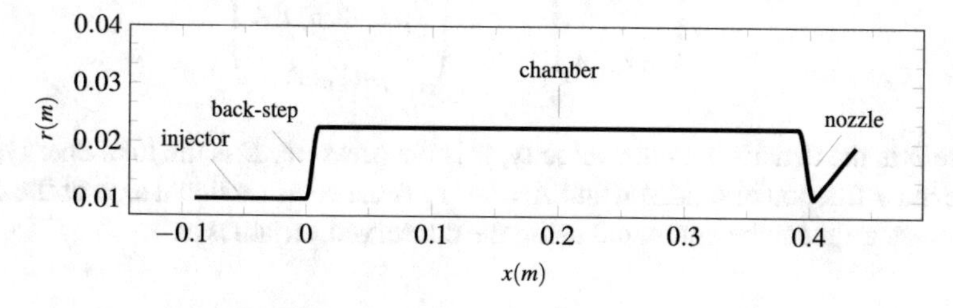

Fig. 12 Geometry of quasi-1D CVRC model

less than one, there is oxidizer left after the combustion. Therefore, only two species need to be considered: oxidizer and combustion products.

The governing equations that describe the conservation of mass, momentum, and energy of the quasi-1D CVRC flow, are the quasi-1D unsteady Euler equations for

Table 1 Geometry parameters of the quasi-1D CVRC with an oxidizer post length $L_{op} = 14$ cm

Section	Oxidizer post		Chamber	Nozzle	
	Injector	Back-step		Converging part	Diverging part
Length (cm)	12.99	1.01	38.1	1.27	3.4
Radius (cm)	1.02	1.02–2.25	2.25	2.25–1.04	1.04–1.95

Table 2 CVRC operating conditions

Parameter	Unit	Value
Fuel mass flow rate, $\dot{m}_f$	kg/s	0.027
Fuel temperature, T_f	K	300
Oxidizer mass flow rate, $\dot{m}_{ox}$	kg/s	0.32
Oxidizer temperature, T_{ox}	K	1030
O_2 mass fraction in oxidizer, Y_{O_2}	–	42.4%
H_2O mass fraction in oxidizer, Y_{H_2O}	–	57.6%
Mean chamber pressure	MPa	1.34
Equivalence ratio, E_r	–	0.8

multiple species, expressed in conservative form as

$$\frac{\partial}{\partial t}v + \frac{\partial}{\partial x}F_v = s_A + s_f + s_q. \tag{50}$$

The conserved variable vector v and the convective flux vector F are

$$v = \begin{pmatrix} \rho A \\ \rho u A \\ \rho E A \\ \rho Y_{ox} A \end{pmatrix}, \; F = \begin{pmatrix} \rho u A \\ \left(\rho u^2 + p\right) A \\ (\rho E + p)\, u A \\ \rho u Y_{ox} A \end{pmatrix}, \tag{51}$$

where ρ is the density, u is the velocity, p is the pressure, E is the total energy, Y_{ox} is the mass fraction of oxidizer, and $A = A(x)$ is the cross sectional area of the duct. The pressure p can be computed using the conserved variables as

$$E = \frac{p}{\rho(\gamma - 1)} + \frac{u^2}{2} - C_p T_{ref}, \tag{52}$$

where T_{ref} is the reference temperature and is set as 298.15 K. The temperature T is recovered from the equation of state $p = \rho R T$. The gas properties C_p, R and γ are computed as $C_p = \sum C_{pi} Y_i$, $R = \sum R_i Y_i$ and $\gamma = C_p/(C_p - R)$, respectively.

The source terms are

$$s_A = \begin{pmatrix} 0 \\ p\dfrac{dA}{dx} \\ 0 \\ 0 \end{pmatrix}, s_f = \begin{pmatrix} \dot{\omega}_f \\ \dot{\omega}_f u \\ \dot{\omega}_f \left(h_0^f + \Delta h_0^{rel}\right) \\ \dot{\omega}_{ox} \end{pmatrix}, s_q = \begin{pmatrix} 0 \\ 0 \\ q' \\ 0 \end{pmatrix}, \tag{53}$$

where $\dot{\omega}_f$ is the depletion rate of the fuel, $\dot{\omega}_{ox}$ is the depletion rate of the oxidizer, h_0^f is the total enthalpy of the fuel, Δh_0^{rel} is the heat of reaction per unit mass of fuel and q' is the unsteady heat release term. s_A accounts for area variations, s_f and s_q are related to the combustion. s_f represents the addition of the fuel and its combustion with the oxidizer, which in turn results in the creation of the combustion products. The depletion rate of the fuel is

$$\dot{\omega}_f = \frac{k_f \dot{m}_f Y_{ox} (1 + sin\xi)}{l_f - l_s}, \tag{54}$$

where

$$\xi = -\frac{\pi}{2} + 2\pi \frac{x - l_s}{l_f - l_s}, \qquad \forall\ l_s < x < l_f. \tag{55}$$

The setting of the fuel injection restricts the combustion to the region $l_s < x < l_f$. The reaction constant k_f is selected to insure that the fuel is consumed within the specified combustion zone. The depletion rate of the oxidizer is computed by

$$\dot{\omega}_{ox} = C_{o/f} \dot{\omega}_f, \tag{56}$$

where $C_{o/f}$ is the oxidizer-to-fuel ratio.

The unsteady heat release term q', also called the combustion response function, models the coupling between acoustics and combustion. Here, we use the combustion response function designed by Frezzotti et al. [15, 16], which is a function of the velocity, sampled at specific abscissa $\hat{x}$ that is almost coincident with the antinode of the first longitudinal modal shape with a time lag t_0, i.e.,

$$q'(x,t) = \alpha g(x) A(x) \left[u\left(\hat{x}, t - t_0\right) - \bar{u}\left(\hat{x}\right)\right]. \tag{57}$$

Here $\bar{u}$ is the time averaged velocity, estimated with the steady-state quasi-1D model assuming $q' = 0$, and $g(x)$ is a Gaussian distribution

$$g(x) = \frac{1}{\sqrt{2\pi\sigma^2}} \exp\left(-\frac{(x-\mu)^2}{2\sigma^2}\right), \tag{58}$$

where μ is the mean and σ is the standard deviation. The amount of heat release due to velocity oscillations is controlled by the parameter α, in (57).

The boundary conditions for the quasi-1D CVRC flow include the fixed mass flow rate and the stagnation temperature at the head-end of the oxidizer injector, and the supersonic outflow at the exit of the nozzle.

Prior to the unsteady simulation, the quasi-1D CVRC needs to be excited, which is achieved by adding a perturbation to the steady-state solution. The perturbation is added by forcing the mass flow rate with a multi-sine signal

$$\dot{m}_{ox}(t) = \dot{m}_{ox,0}\left[1 + \delta \sum_{k=1}^{K} sin\left(2\pi k \Delta f t\right)\right], \tag{59}$$

where $\dot{m}_{ox,0}$ is the oxidizer mass flow rate in Table 2, Δf is the frequency resolution and K is the number of frequencies. In this paper, $\Delta f = 50$ Hz and $K = 140$, resulting in a minimal frequency of 50 Hz and a maximal frequency of 7000 Hz. δ is required to be small to control the amplitude of the perturbation and is set as 0.1%.

The procedure of the unsteady simulation of the quasi-1D CVRC flow includes three steps:

1. Compute the steady-state solution by setting $\dot{m}_{ox} = \dot{m}_{ox,0}$ and $q' = 0$.
2. Excite the system by adding a perturbation to the oxidizer mass flow rate according to (57) and setting $q' = 0$.
3. Perform the unsteady simulation by turning on the combustion response function q' in (53) and turning off the oxidizer mass flow rate perturbation by setting $\dot{m}_{ox} = \dot{m}_{ox,0}$.

Introduction of an artificial viscosity is essential for a robust and long time-integration of (53). Common discretization schemes for (53) are often dissipative, e.g., the Lax–Friedrich scheme used in [41]. Since the skew-symmetric discretization is non-dissipative, we modify (53) as

$$\frac{\partial}{\partial t}v + \frac{\partial}{\partial x}F = s_A + s_f + s_q + d, \quad d = \left(0, \frac{\partial}{\partial x}\tau, 0, 0\right)^T, \tag{60}$$

with $\tau = \mu\partial(uA)/\partial x$, and $\mu = 6 \times 10^{-5}$. This type of artificial viscosity is chosen for its simplicity. This, however, can be replaced with a more moderate and sophisticated method.

Note that the right hand side in (60) suggests that, in general, mass, momentum, and energy is not conserved. Furthermore, the complex coupling of the variables in (53) and the non-constant adiabatic gas index prohibits the application of complex and implicit time integration schemes. Therefore, a quasi-skew-symmetric form, introduced in (17), is used for (53). It is straight-forward to check [36], for $t, s \in \mathbb{R}^N$

$$\frac{1}{2}\delta_x(st)_j + \frac{1}{2}s_j\delta_x(t)_j + \frac{1}{2}t_j\delta_x(s)_j = \frac{1}{4}\delta_x^+(s_j + s_{j-1})(t_j + t_{j-1}). \tag{61}$$

where $\delta_x(v)_j = \left(v_{j+1} - v_{j-1}\right)/\Delta x$ is centered finite difference approximation of the space derivative and $\delta_x^+(v_j) = \left(v_{j+1} - v_j\right)/\Delta x$, for some $v \in \mathbb{R}^N$. Therefore,

$$F_{i+1/2}^{\Delta}(s_j t_j, s_{j+1} t_{j+1}) = (s_j + s_{j-1})(t_j + t_{j-1}), \tag{62}$$

can be interpreted as an approximation of a quadratic flux function at the boundary of two adjacent finite volume cells. A better approximation of the flux in (62) corresponds to a higher order skew-symmetric form for a quadratic variable st in (61). We discretize the real line into N uniform cells of size Δx. A quasi-skew-symmetric form for (60) now takes the form

$$\begin{aligned}\frac{d}{dt} q_j^i + \delta^+ F_{i+1/2}^{\Delta}(q_j^i r_j^i, q_{j+1}^i r_{j+1}^i) - \delta^+ F_d^{\Delta}(d_j^i, d_{j+1}^i) + \delta^+ F_p^{\Delta}(p_j, p_{j+1})\\ = \int_{c_j} s_A + s_f + s_q \, dx.\end{aligned} \tag{63}$$

for $j = 1, \ldots, N$. Here, c_j is the jth cell, $q_j^i = \int_{c_j} v^i \, dx$ is the cell average of the ith component of v, F_p^{Δ} is the flux approximation of the pressure term, F_d^{Δ} is the flux approximation for the viscous term and $r = (u, u, u, u)^T$.

The three-stage Runge–Kutta (SSP RK3) [25] is used to integrate (60) in time. The pressure profile for the steady state, with $q' = 0$, and the pressure oscillatory mode in the unsteady phase is presented in Fig. 13a, b, respectively.

The discontinuities that appear in the solution of (60) suggests that a relatively large basis is required to resolve fine structures in the solution. Here, a POD basis is generated with $k = 200$, $k = 300$ and $k = 400$ number of basis vectors. To avoid basis changes in the reduced system, only one POD basis is considered for ρ, ρu ρE and ρY_{ox}. The explicit SSP RK3 is then used to integrated the reduced system in time, for the unsteady system. The source terms are evaluated in the high-fidelity space and projected onto the reduced space. However, in principle, the DEIM can be applied to accelerate the evaluation this component.

Figure 13c shows the approximation error of the pressure, due to MOR. It is observed that the approximation is consistently improved as the number of basis vectors increases. Furthermore, the approximate solution maintains high accuracy over a relatively long time-integration. The oscillations of pressure is demonstrated in Fig. 13d. The overall behaviour of pressure is well approximated by the reduced system. Similar results are obtained for a POD basis with higher number of modes.

We note that the discrete form of (60) is not in the full skew-symmetric form. Nonetheless, the quasi-skew-symmetric discretization offers remarkable stability preservation.

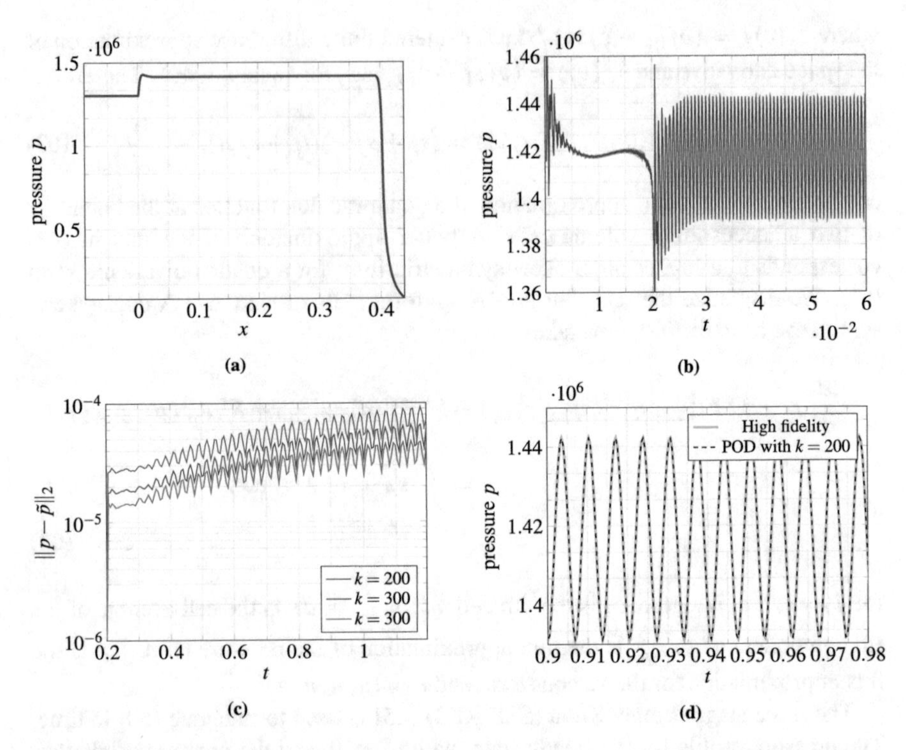

Fig. 13 (**a**) Pressure profile of the steady state. (**b**) Oscillatory mode of pressure located at $x = 0.36$ for the unsteady flow. (**c**) Relative error between the high-fidelity and approximated pressure. (**d**) Approximation of the oscillations

6 Conclusions

Conservation of nonlinear invariants are not, in general, guaranteed with conventional model reduction techniques. The violation of such invariants often result in a qualitatively wrong or unstable reduced system, even when the high-fidelity system is stable. This is particularly important for fluid flow, where conservation of the energy, as a nonlinear invariant of the system, is crucial for a correct numerical evaluation.

In this paper, we discuss that conservative properties of the skew-symmetric form for fluid flow can naturally be extended to the reduced system. Conventional MOR techniques preserves the skew-symmetry of differential operator which result in the conservation of quadratic invariants at the level of the reduced system. Furthermore, the reduced system also contains quadratic invariants with respect to the reduced variables that approximates the invariants of the high-fidelity system. This results in the construction of a physically meaningful reduced system, rather than a mere coupled system of differential equations.

Numerical experiments for the incompressible and compressible Euler equation confirms conservation of mass, momentum and energy for the reduced model with the skew-symmetric discretization. In contrast, when a non-skew-symmetric form, e.g. divergence form or advective form, is considered, MOR does not necessarily yield a stable reduced system. On the other hand the skew-symmetric form consistently yields a robust reduced system over long time-integration, even when the reduced space does not represent the high-fidelity solution accurately.

Finally, a MOR of a quasi-skew-symmetric form for the CVRC model is presented. Although this model is not in a full skew-symmetric form and an explicit Runge–Kutta method used for time-integration, we still recover a reduced model with excellent stability properties.

Acknowledgments The work was partially supported by AFOSR under grant FA9550-17-1-9241 and by SNSF under the grant number P1ELP2-175039. We also thank prof. Karen E. Willcox who provided insight and expertise that greatly assisted this work.

References

1. Ballarin, F., Manzoni, A., Quarteroni, A., Rozza, G.: Supremizer stabilization of POD-Galerkin approximation of parametrized steady incompressible Navier–Stokes equations. Int. J. Numer. Methods Eng. **102**(5), 1136–1161 (2015)
2. Barone, M.F., Kalashnikova, I., Segalman, D.J., Thornquist, H.K.: Stable Galerkin reduced order models for linearized compressible flow. J. Comput. Phys. **228**(6), 1932–1946 (2009)
3. Beattie, C., Gugercin, S.: Structure-preserving model reduction for nonlinear port-Hamiltonian systems. In 2011 50th IEEE Conference on Decision and Control and European Control Conference (CDC-ECC), pp. 6564–6569. IEEE (2011)
4. Benner, P., Breiten, T.: Interpolation-based $\mathcal{H}_2$-model reduction of bilinear control systems. SIAM J. Matrix Anal. Appl. **33**(3), 859–885 (2012)
5. Benner, P., Goyal, P.: An Iterative Model Reduction Scheme for Quadratic-Bilinear Descriptor Systems with an Application to Navier–Stokes Equations, pp. 1–19. Springer, Cham, (2018)
6. Blaisdell, G.A.: Numerical simulations of compressible homogeneous turbulence. Ph.D. thesis, Stanford University (1991)
7. Carlberg, K., Choi, Y., Sargsyan, S.: Conservative model reduction for finite-volume models. J. Comput. Phys. **371**, 280–314 (2018)
8. Carlberg, K., Tuminaro, R., Boggs, P.: Preserving Lagrangian structure in nonlinear model reduction with application to structural dynamics. SIAM J. Sci. Comput. **37**(2), B153–B184 (2015)
9. Chandrasekhar, S.: Hydrodynamic and Hydromagnetic Stability. Courier Corporation, Chelmsford (2013)
10. Chaturantabut, S., Beattie, C., Gugercin, S.: Structure-preserving model reduction for nonlinear port-Hamiltonian systems. SIAM J. Sci. Comput. **38**(5), B837–B865 (2016)
11. Desjardins, O., Blanquart, G., Balarac, G., Pitsch, H.: High order conservative finite difference scheme for variable density low Mach number turbulent flows. J. Comput. Phys. **227**(15), 7125–7159 (2008)
12. Dritschel, D.G., Zabusky, N.J.: On the nature of vortex interactions and models in unforced nearly inviscid two dimensional turbulence. Phys. Fluids **8**(5), 1252–1256 (1996)
13. Edsberg, L.: Introduction to Computation and Modeling for Differential Equations. Wiley-Interscience, New York (2008)

14. Farhat, C., Chapman, T., Avery, P.: Structure-preserving, stability, and accuracy properties of the energy-conserving sampling and weighting method for the hyper reduction of nonlinear finite element dynamic models. Int. J. Numer. Methods Eng. **102**(5), 1077–1110 (2015)
15. Frezzotti, M.L., Nasuti, F., Huang, C., Merkle, C.L., Anderson, W.E.: Quasi-1D modeling of heat release for the study of longitudinal combustion instability. Aerosp. Sci. Technol. **75**, 261–270 (2018)
16. Frezzotti, M.L., D'Alessandro, S., Favini, B., Nasuti, F.: Numerical issues in modeling combustion instability by quasi-1D Euler equations. Int. J. Spray Combust. Dyn. **9**(4), 349–366 (2017)
17. Frezzotti, M.L., Nasuti, F., Huang, C., Merkle, C., Anderson, W.E.: Determination of heat release response function from 2D hybrid RANS-LES data for the CVRC combustor. In: 51st AIAA/SAE/ASEE Joint Propulsion Conference, p. 3841 (2015)
18. Garby, R.: Simulations of flame stabilization and stability in high-pressure propulsion systems. Ph.D. thesis, INPT (2013)
19. Gugercin, S., Polyuga, R.V., Beattie, C., Van Der Schaft, A.: Structure-preserving tangential interpolation for model reduction of port-Hamiltonian systems. Automatica **48**(9), 1963–1974 (2012)
20. Haasdonk, B.: Convergence rates of the POD-Greedy method. ESAIM: Math. Model. Numer. Anal. **47**(3), 859–873 (2013)
21. Hairer, E., Lubich, C., Wanner, G.: Geometric Numerical Integration: Structure-Preserving Algorithms for Ordinary Differential Equations, vol. 31. Springer, Berlin (2006)
22. Hesthaven, J.S., Rozza, G., Stamm, B.: Certified Reduced Basis Methods for Parametrized Partial Differential Equations. SpringerBriefs in Mathematics. Springer, Berlin (2015)
23. Honein, A.E., Moin, P.: Higher entropy conservation and numerical stability of compressible turbulence simulations. J. Comput. Phys. **201**(2), 531–545 (2004)
24. Honein, A.E.: Numerical aspects of compressible turbulence simulations (2005)
25. Jiang, G.-S., Shu, C.-W.: Efficient implementation of weighted ENO schemes. J. Comput. Phys. **126**(1), 202–228 (1996)
26. Kalashnikova, I., van Bloemen Waanders, B., Arunajatesan, S., Barone, M.: Stabilization of projection-based reduced order models for linear time-invariant systems via optimization-based eigenvalue reassignment. Comput. Methods Appl. Mech. Eng. **272**, 251–270 (2014)
27. Kevlahan, N.K.R., Farge, M.: Vorticity filaments in two-dimensional turbulence: creation, stability and effect. J. Fluid Mech. **346**, 49–76 (1997)
28. Maboudi Afkham, B., Hesthaven, J.S.: Structure preserving model reduction of parametric Hamiltonian systems. SIAM J. Sci. Comput. **39**(6), A2616–A2644 (2017)
29. Maboudi Afkham, B., Hesthaven, J.S.: Structure-preserving model-reduction of dissipative Hamiltonian systems. J. Sci. Comput. **81**(1), 3–21 (2019)
30. Morinishi, Y.: Skew-symmetric form of convective terms and fully conservative finite difference schemes for variable density low-Mach number flows. J. Comput. Phys. **229**(2), 276–300 (2010)
31. Morinishi, Y., Lund, T.S., Vasilyev, O.V., Moin, P.: Fully conservative higher order finite difference schemes for incompressible flow. J. Comput. Phys. **143**(1), 90–124 (1998)
32. Morinishi, Y., Tamano, S., Nakabayashi, K.: A DNS algorithm using B-spline collocation method for compressible turbulent channel flow. Comput. Fluids **32**(5), 751–776 (2003)
33. Peng, L., Mohseni, K.: Symplectic model reduction of Hamiltonian systems. SIAM J. Sci. Comput. **38**(1), A1–A27 (2016)
34. Quarteroni, A., Manzoni, A., Negri, F.: Reduced Basis Methods for Partial Differential Equations: An Introduction. UNITEXT. Springer, Berlin (2015)
35. Reiss, J., Sesterhenn, J.: A conservative, skew-symmetric finite difference scheme for the compressible Navier–Stokes equations. Comput. Fluids **101**, 208–219 (2014)
36. Sjögreen, B., Yee, H.C.: On skew-symmetric splitting and entropy conservation schemes for the Euler equations. In: Numerical Mathematics and Advanced Applications 2009, pp. 817–827. Springer, Berlin (2010)

37. Smith, R., Ellis, M., Xia, G., Sankaran, V., Anderson, W., Merkle, C.L.: Computational investigation of acoustics and instabilities in a longitudinal-mode rocket combustor. AIAA J. **46**(11), 2659–2673 (2008)
38. Tadmor, E.: Skew-selfadjoint form for systems of conservation laws. J. Math. Anal. Appl. **103**(2), 428–442 (1984)
39. Thompson, W.J.: Fourier series and the Gibbs phenomenon. Am. J. Phys. **60**(5), 425–429 (1992)
40. Trefethen, L.N., Bau, D.: Numerical Linear Algebra. SIAM (1997)
41. Wang, Q., Hesthaven, J.S., Ray, D.: Non-intrusive reduced order modeling of unsteady flows using artificial neural networks with application to a combustion problem. J. Comput. Phys. **384**, 289–307 (2019)
42. Yu, J., Hesthaven, J.S.: A Comparative Study of Shock Capturing Models for the Discontinuous Galerkin Method. No. EPFL-ARTICLE-231188. Elsevier, Amsterdam (2017)
43. Yu, Y., Koeglmeier, S., Sisco, J., Anderson, W.: Combustion instability of gaseous fuels in a continuously variable resonance chamber (CVRC). In: 44th AIAA/ASME/SAE/ASEE Joint Propulsion Conference and Exhibit, p. 4657 (2008)

Piecewise Polynomial Approximation of Probability Density Functions with Application to Uncertainty Quantification for Stochastic PDEs

Giacomo Capodaglio and Max Gunzburger

Abstract The probability density function (PDF) associated with a given set of samples is approximated by a piecewise-linear polynomial constructed with respect to a binning of the sample space. The kernel functions are a compactly supported basis for the space of such polynomials, i.e. finite element hat functions, that are centered at the bin nodes rather than at the samples, as is the case for the standard kernel density estimation approach. This feature naturally provides an approximation that is scalable with respect to the sample size. On the other hand, unlike other strategies that use a finite element approach, the proposed approximation does not require the solution of a linear system. In addition, a simple rule that relates the bin size to the sample size eliminates the need for bandwidth selection procedures. The proposed density estimator has unitary integral, does not require a constraint to enforce positivity, and is consistent. The proposed approach is validated through numerical examples in which samples are drawn from known PDFs. The approach is also used to determine approximations of (unknown) PDFs associated with outputs of interest that depend on the solution of a stochastic partial differential equation.

1 Introduction

The problem of estimating a probability density function (PDF) associated with a given set of samples is of major relevance in a variety of mathematical and statistical applications; see, e.g., [1, 5, 6, 9, 11, 14, 20, 27, 31]. Histograms are perhaps the most popular means used in practice for this purpose. A histogram is a piecewise-constant approximation of an unknown PDF that is based on the subdivision of the sample space into subdomains that are commonly referred as bins. For simplicity, consider

G. Capodaglio · M. Gunzburger (✉)
Department of Scientific Computing, Florida State University, Tallahassee, FL, USA
e-mail: gcapodaglio@fsu.edu; mgunzburger@fsu.edu

M. D'Elia et al. (eds.), *Quantification of Uncertainty: Improving Efficiency and Technology*, Lecture Notes in Computational Science and Engineering 137,
https://doi.org/10.1007/978-3-030-48721-8_5

the case in which all bins have equal volume. The value of the histogram on any bin is given by the number of samples that lie within that bin along with a global scaling applied to ensure that the histogram has unitary integral. As an example, assume that a bounded one-dimensional sample domain Γ has been discretized into a set $\mathcal{B}^\delta = \{\mathcal{B}_\ell\}_{\ell=1}^{N_{bins}}$ of N_{bins} non-overlapping, covering, bins (i.e., intervals) of length δ. Assume also that one has in hand M samples $\{Y_m\}_{m=1}^M$ of an unknown PDF $f(Y)$. For every $\mathcal{B}_\ell \in \mathcal{B}^\delta$, the value of the histogram function $f_{\delta,M}^{histo}(Y)$ on $\mathcal{B}_\ell$ is given by

$$f_{\delta,M}^{histo}(Y)\big|_{\mathcal{B}_\ell} = \frac{1}{\delta M}\sum_{m=1}^{M} \mathcal{X}_{\mathcal{B}_\ell}(Y_m), \quad \ell = 1, \dots, N_{bins}, \tag{1}$$

where $\mathcal{X}_{\mathcal{B}_\ell}(Y)$ denotes the indicator function for $\mathcal{B}_\ell$. The above formula easily generalizes to higher-dimensional sample domains. Although the histogram is probably the easiest, with regards to implementation, means for estimating a PDF, it suffers from some limitations. For instance, being only a piecewise-constant approximation, it is discontinuous across bin boundaries and also the use of piecewise-constant approximations severely limits the discretization accuracy one can achieve.

A popular alternative method capable of overcoming the differentiability issue is *kernel density estimation* (KDE) for which the PDF is approximated by a sum of *kernel* functions $\mathcal{K}(\cdot)$ centered at the samples so that a desired smoothness of the approximation can be obtained [20]. Considering again a one-dimensional sample domain Γ, the KDE approximation is defined as

$$f_{b,M}^{kde}(Y) = \frac{1}{bM}\sum_{m=1}^{M} \mathcal{K}\Big(\frac{Y - Y_m}{b}\Big), \tag{2}$$

where b, which is referred to as the bandwidth, usually governs the decay rate of the kernel as $|Y - Y_m|$ increases. KDE approximations have been shown to be effective in a variety of applications; see, e.g., [14, 22, 29, 30]. A limitation of KDE is that the choice of the bandwidth strongly affects the accuracy [19, 28] of the approximation. More important, the naive KDE method of (2) does not scale well with the dimension M of the sample data set, i.e., for a given Y, the evaluation of the KDE approximation (2) requires M kernel evaluations so that clearly evaluating (2) becomes more expensive as the value of M grows. A way to overcome this issue is performing an appropriate binning of the sample set, as for the histogram, and appropriately transferring the information from the samples to the grid points. The kernel functions are then centered at the grid points and scalability with respect to the sample size can be achieved [12, 17]. Another method related to KDE is the one presented in [18], which is based on spline smoothing and on a finite element discretization of the estimator. This method is related to our approach because the PDF is also approximated by a finite element function. Our method presents several advantages compared to that of [18]. First, to determine the coefficients, the solution of a linear system is not required. Second, our method only involves the binning size as a smoothing parameter, whereas the one in [18] also requires the treatment of an additional smoothing parameter λ. Third, no special constraints have to be introduced to ensure positivity of the PDF approximation. In turn, the method in [18] has features that our approach does not provide such as the ability to match the sample moments up to a certain degree and the possibility of allowing aggregate data. We note that in [25], sparse-grid basis functions are substituted for the standard finite element basis functions in the method of [18], allowing for consideration of larger dimension N_Γ.

In our approach, the unknown PDF is approximated by a piecewise-polynomial function, specifically a piecewise-linear polynomial, that is defined, as are histograms, with respect to a subdivision of the sample domain Γ into bins. The piecewise-linear approximation we propose is a finite element function obtained as a linear combination of hat functions. The procedure to determine the coefficients of the linear combination is purely algebraic and can be efficiently carried out. The approach is consistent in the sense that the approximate PDF converges to the exact PDF with respect to the $L^2(\Gamma)$ norm as the number of samples tends to infinity and the volume of the bins tend to zero. Moreover, the only smoothing parameter involved is the bin size that can be related heuristically to the sample size by a simple rule.

The paper is structured as follows. In Sect. 2, the mathematical foundation of our approach is laid out; there, it is shown analytically that the proposed approximation satisfies several requirements needed for it to be considered as a PDF estimator. A numerical investigation of the accuracy and computational costs incurred by our approach is then provided. First, in Sect. 3, our method is tested and validated using sample sets associated with different types of known PDFs so that exact errors can be determined. Then, in Sect. 4, the method is applied to the estimation of *unknown* PDFs of outputs of interest associated with the solution of a stochastic partial differential equation. Finally, concluding remarks and future work are discussed in Sect. 5.

2 Piecewise-Linear Polynomial Approximations of PDFs

Let $\boldsymbol{Y}$ denote a multivariate random variable belonging to a closed, bounded parameter domain $\Gamma \subset \mathbb{R}^{N_\Gamma}$ which, for simplicity, we assume is a polytope in $\mathbb{R}^{N_\Gamma}$. The probability density function (PDF) $f(\boldsymbol{Y})$ corresponding to $\boldsymbol{Y}$ is not known. However, we assume that we have in hand a data set of M samples $\boldsymbol{Y}_m \in \Gamma$, $m = 1, \ldots, M$, of $\boldsymbol{Y} \in \Gamma$. The goal is to construct an estimator for $f(\boldsymbol{Y})$ using the given data set $\{\boldsymbol{Y}_m\}_{m=1}^M$ of samples. To this end, we use an approximating space that is popular in the finite element community [7, 10].

Let $\mathcal{B}^\delta = \{\mathcal{B}_\ell\}_{\ell=1}^{N_{bins}}$ denote a covering, non-overlapping subdivision of the sample domain Γ into N_{bins} bins.[1] Here, δ parametrizes the subdivision and may be taken as, e.g, the largest diameter of any of the bins $\{\mathcal{B}_\ell\}$. The bins are chosen to be hyper-quadrilaterals; for example, if $N_\Gamma = 2$, they would be quadrilaterals. It is also assumed that the faces of the bins are either also complete faces of abutting bins or part of the boundary of Γ. From a practical point of view, our considerations are limited to relatively small N_Γ because $N_{bins} = O(1/\delta^{N_\Gamma})$. Detailed discussion about the subdivisions we use can be found in, e.g., [7, 10].

Let $\{\widehat{\boldsymbol{Y}}_j\}_{j=1}^{N_{nodes}}$ denote the set of nodes, i.e., vertices, of $\mathcal{B}^\delta$ with N_{nodes} denoting the number of nodes of $\mathcal{B}^\delta$. Note that we also have that $N_{nodes} = O(1/\delta^{N_\Gamma})$. Based on the subdivision $\mathcal{B}^\delta$, we define the *space of continuous piecewise polynomials*

$$\mathcal{V}_\delta = \left\{ v \in C(\Gamma) \; : \; v|_{\mathcal{B}_\ell} \in \mathcal{P}_1(\mathcal{B}_\ell) \quad \text{for } \ell = 1, \ldots, N_{bins} \right\},$$

where, for hyper-quadrilateral elements, $\mathcal{P}_1(\cdot)$ denotes the space of N_Γ*-linear* polynomials, e.g., bilinear and trilinear polynomials in two and three dimensions, respectively.

A basis $\{\phi_j(\boldsymbol{Y})\}_{j=1}^{N_{nodes}}$ for $\mathcal{V}_\delta$ is given by, for $j = 1, \ldots, N_{nodes}$,

$$\phi_j(\boldsymbol{Y}) = \left\{ \phi_j(\boldsymbol{Y}) \in \mathcal{V}_\delta \; : \; \phi_j(\widehat{\boldsymbol{Y}}_{j'}) = \delta_{jj'} \quad \text{for } j' = 1, \ldots, N_{nodes} \right\},$$

where $\delta_{jj'}$ denotes the Kronecker delta function. In detail, we have that $\{\phi_j(\boldsymbol{Y})\}_{j=1}^{N_{nodes}}$ denotes the continuous piecewise-linear or piecewise N_Γ-linear Lagrangian FEM basis corresponding to $\mathcal{B}^\delta$, i.e, we have that, for $j = 1, \ldots, N_{nodes}$,

- for hyper-quadrilateral bins, $\phi_j(\boldsymbol{Y})$ is an N_Γ-linear function on each bin $\mathcal{B}_\ell$, $\ell = 1, \ldots, N_{bins}$, e.g., for $N_\Gamma = \{1, 2, 3\}$, a linear, bilinear, or trilinear function, respectively;
- $\phi_j(\boldsymbol{Y})$ is continuous on Γ;
- $\phi_j(\widehat{\boldsymbol{Y}}_j) = 1$ at the j-th node $\widehat{\boldsymbol{Y}}_j$ of the subdivision $\mathcal{B}^\delta$; and
- if $j' \neq j$, $\phi_j(\widehat{\boldsymbol{Y}}_{j'}) = 0$ at the j'-th node $\widehat{\boldsymbol{Y}}_{j'}$ of the subdivision $\mathcal{B}^\delta$.

[1] In the partial differential equation (PDE) setting, what we refer to as bins are often referred to as grid cells or finite elements or finite volumes. We instead refer to the subdomains $\{\mathcal{B}_\ell\}_{\ell=1}^{N_{bins}}$ as *bins* because that is the notation in common use for histograms which we use to compare to our approach. Furthermore, in Sect. 4, we also use finite element grids for spatial discretization of partial differential equations, so that using the notation "bins" for parameter domain subdivisions helps us differentiate between subdivisions of parameter and spatial domains. For the same reason, we use δ instead of h to parametrize parameter bin sizes because h is in common use to parametrize spatial grid sizes.

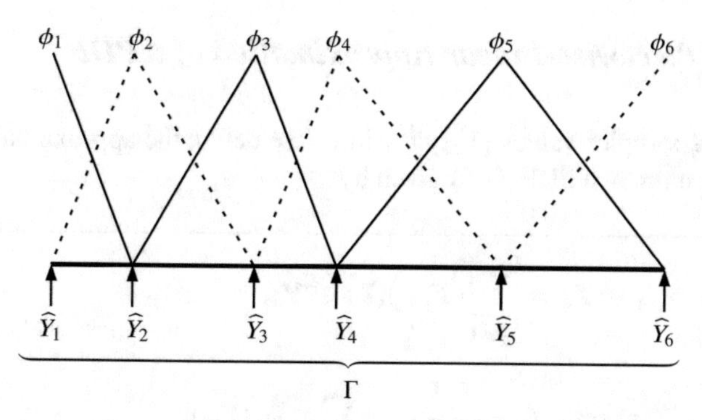

Fig. 1 The set of basis function $\{\phi_j(\boldsymbol{Y})\}_{j=1}^6$ for the case of $N_{nodes} = 6$ in one dimension. Note that the support of the basis functions is limited to the two intervals that contain the corresponding node. Here, the number of bins is $N_{bins} = 5$ and the number of bins $N_{bins,j}$ in the support of the basis functions $\phi_j(Y)$ is one for $j = 1, 6$ and two for $j = 2, 3, 4, 5$

For $j = 1, \ldots, N_{nodes}$, let $S_j(\boldsymbol{Y}) = support\{\phi_j(\boldsymbol{Y})\} \subset \Gamma$ and let $V_j = volume\{S_j(\boldsymbol{Y})\}$; note that $S_j(\boldsymbol{Y})$ consists of the union of the bins $\mathcal{B}_\ell \in \mathcal{B}^\delta$ having the node $\widehat{\boldsymbol{Y}}_j$ as one of its vertices. Thus, the basis functions have compact support with respect to Γ. An illustration of the basis functions in one dimension is given in Fig. 1. We further let $N_{bins,j}$, for $j = 1, \ldots, N_{nodes}$, denote the number of bins in $S_j(\boldsymbol{Y})$, i.e., the number of bins that share the vertex $\widehat{\boldsymbol{Y}}_j$.

Note that the approximating space $\mathcal{V}_\delta$ and the basis $\{\phi_j(\boldsymbol{Y})\}_{j=1}^{N_{nodes}}$ are in common use for the finite element discretization or partial differential equations. Details about the geometric subdivision $\mathcal{B}^\delta$, the approximation space $\mathcal{V}_\delta$, and the basis functions $\{\phi_j(\boldsymbol{Y})\}$ and their properties may be found in, e.g., [7, 10].

Below we make use of two properties of the basis $\{\phi_j(\mathbf{Y})\}_{j=1}^{N_{nodes}}$. First, we have the well-known relation

$$\sum_{j=1}^{N_{nodes}} \phi_j(\mathbf{Y}) = 1 \quad \forall \mathbf{Y} \in \Gamma. \tag{3}$$

We also have that

$$C_j = \int_\Gamma \phi_j(\boldsymbol{Y}) d\boldsymbol{Y} = \int_{S_j(\boldsymbol{Y})} \phi_j(\boldsymbol{Y}) d\boldsymbol{Y} = \sum_{\mathcal{B}_\ell \in S_j(\boldsymbol{Y})} \int_{\mathcal{B}_\ell} \phi_j(\boldsymbol{Y}) d\boldsymbol{Y}. \tag{4}$$

Note that, in general, C_j is proportional to V_j.

2.1 The Piecewise-Linear Approximation of a PDF

Given the M samples values $\{\boldsymbol{Y}_m\}_{m=1}^{M}$ in Γ, we define the approximation $f_{\delta,M}(\boldsymbol{Y}) \in \mathcal{V}_\delta$ of the unknown PDF $f(\boldsymbol{Y})$ given by

$$
\boxed{
\begin{aligned}
&f(\boldsymbol{Y}) \approx f_{\delta,M}(\boldsymbol{Y}) = \sum_{j=1}^{N_{nodes}} F_j \phi_j(\boldsymbol{Y}) \in \mathcal{V}_\delta,\\
&\qquad \text{where} \quad F_j = \frac{1}{MC_j} \sum_{\boldsymbol{Y}_m \in S_j(\boldsymbol{Y})} \phi_j(\boldsymbol{Y}_m),\\
&\qquad \text{with} \quad C_j = \int_{S_j(\boldsymbol{Y})} \phi_j(\boldsymbol{Y}) d\boldsymbol{Y}, \;\; j = 1, \ldots, N_{nodes}.
\end{aligned}
}
\tag{5}
$$

Note that only the samples $\boldsymbol{Y}_m \in S_j(\boldsymbol{Y})$, i.e., only the samples in the support $S_j(\boldsymbol{Y})$ of the basis function $\phi_j(\boldsymbol{Y})$, are used to determine F_j. We observe that the proposed estimator can be regarded as a kernel density estimator with linear binning [17], where the binning kernel and the kernel associated with a given grid point are equal to the same hat function. With this choice, the smoothing parameter of the kernel becomes the binning parameter δ, so no additional tuning of the bandwidth is necessary.

Of course, the approximate PDF (5) should be a PDF in its own right. That it is indeed a PDF is shown in the following lemma.

Lemma 1 *$f_{\delta,M}(\boldsymbol{Y}) \geq 0$ for all $\boldsymbol{Y} \in \Gamma$ and $\int_\Gamma f_{\delta,M}(\boldsymbol{Y}) d\boldsymbol{Y} = 1$.*

Proof Clearly $f_{\delta,M}(\boldsymbol{Y})$ is non-negative because it is a linear combination of non-negative functions with non-negative coefficients.

$$
\begin{aligned}
\int_\Gamma f_{\delta,M}(\mathbf{Y}) d\mathbf{Y} &= \sum_{j=1}^{N_{nodes}} F_j \int_\Gamma \phi_j(\mathbf{Y}) d\mathbf{Y}\\
&= \sum_{j=1}^{N_{nodes}} \frac{1}{M\,C_j} \sum_{m=1}^{M} \phi_j(\mathbf{Y}_m) \int_\Gamma \phi_j(\mathbf{Y}) d\mathbf{Y}\\
&= \frac{1}{M} \sum_{j=1}^{N_{nodes}} \sum_{m=1}^{M} \phi_j(\mathbf{Y}_m) = \frac{1}{M} \sum_{m=1}^{M} 1 = 1.
\end{aligned}
\tag{6}
$$

The third and fourth equalities hold because of (4) and (3), respectively. □

The next lemma is useful to prove the convergence of our approximation.

Lemma 2 *Let $f \in C^2(\Gamma)$ with $f|_{\partial\Gamma} = 0$ and let $\mathbb{E}[F_j]$ denote the expectation of F_j with respect to f. Then*

$$\left| f(\widehat{\mathbf{Y}}_j) - \mathbb{E}[F_j] \right| \leq C\delta^{\alpha}, \tag{7}$$

where the constant C does not depend on either δ or M, and α is a positive integer. If $[-1, 1]^{N_\Gamma} \cap \Gamma = [-c_1, c_1]^{N_\Gamma}$ for some positive constant c_1, then $\alpha = 2$. Otherwise $\alpha = 1$.

Proof Let $\chi_{[-1,1]}$ be the characteristic function of $[-1, 1]$ and define $\phi(Y) := (1 - |Y|)\chi_{[-1,1]}$. Let $\phi(\mathbf{Y}) := \prod_{n=1}^{N_\Gamma} \phi(Y_n)$ be defined in the usual tensor product fashion. Assuming $f|_{\partial\Gamma} = 0$, we have $C_j = \delta^{N_\Gamma}$ for all j. Then

$$\phi_j(\mathbf{Y}_m) = \phi\Big(\frac{\widehat{\mathbf{Y}}_j - \mathbf{Y}_m}{\delta}\Big), \qquad F_j = \frac{1}{\delta^{N_\Gamma} M} \sum_{m=1}^{M} \phi\Big(\frac{\widehat{\mathbf{Y}}_j - \mathbf{Y}_m}{\delta}\Big). \tag{8}$$

F_j is the value of a naive kernel density estimator of f evaluated at $\widehat{Y}_j$, with the function ϕ as a kernel. Using a standard argument for the bias of kernel density estimators we have that

$$\begin{aligned} \left| f(\widehat{\mathbf{Y}}_j) - \mathbb{E}[F_j] \right| \leq & \delta \left| \frac{\partial f(\widehat{\mathbf{Y}}_j)}{\partial \mathbf{Y}} \int_{[-1,1]^{N_\Gamma} \cap \Gamma} \phi(\mathbf{Y}')\mathbf{Y}' d\mathbf{Y}' \right| \\ & + \frac{\delta^2}{2} \left| \int_{[-1,1]^{N_\Gamma} \cap \Gamma} \mathbf{Y}'^T \frac{\partial^2 f(\widehat{\mathbf{Y}}_j)}{\partial \mathbf{Y}^2} \mathbf{Y}' d\mathbf{Y}' \right| + O(\delta^2). \end{aligned} \tag{9}$$

The above inequality proves the result for $\alpha = 1$. Thanks to the symmetry of ϕ, if $[-1, 1]^{N_\Gamma} \cap \Gamma = [-c_1, c_1]^{N_\Gamma}$ for some positive constant c_1, then $\int_{[-1,1]^{N_\Gamma} \cap \Gamma} \phi(\mathbf{Y}')\mathbf{Y}' d\mathbf{Y}' = 0$, hence the result follows with $\alpha = 2$. □

The next theorem shows that the approximate PDF obtained with our method converges to the exact PDF with respect to the $L^2(\Gamma)$ norm.

Theorem 1 *Let Γ be a polytope in $\mathbb{R}^{N_\Gamma}$ and $f \in C^2(\Gamma)$ with $f|_{\partial\Gamma} = 0$. If $f_{\delta,M}$ is the approximation of f given in* (5)*, then:*

$$\lim_{\delta \to 0} \lim_{M \to \infty} \|f - f_{\delta,M}\|_{L^2(\Gamma)} = 0.$$

Moreover, if $[-1, 1]^{N_\Gamma} \cap \Gamma = [-c_1, c_1]^{N_\Gamma}$ for some positive constant c_1, then

$$\lim_{M \to \infty} \|f - f_{\delta,M}\|_{L^2(\Gamma)} \leq C\delta^2,$$

where C is a constant that does not depend on δ or M.

Proof Let $\mathcal{I}_\delta f = \sum_{j=1}^{N_{nodes}} f(\widehat{\mathbf{Y}}_j)\phi_j$ be the finite element nodal interpolant of f, then

$$\begin{aligned}\|f - f_{\delta,M}\|_{L^2(\Gamma)} &\leq \|f - \mathcal{I}_\delta f\|_{L^2(\Gamma)} + \|\mathcal{I}_\delta f - f_{\delta,M}\|_{L^2(\Gamma)} \\ &\leq C_1\delta^2 + \|\mathcal{I}_\delta f - f_{\delta,M}\|_{L^2(\Gamma)},\end{aligned} \tag{10}$$

where C_1 is a constant that does not depend on δ [7]. Considering the second term in the above inequality, we have

$$\begin{aligned}\|\mathcal{I}_\delta f - f_{\delta,M}\|_{L^2(\Gamma)} &\leq \sqrt{\int_\Gamma \Big(\sum_{j=1}^{N_{nodes}} |f(\widehat{\mathbf{Y}}_j) - F_j|\phi_j\Big)^2} \\ &\leq \sqrt{\int_\Gamma \Big[\Big(\sum_{j=1}^{N_{nodes}} |f(\widehat{\mathbf{Y}}_j) - \mathbb{E}[F_j]|\phi_j\Big) + \Big(\sum_{j=1}^{N_{nodes}} |\mathbb{E}[F_j] - F_j|\phi_j\Big)\Big]^2} \\ &\leq \sqrt{\int_\Gamma \Big[C_2\delta^\alpha + \Big(\sum_{j=1}^{N_{nodes}} |\mathbb{E}[F_j] - F_j|\phi_j\Big)\Big]^2}.\end{aligned} \tag{11}$$

The last inequality is obtained using Lemma 2 and (3). Considering that $\mathbb{E}[F_j] = \mathbb{E}\Big[\frac{\phi_j}{\delta^{N_\Gamma}}\Big] = \frac{1}{\delta^{N_\Gamma}}\mathbb{E}\Big[\phi_j\Big]$, we have

$$\begin{aligned}|\mathbb{E}[F_j] - F_j| &= \Big|\mathbb{E}\Big[\frac{\phi_j}{\delta^{N_\Gamma}}\Big] - \frac{1}{M}\sum_{m=1}^{M}\frac{\phi_j(\mathbf{Y}_m)}{\delta^{N_\Gamma}}\Big| \\ &= \frac{1}{\delta^{N_\Gamma}}\Big|\mathbb{E}\Big[\phi_j\Big] - \frac{1}{M}\sum_{m=1}^{M}\phi_j(\mathbf{Y}_m)\Big| \leq \frac{\sigma(\phi_j)}{\delta^{N_\Gamma}\sqrt{M}},\end{aligned} \tag{12}$$

where $\sigma(\phi_j) = \sqrt{\mathbb{E}[\phi_j^2] - \mathbb{E}[\phi_j]^2} \leq C_3\sqrt{\delta^{N_\Gamma}}$ for all j, with C_3 independent of both δ and M. Hence

$$\|f - f_{\delta,M}\|_{L^2(\Gamma)} \leq C_1\delta^2 + C_2\delta^\alpha + \frac{C_3}{\sqrt{\delta^{N_\Gamma}}\sqrt{M}}, \tag{13}$$

so the first result is obtained. If $[-1,1]^{N_\Gamma} \cap \Gamma = [-c_1, c_1]^{N_\Gamma}$ for some positive constant c_1, then $\alpha = 2$ in (13), so the second result also follows taking the limit as $M \to \infty$. □

We note that the numerical examples considered below show that convergence can be obtained even for cases where the PDF is not in $C^2(\Gamma)$, even when the PDF is not differentiable or even continuous.

2.2 Numerical Illustrations

In Sect. 3, we validate our approach by approximating known joint PDFs $f(\boldsymbol{Y})$. Of course, in comparing approximations to an exact known PDF, we pretend that we have no or very little knowledge about the latter except that we have available M samples $\{\boldsymbol{Y}_m\}_{m=1}^M$ at M points $\boldsymbol{Y}_m \in \Gamma$, $m = 1, \ldots, M$. For the rest of the paper, whenever $M_1 < M_2$, then $M_1 \subset M_2$, meaning that smaller sample sets are obtained as subsets of a larger sample set. Comparing with known PDFs allows us to precisely determine errors in the approximation of the PDF determined using our method. Then, in Sect. 4, we use our method to approximate the PDFs of outputs of interest associated with the solution of a stochastic partial differential equation; in that case, the PDF is not known. All computations were performed on a Dell Inspiron 15, 5000 series laptop with the CPU {Intel(R) Core(TM) i3-4030U CPU1.90 GHz, 1895 MHz} and 8 GB of RAM.

Note that in all the numerical examples, Γ denotes a sampling domain, i.e., all samples $\{\boldsymbol{Y}_m\}_{m=1}^M$ lie within Γ. For most cases, Γ is also the support domain for the PDF. However, we also consider the case in which the support of the PDF is not known beforehand so that the sampling domain Γ is merely assumed to contain, but not be the same as, the support domain.

For simplicity, the sample space is assumed to be a bounded N_Γ-dimensional box $\Gamma = [a, b]^{N_\Gamma}$ with $a < b$. We subdivide the parameter domain Γ into a congruent set of bins $\mathcal{B}^\delta = \{\mathcal{B}_\ell\}_{k=1}^{N_{bins}}$ consisting of N_Γ-dimensional hypercubes of side $\delta = (b - a)/N_\delta$, where N_δ denotes the number of intervals in the subdivision $\mathcal{B}^\delta$ of Γ along each of the N_Γ coordinate directions. We then have that the number of bins is given by $N_{bins} = N_\delta^{N_\Gamma}$ and the number of nodes is given by $N_{nodes} = (N_\delta+1)^{N_\Gamma}$. For simplicity, we assume throughout that the components of the random variable $\boldsymbol{Y}$ are independently distributed so that the joint PDFs are given as the product of univariate PDFs; our method can also be applied in a straightforward way to cases in which the components of $\boldsymbol{Y}$ are correlated.

3 Validation Through Comparisons with Known PDFs

In this section, we assume that we have available M samples $\{\boldsymbol{Y}_m\}_{m=1}^M$ drawn from a *known* PDF $f(\boldsymbol{Y})$. The error incurred by any approximation $f_{approx}(\boldsymbol{Y})$ of the exact

PDF $f(\boldsymbol{Y})$ is measured by

$$\mathcal{E}_{f^{approx}} = \Big(\frac{1}{M}\sum_{m=1}^{M}\big(f(\boldsymbol{Y}_m) - f^{approx}(\boldsymbol{Y}_m)\big)^2\Big)^{1/2}. \tag{14}$$

In particular, we use this error measure for our approximation $f_{\delta,M}(\boldsymbol{Y})$ defined in (5).

The accuracy of approximations of a PDF, be they by histograms or by our method, depends on both M (the number of samples available) and δ (the length of the bin edges). Thus, M and δ should be related to each other in such a way that errors in (14) due to sampling and bin size are commensurate with each other. Thus, if the bin size errors in (14) are of $O(\delta^r)$ and the sampling error is of $O(M^{-1/2})$, we set

$$M = (b-a)^{2r}\delta^{-2r} = N_\delta{}^{2r}. \tag{15}$$

Thus, once a and b are specified, one can choose N_δ (or equivalently δ) and the value of M is determined by (15) or vice versa. Clearly, M increases as δ decreases.

For most of the convergence rate illustrations given below, we

$$\text{choose} \quad N_\delta = 2^{(3-r)k},\ k = 1, 2, \ldots, \quad \text{so that} \quad \delta = \frac{(b-a)}{2^{(3-r)k}} \quad \text{and} \quad M = 2^{2r(3-r)k}. \tag{16}$$

Note that neither (15) or (16) depend on the dimension N_Γ of the parameter domain but, of course, $N_{bins} = N_\delta^{N_\Gamma}$ and $N_{nodes} = (N_\delta + 1)^{N_\Gamma}$ do. If the variance of the PDF is large, one may want to increase the size of M by multiplying the term δ^{-2r} in (15) by the variance.

The computation of the coefficients F_j defined in (5) may be costly if M is large and consequently δ is small. To improve the computational efficiency, we evaluate a basis function $\phi_j(\boldsymbol{Y})$ at a sample point $\boldsymbol{Y}_m$ only if the point is within the support of $\phi_j(\boldsymbol{Y})$. However, the determination of the bin $\mathcal{B}_\ell \in \mathcal{B}^\delta$ such that $\boldsymbol{Y}_m \in \mathcal{B}_\ell$ may be expensive in case of large M and $N_\Gamma > 2$. For this task, we employ the efficient point locating algorithm described in [8].

3.1 A Smooth PDF with Known Support

For the first example we consider, we ignore the fact that we know the exact PDF we are trying to approximate. However, we assume we know the support of the PDF so the sampling domain Γ is also the support domain. We also use this example to illustrate that the number of Monte Carlo samples needed is independent of the dimension N_Γ of the parameter space.

We set $\Gamma = [-5.5, 5.5]^{N_\Gamma}$ so that $(b-a) = 11$ and assume that the components of the random vector $\boldsymbol{Y} = (Y_1, \ldots, Y_{N_\Gamma})$ are independently and identically distributed according to a truncated standard Gaussian PDF so that the joint PDF is given by

$$f(\boldsymbol{Y}) = \prod_{n=1}^{N_\Gamma} \frac{1}{\sqrt{2\pi}C_G} \exp\Big(-\frac{Y_n^2}{2}\Big) \qquad \text{for } \boldsymbol{Y} \in \Gamma = [-5.5, 5.5]^{N_\Gamma}, \tag{17}$$
$$\text{with} \quad C_G = \frac{1}{2}\big(\mathrm{erf}(5.5/\sqrt{2}) - \mathrm{erf}(-5.5/\sqrt{2})\big).$$

The scaling factor C_G is introduced to ensure that we indeed have a PDF, i.e., that the integral of $f(\boldsymbol{Y})$ over Γ is unity. Note that because the standard deviation of the underlying standard Gaussian PDF is unity, near the faces of the box $\Gamma = [-5.5, 5.5]^{N_\Gamma}$ the values of the truncated Gaussian distribution (17) are very small so that the results of this example are given to a precision such that they would not change if one considers instead the (non-truncated) standard Gaussian distribution. Also note that because the second moment of the standard Gaussian distribution is unity, the absolute error (14) is also very close to the error relative to the given PDF.

Before we use the formula (16) to relate M and δ, we first separately examine the convergence rates with respect to δ and M. To this end, to illustrate the convergence with respect to δ, we set

$$M = 10^7 \quad \text{and} \quad \delta = 11/2^k \quad \text{for} \qquad k = 3, 4, 5, 6, \tag{18}$$

so that the error due to sampling is relatively negligible. For illustrating the convergence with respect to M, we set

$$\delta = 11/2^8 \quad \text{and} \quad M = 10^k \qquad \text{for} \qquad k = 3, 4, 5, 6, \tag{19}$$

so that the error due to the bin size is relatively negligible. The plots for $N_\Gamma = 1$ in Fig. 2 illustrate a second-order convergence rate with respect to δ and a half-order convergence rate with respect to $1/M$.

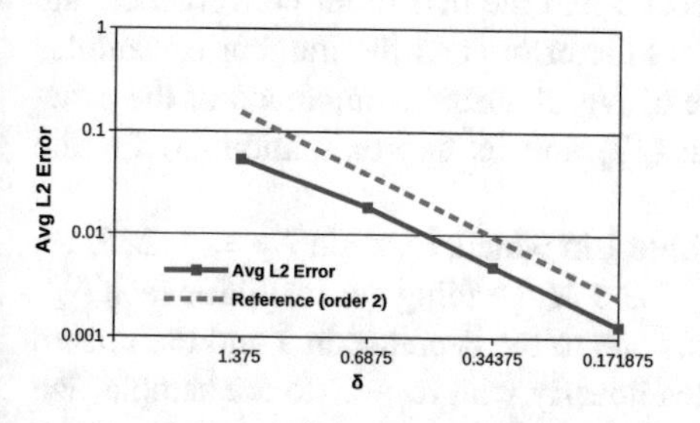

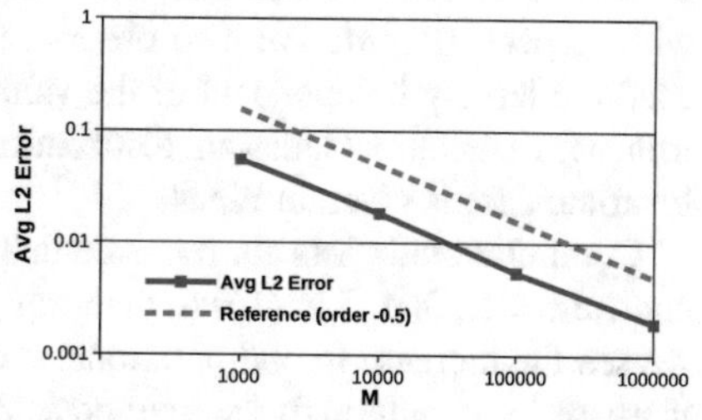

Fig. 2 Errors and convergence rates for the approximation (5) for the one-dimensional truncated standard Gaussian PDF (17). Left: second-order convergence rate with respect to δ with $M = 10^7$. Right: half-order convergence rate with respect to M with $\delta = 11/2^8$

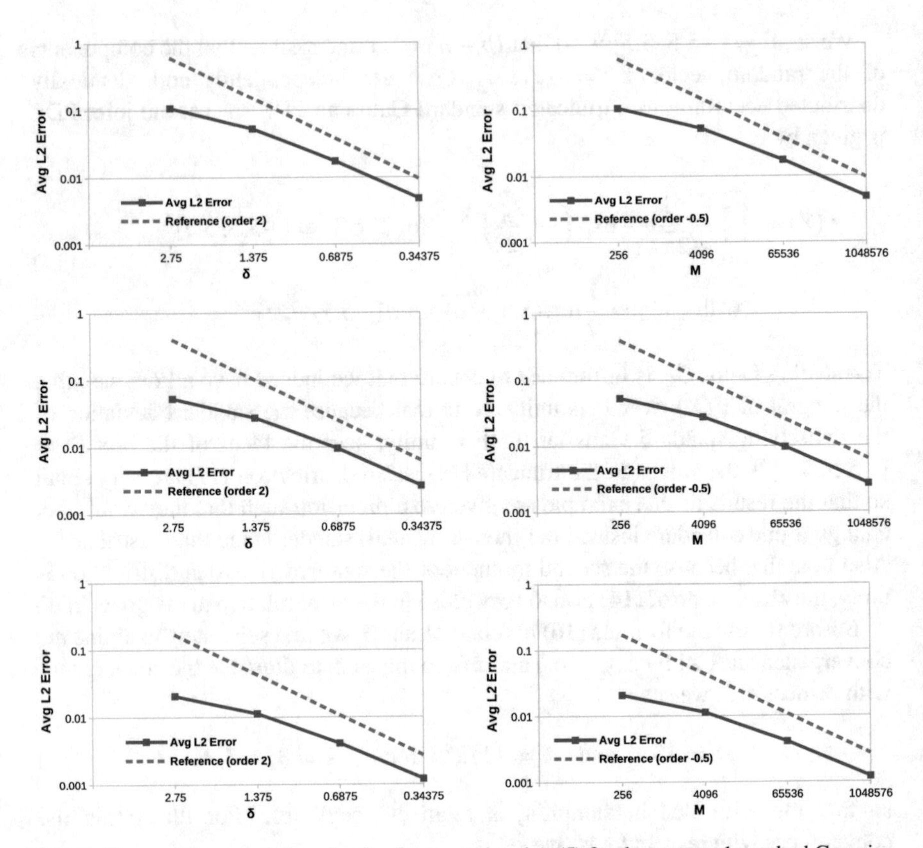

Fig. 3 Errors and convergence rates for the approximation (5) for the truncated standard Gaussian PDF (17) with M and δ related through (16). Left: convergence rates with respect to δ. Right: convergence rates with respect to M. Top to bottom: $N_\Gamma = 1, 2, 3$

We now turn to relating M and δ using the formula (16). We consider the multivariate truncated standard Gaussian PDF (17) for $N_\Gamma = 1, 2, 3$. Plots of the error vs. both δ and M are given in Fig. 3 from which we observe, in all cases, the second-order convergence rate with respect δ and the half-order convergence rate with respect to $1/M$. We also observe that the errors and the number of samples used are largely independent of the value of N_Γ. A visual comparison of the exact truncated standard Gaussian distribution (17) and its approximation (5) for the bivariate case is given in Fig. 4.

Computational costs are reported in Table 1 in which, for each $N_\Gamma = 1, 2, 3$, we choose $k = 2, 3, 4, 5$ in (16) to determine δ and M. Reading vertically for each N_Γ, we see the increase in computational costs due to the decrease in δ and the related increase in M, although the method scales linearly with respect to the sample size M. Reading horizontally so that δ and M are fixed, the increase in costs is due to the increasing number of bins and nodes as N_Γ increases. We note that our method is amenable to highly scalable parallelization not only as δ decreases and M increases,

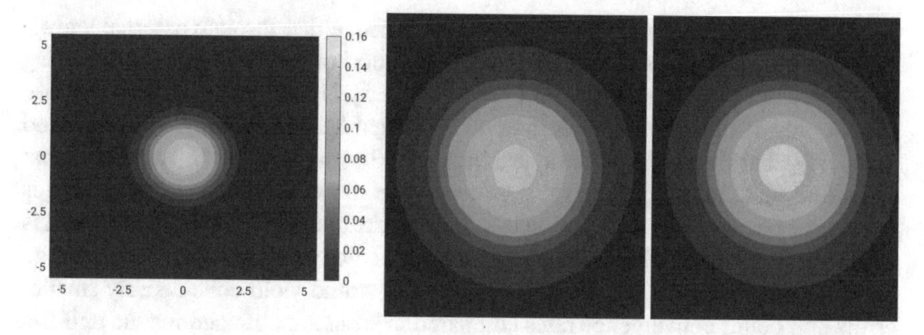

Fig. 4 Left: the approximation (5) of the bivariate truncated standard Gaussian PDF (17). Center: a zoom-in of the approximate PDF. Right: a zoom-in of the exact PDF. For these plots, $\delta = 0.34375 = 11/2^5$ and $M = 1048576 = 16^5$

Table 1 Computational time (in seconds) for determining the approximation (5) of the truncated standard Gaussian PDF (17).

				Computation time in seconds					
k	N_δ	$\delta = \frac{11}{N_\delta}$	$M = 2^{4k}$	N_{bins}	$N_\Gamma = 1$	N_{bins}	$N_\Gamma = 2$	N_{bins}	$N_\Gamma = 3$
2	4	2.75	256	4	7.710e−04	16	3.777e−02	64	1.190e−01
3	8	1.375	4096	8	1.356e−02	64	5.114e−01	512	1.955e−00
4	16	0.6875	65,536	16	2.796e−01	256	8.106e−00	4096	3.545e+01
5	32	0.34375	1,048,576	32	5.870e−00	1024	1.281e+02	32,768	1.013e+03

but also as N_Γ increases so that, through parallelization, our method may prove to be useful in dimensions higher than those considered here. When developing a parallel implementation of our method, using a point locating algorithm such as that of [8] to locate a sample on a finite element grid shared by several processors would be crucial to realize the gains in efficiency due to parallelization.

3.2 A Smooth PDF with Unknown Support

Still considering a known PDF, we now consider a case for which we not only pretend we do not know the PDF, but, in addition, we pretend we do not know its support. Specifically, we consider the uniform distribution $f(Y) = 0.5$ on $[-1, 1]$. A univariate distribution suffices for the discussions of this case; multivariate distributions can be handled by the obvious extensions of what is said here about the univariate case. We assume that we know that the support of the known PDF lies within a larger interval Γ. Of course, we may be mistaken about this so that once we examine the sample set $\{\boldsymbol{Y}_m\}_{m=1}^M$, we may observe that some of the samples fall outside of Γ. In this case we can enlarge the interval Γ until we observe that the interval spanned by smallest to largest sample values is contained within the new Γ.

We first simply assume that we have determined, either through external knowledge or by the process just described, that the support of the PDF we are considering lies somewhere within the interval $\Gamma = [-1.5, 1.5]$. Not knowing the true support, we not only sample in the larger interval Γ (so that here we have $(b-a) = 3$ and $\delta = 3/N_\delta$), but we also build the approximate PDF with respect to Γ. We remark that a uniform distribution provides a stern test when the support of the distribution is not known because that distribution is as large at the boundaries of its support as it is in the interior. Distributions that are small near the boundaries of their support, e.g., the truncated Gaussian distribution of Sect. 3.1, would yield considerably smaller errors and better convergence rates compared to what are obtained for the uniform distribution. Choosing $k = 2, 3, 4, 5$ and $r = 2$ in (16), we obtain the errors plotted in Fig. 5. Clearly, the convergence rates are nowhere near optimal. Of course, the reason for this is that by building the approximation with respect to Γ, we are not approximating the uniform distribution on $[-1, 1]$, but instead we are approximating the *discontinuous* distribution

$$f_{[-1.5,1.5]}(Y) = \begin{cases} 1 & \text{for } Y \in [-1, 1] \\ 0 & \text{for } Y \in [-1.5, -1) \text{ and } Y \in (1, 1.5]. \end{cases}$$

For comparison purposes we provide, in Fig. 6, results for the case where we use the support interval $[-1, 1]$ for both sampling and for approximation construction. Because now the PDF is smooth, in fact constant, throughout the interval in which the approximation is constructed, we obtain optimal convergence rates.

One can improve on the results of Fig. 5, even if one does not know the support of the PDF one is trying to approximate, by taking advantage of the fact that the samples obtained necessarily have to belong to the support of the PDF and therefore provide an estimate for that support. For instance, for the example we are considering, one could proceed as follows.

1. For a chosen M, sample $\{Y_m\}_{m=1}^M$ over $[-1.5, 1.5]$.
2. Determine the minimum and maximum values Y_{min} and Y_{max}, respectively, of the sample set $\{Y_m\}_{m=1}^M$.

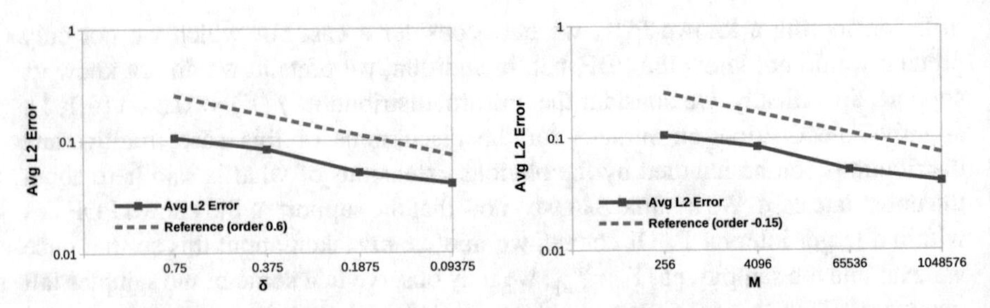

Fig. 5 For M and δ related through (16) with $r = 2$, convergence rates with respect to δ (left) and M (right) for the uniform distribution on $[-1, 1]$ but for an approximation built with respect to the larger interval $[-1.5, 1.5]$

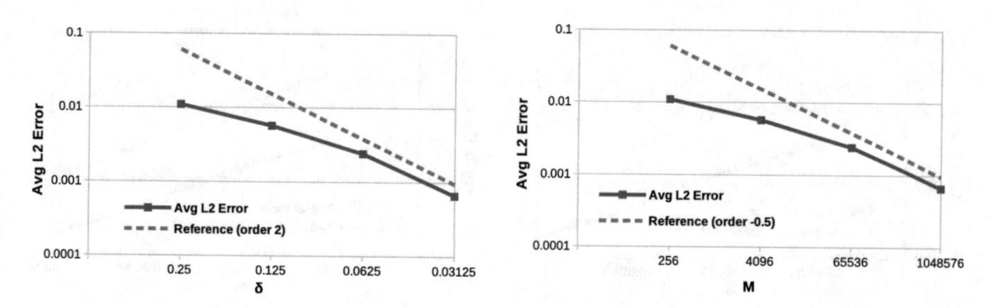

Fig. 6 For M and δ related through (16) with $r = 2$, convergence rates with respect to δ (left) and M (right) for the uniform distribution on $[-1, 1]$ for an approximation built with respect to the same interval

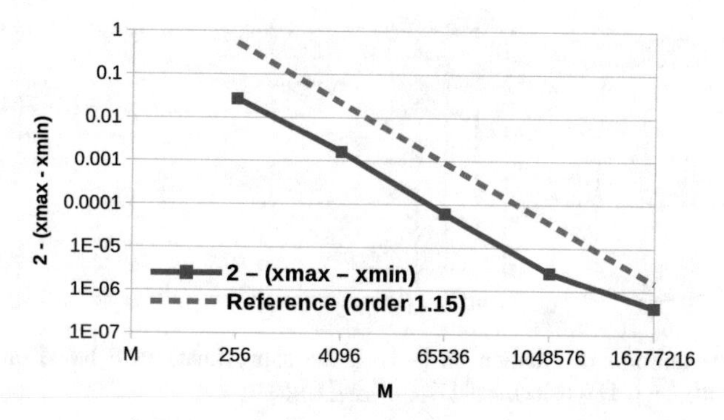

Fig. 7 For the uniform distribution and for M and δ related through (16) with $r = 2$, convergence of the approximate support interval $[Y_{min}, Y_{max}]$ to the exact support interval $[-1, 1]$

3. Choose the number of bins N_{bins} and set $\delta = (Y_{max} - Y_{min})/N_{bins}$.
4. Build the approximation over the interval $[Y_{min}, Y_{max}]$ with a bin size δ.

It is reasonable to expect that as M increases, the interval $[Y_{min}, Y_{max}]$ becomes a better approximation to the true support interval $[-1, 1]$. Figure 7 illustrates the convergence of $[Y_{min}, Y_{max}]$ to $[-1, 1]$. Note that because $[Y_{min}, Y_{max}] \subset [-1, 1]$, the exact PDF is continuous within $[Y_{min}, Y_{max}]$. Thus, it is also reasonable to expect that because the approximate PDF is built with respect to an interval which is contained within the support of the exact PDF, that there will be an improvement in the accuracy of that approximation compared to that reported in Fig. 5 and, in particular, that as one increases N_{bins} so that δ decreases and M increases, better rates of convergence will be obtained. Figure 8 corresponds to the application of this procedure and shows the substantially smaller errors and substantially higher convergence rates compared to that reported in Fig. 5.

A visual comparisons of the approximations obtained using the smallest δ/largest M pairing corresponding to Figs. 5, 6, and 8 are given in Fig. 9. The defects resulting from the use of the interval $[-1.5, 1.5]$ for constructing the approximation of a uniform PDF that has support on the interval $[-1, 1]$ are clearly evident. On

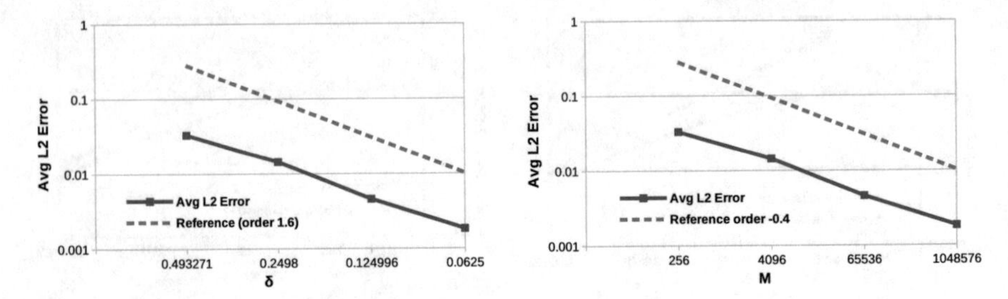

Fig. 8 For M and δ related through (16) with $r = 2$, convergence rates with respect to δ (left) and M (right) for the uniform distribution on $[-1, 1]$ with approximations built with respect to the approximate support interval $[Y_{min}, Y_{max}]$

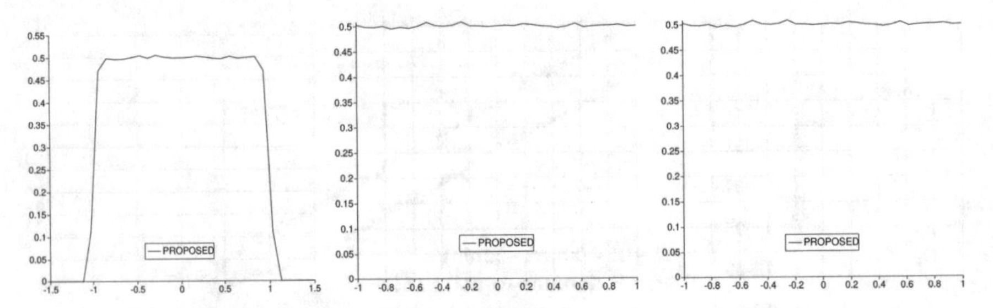

Fig. 9 For the uniform distribution on $[-1, 1]$, the approximate PDF based on sampling in $[-1.5, 1.5]$ (left), $[-1, 1]$ (center), and $[Y_{min}, Y_{max}]$ (right)

the other hand, using the support interval approximation process outlined above results in a visually identical approximation as that obtained using the correct support interval $[-1, 1]$. Note that for the smallest value of δ, we have that Y_{min} approximates -1 and Y_{max} approximates 1 to seven decimal places.

3.3 A Non-smooth PDF

We next consider the approximation of a non-smooth PDF. Specifically, we consider the centered truncated Laplace distribution

$$f(Y) = \frac{1}{3C_L} \exp\left(\frac{-|Y|}{1.5}\right) \tag{20}$$

over $\Gamma = [-5.5, 5.5]$, where $C_L = 1 - \exp(-5.5/1.5)$ is a scaling factor that ensures a unitary integral of the PDF over Γ. Here, the support domain and sampling domain are the same. This distribution is merely continuous. i.e., its derivative is discontinuous at $Y = 0$, so one cannot expect optimally accurate approximations. However, as illustrated in Fig. 10, it seems the approximation does converge, but

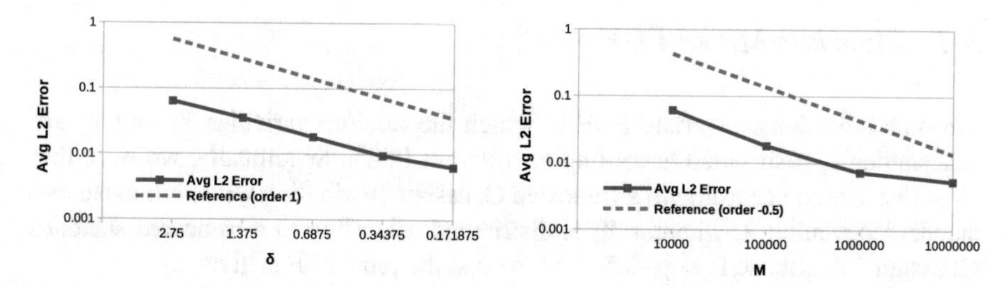

Fig. 10 Errors and convergence rates for the approximation (5) of the Laplace distribution (20). Left: convergence with respect to δ with $M = 10^7$ is fixed. Right: convergence with respect to M with $h = 11/2^{12}$ fixed

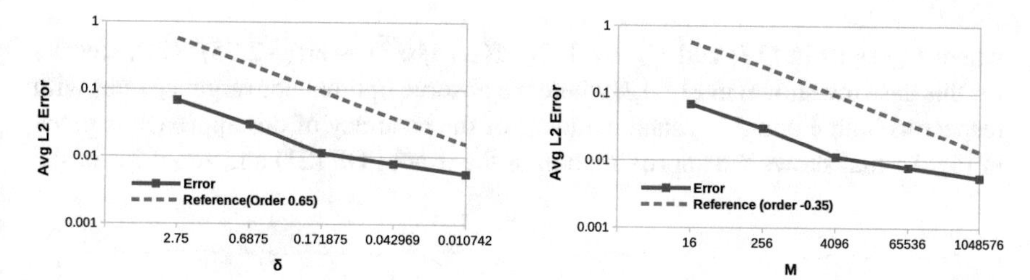

Fig. 11 For M and δ related through (16) with $r = 1$, convergence rates with respect to δ (left) and M (right) for the Laplace distribution with approximations built with respect to the approximate support interval $\Gamma = [-5.5, 5.5]$

at a lower rate with respect to δ and at the optimal rate with respect to M. The latter is not surprising because Monte Carlo sampling is largely impervious to the smoothness or lack thereof of the function being approximated.

Whenever there is any information about the smoothness of the PDF, one can choose an appropriate value of r in (16). Alternately, possibly through a preliminary investigation, one can estimate the convergence rate of the approximation (5). In the case of the Laplace distribution which is continuous but not continuously differentiable, one cannot expect a convergence rate greater than one. Selecting $r = 1$ in (16) to relate M and δ, we obtain the results given in Fig. 11 which depicts rates somewhat worse that we should perhaps expect.

The Laplace distribution, although not globally C^2, is piecewise smooth, with failure of smoothness only occurring at the symmetry point of the distribution. For example, for the particular case of the centered distribution (20), the distribution is smooth for $Y > 0$ and $Y < 0$. Thus, in general, one could build two separate, optimally accurate approximations, one for the right of the symmetry point and the other for the left of that point. Of course, doing so requires knowledge of where that point is located. If this information is not available, then one can estimate the location of that point by a process analogous to what we described in Sect. 3.2 for distributions whose support is not known a priori. Such a process can be extended to distributions with multiple points at which smoothness is compromised.

3.4 Bivariate Mixed PDF

We now consider a bivariate PDF in which the random variables Y_1 and Y_2 are independently distributed according to different PDFs. Specifically, we have that Y_1 is distributed according to a truncated Gaussian distribution with zero mean and standard deviation 2, whereas Y_2 is distributed according to a truncated standard Gaussian. We choose $\Gamma = [-5.5, 5.5]^2$ so that the joint PDF is given by

$$f(\mathbf{Y}) = \frac{1}{\sqrt{8\pi}C'_G} \exp\Big(-\frac{Y_1^2}{8}\Big)\frac{1}{\sqrt{2\pi}C_G} \exp\Big(-\frac{Y_2^2}{2}\Big), \tag{21}$$

where C_G is as in (17) and $C'_G = 1/2(\text{erf}(2.75/\sqrt{2}) - \text{erf}(-2.75/\sqrt{2}))$. Results for this case are shown in Fig. 12, where we observe optimal convergence rates with respect to both δ and M. Visual evidence of the accuracy of our approach is given in Fig. 13 that shows the approximation of the exact PDF (21) and zoom-ins of the

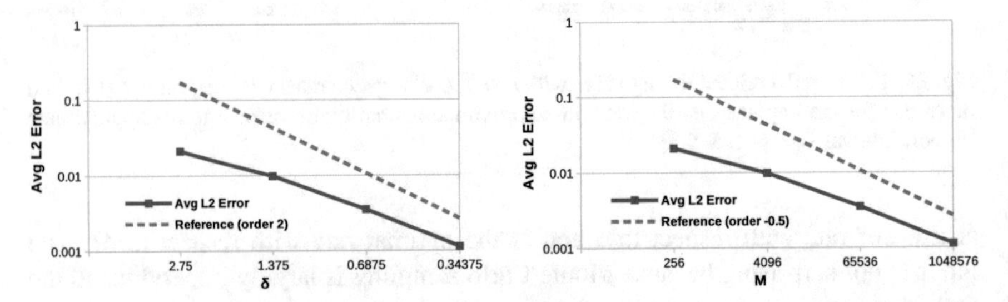

Fig. 12 Errors and convergence rates for the approximation (5) of the bivariate mixed-distribution PDF (21). Left: convergence with respect to δ with $M = 10^7$ is fixed. Right: convergence with respect to M with $h = 11/2^8$ fixed

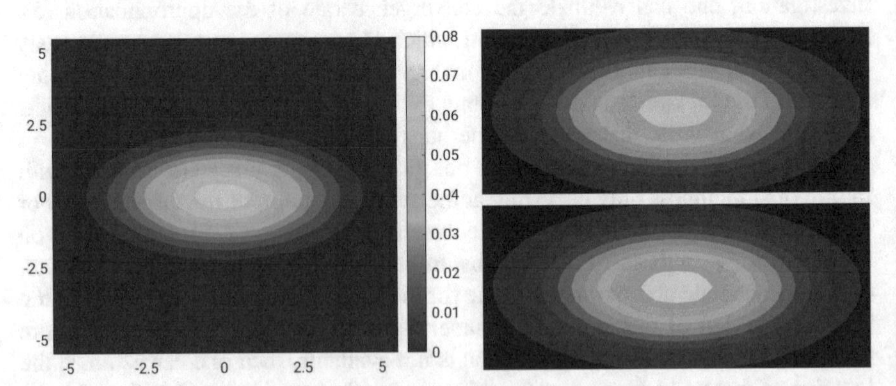

Fig. 13 Left: the approximation (5) of the bivariate mixed-distribution PDF (21). Right-top: a zoom-in of the exact PDF. Right-bottom: a zoom-in of the approximate PDF. For these plots, $\delta = 0.34375 = 11/2^5$ and $M = 1048576 = 16^5$

approximate and approximate PDFs. Computational times are very similar to those for $N_\Gamma = 2$ in Table 1 so that they are not provided here.

4 Application to an Unknown PDFs Associated with a Stochastic PDE

In this section, we consider the construction of approximations of the PDF of outputs of interest that depend on the solution of a stochastic PDE. In general, such PDFs are unknown a priori.

The boundary value problem considered is the stochastic Poisson problem

$$\begin{cases} -\nabla\cdot\big(\kappa(\mathbf{x},\mathbf{Z})\,\nabla u(\mathbf{x},\mathbf{Z})\big) = 1 & \text{for } \mathbf{x}\in D,\ \mathbf{Z}\in\Gamma_{input} \\ u(\mathbf{x},\mathbf{Z}) = 0 & \text{for } \mathbf{x}\in\partial D,\ \mathbf{Z}\in\Gamma_{input}, \end{cases} \tag{22}$$

where $D \subset \mathbb{R}^d$ denotes a spatial domain with boundary ∂D and $\Gamma_{input} \subset \mathbb{R}^{N_{\Gamma_{input}}}$ is the sample space for the *input* random vector variable $\mathbf{Z}$ which we assume is distributed according to a *known input joint PDF* $f_{input}(\mathbf{Z})$.

For the coefficient function $\kappa(\mathbf{x},\mathbf{Z})$, we assume that there exists a positive lower bound $\kappa_{\min}$ almost surely on Γ_{input} for all $\mathbf{x}\in D$. We also assume that $\kappa(\mathbf{x},\mathbf{Z})$ is measurable with respect to $\mathbf{Z}$. It is then known that the system (22) is well posed almost surely for $\mathbf{Z}\in\Gamma_{input}$; see, e.g., [4, 16, 23, 24] for details.

Stochastic Galerkin Approximation of the Solution of the PDE

We assume that we have in hand an approximation $u_{approx}(\mathbf{x},\mathbf{Z})$ of the solution $u(\mathbf{x},\mathbf{Z})$ of the system (22). Specifically, spatial approximation is effected via a piecewise-quadratic finite element method [7, 10]. Because this aspect of our algorithm is completely standard, we do not give further details about how we effect spatial approximation. For stochastic approximation, i.e., for approximation with respect to the parameter domain Γ_{input}, we employ a spectral method. Specifically, we approximate using global orthogonal polynomials, where orthogonality is with respect to Γ_{input} and the known PDF $f_{input}(\mathbf{Z})$. Thus, if $\{\Phi_j(\mathbf{x})\}$ denotes a basis for the finite element space used for spatial approximation and $\{\Psi_i(\mathbf{Z})\}$ denotes a basis for the spectral space used for approximation with respect to $\mathbf{Z}$, with the stochastic Galerkin method (SGM) we obtain an approximation of the form

$$\begin{aligned} u_{approx}(\mathbf{x},\mathbf{Z}) &= \sum_i\sum_j U_{i,j}\,\Phi_j(\mathbf{x})\Psi_i(\mathbf{Z}) \\ &= \sum_i u_i(\mathbf{x})\Psi_i(\mathbf{Z}), \qquad \text{where} \quad u_i(\mathbf{x}) = \sum_j U_{i,j}\Phi_j(\mathbf{x}). \end{aligned} \tag{23}$$

Note that once the SGM approximation (23) is constructed, it may be evaluated at any point $\mathbf{x} \in D$ and for any parameter vector $\mathbf{Z} \in \Gamma_{input}$. Having chosen the types of spatial and stochastic approximations we use, the approximate solution $u_{approx}(\mathbf{x}, \mathbf{Z})$, i.e., the set of coefficients $\{U_{i,j}\}$, is determined by a stochastic Galerkin projection, i.e., we determine an approximation to the Galerkin projection of the exact solution with respect to both the spatial domain D and the parameter domain Γ_{input}. This approach is well documented so that we do not dwell on it any further; one may consult, e.g., [2, 3, 9, 15, 16], for details. Note that once the surrogate (23) for the solution of the PDE is built, it can be used through direct evaluation to cheaply determine an approximation of the solution of the PDE for any $\mathbf{x} \in D$ and any $\mathbf{Z} \in \Gamma_{input}$ instead of having to do a new approximate PDE solve for any new choice of $\mathbf{x}$ and $\mathbf{Z}$.

The error in $u_{approx}(\mathbf{x}, \mathbf{Z})$ depends on the grid-size parameter h used for spatial approximation and the degree of the orthogonal polynomials used for approximation with respect to the input parameter vector $\mathbf{Z}$. In practice, these parameters should be chosen so that the two errors introduced are commensurate. However, here, because our focus is on stochastic approximation and because throughout we use the same finite element method for spatial approximation, we use a small enough spatial grid size so that the error due to spatial approximation is, for all practical purposes, negligible compared to the errors due to stochastic approximation.

Outputs of Interest Depending on the Solution of the PDE

In our context, outputs of interest are spatially-independent functionals of the solution $u(\mathbf{x}, \mathbf{Z})$ of the system (22). Here, we consider the two specific functionals

$$Y(\mathbf{Z}) = \sum_i \Big(\frac{1}{|D|} \int_D u_i(\mathbf{x}) d\mathbf{x} \Big) \Psi_i(\mathbf{Z}), \tag{24}$$

or

$$Y(\mathbf{Z}) = \sum_i \Big(\int_D u_i^2(\mathbf{x}) d\mathbf{x} \Big) \Psi_i(\mathbf{Z}), \tag{25}$$

i.e., the average of the approximate solution and a functional of the integral of the square u_i, respectively, over the spatial domain D. The output of interest Y is a random variable that depends on the random input vector $\mathbf{Z}$. Note that although we consider scalar outputs of interest, the extension to vector-valued outputs of interest is straightforward. Note that throughout, all outputs of interest are standardized, namely, they are translated by their mean and scaled by their standard deviation. Hence, we seek approximations of the PDFs of standardized outputs of interest.

It is important to keep in mind that we are dealing with two random variables. First, we have the *input* random variable $\mathbf{Z}$ having a known PDF $f_{input}(\mathbf{Z})$ supported over the known parameter domain Γ_{input}. Second, we have the *output* random variable Y having an unknown PDF $f_{output}(Y)$ supported over an unknown parameter domain Γ_{output}. Although we do not know the output PDF, that is what we want to construct, so that further samples of the output of interest Y can be obtained by simple direct sampling of the PDF $f_{output}(Y)$ for Y.

Thus, the task at hand is

given the known PDF $f_{input}(\mathbf{Z})$ of the random input $\mathbf{Z} \in \Gamma_{input}$, determine an approximation of the unknown *PDF $f_{output}(Y)$ of an output of interest $Y(\mathbf{Z})$.*

To deal with this task, one simply follows the recipe:

1. construct the stochastic Galerkin approximation $u_{approx}(\mathbf{x}, \mathbf{Z})$ given in (23) of the solution $u(\mathbf{x}, \mathbf{Z})$ of the PDE (22);
2. choose M samples $\{\mathbf{Z}_m\}_{m=1}^{M}$ of the input random vector $\mathbf{Z}$ according to the known given input PDF $f_{input}(\mathbf{Z})$;
3. determine M samples of the approximate solution $\{u_{approx}(\mathbf{x}, \mathbf{Z}_m)\}_{m=1}^{M}$ of the PDE (22) by evaluating (23) at each of the samples $\mathbf{Z}_m$ chosen in step 2;
4. use the approximate solution samples obtained in step 3 to determine M samples $\{Y_m = Y(\mathbf{Z}_m)\}_{m=1}^{M}$ of an output of interest from, e.g., (24) or (25);
5. use the output of interest samples $\{Y_m\}_{m=1}^{M}$ obtained in step 4 to determine, using (5), an approximation to the output PDF $f_{output}(Y) = f_{output}\big(Y(\mathbf{Z})\big)$.

Of course, because the exact PDF $f_{output}(Y)$ is not known, we cannot use (14) to compute errors. Thus, as a surrogate for the exact PDF, we use a histogram approximation $f_{\widehat{M},\widehat{\delta}}^{histo}(Y)$ obtained with a large number of bins (and therefore a very small $\widehat{\delta}$ and a large number of samples $\widehat{M}$), where "large" is relative to what is used in obtaining approximations using, e.g., (5). Thus, we now use

$$\mathcal{E}_{f_{approx}} = \left(\frac{1}{M}\sum_{m=1}^{M}\left(f_{\widehat{M},\widehat{\delta}}^{histo}(Y_m) - f_{approx}(Y_m)\right)^2\right)^{1/2} \tag{26}$$

as a measure of the error in any approximation $f_{approx}(Y)$ of $f_{output}(Y)$ that involves $M \ll \widehat{M}$ samples and a bin width $\delta \gg \widehat{\delta}$.

Illustrative Results for a Specific Choice for the Coefficient of the PDE

For the coefficient function in the PDE (22), we choose

$$\kappa(\mathbf{x}, \mathbf{Z}) = \kappa_{\min} + \exp\big(\gamma(\mathbf{x}, \mathbf{Z})\big) \quad \text{with} \quad \gamma(\mathbf{x}, \mathbf{Z}) = \mu + \sum_{n=1}^{N_{input}} \sqrt{\lambda_n}\psi_n(\mathbf{x})Z_n, \tag{27}$$

where $\{\lambda_n, \psi_n(\mathbf{x})\}$ are the eigenpairs of a given covariance function, with the eigenvalues arranged in non-increasing order and the random variables $\{Z_n\}_{n=1}^{N_{input}}$ are independent and identically distributed standard Gaussian variables. One recognizes that $\gamma(\mathbf{x}, \mathbf{Z})$ is a truncated Karhunen-Loève (KL) expansion corresponding to a correlated Gaussian random field with mean μ [13, 21, 26].

Here, we consider the specific covariance function

$$C_\gamma(\mathbf{x}, \mathbf{x}') = \sigma_\gamma^2 \exp\Big[-\frac{1}{L}\Big(\sum_{i=1}^{d} |x_i - x_i'|\Big)\Big], \tag{28}$$

where σ_γ^2 denotes a variance and $0 < L \leq \mathrm{diam}(D)$ a correlation length. The eigenpairs satisfy the generalized eigenvalue problem

$$\int_D C_\gamma(\mathbf{x}, \mathbf{x}')\psi_n(\mathbf{x}')d\mathbf{x}' = \lambda_n \psi_n(\mathbf{x}). \tag{29}$$

The eigenpairs $\{\lambda_n, \psi_n\}$ are approximately determined by means of a Galerkin projection as in [9, 16].

The N_{input} components of the input stochastic variable $\mathbf{Z}$ are independent and are all distributed according to a standard Gaussian PDF. Thus, in the spectral method discretization with respect to the input stochastic variable $\mathbf{Z}$, we use tensor products of Hermite polynomials as the orthonormal basis $\{\Psi_j(\mathbf{Z})\}$. Due to the orthogonality of these polynomials with respect to the Gaussian PDF, this choice results in very substantial savings in both the assembly and solution aspects of the discretized SGM system. Details can be found in, e.g., [9, 16].

For the numerical tests, we choose the spatial domain D to be the unit square $[0, 1]^2$, the correlation length $L = 0.1$, $\kappa_{\min} = 0.01$, and $\mu = 0$.

Output of Interest (24) Here, we consider the approximation (5) of the PDF $f_{output}(Y)$ of the standardized output of interest Y given by (24).

We consider $\Gamma_{output} = [-5, 3]$, and in (27) and (28) we set $N_{input} = 2$, $\sigma_\gamma = 1.4$. Because $\mathbf{Z} = (Z_1, Z_2)$ with Z_1 and Z_2 being standard Gaussian variables, $\Gamma_{input} = (-\infty, \infty)^{N_{input}}$. A plot of the output of interest (24) as a function of the input variables $\mathbf{Z} = (Z_1, Z_2)$ is given, for $\mathbf{Z} \in [-3.5.3.5]^2$, in Fig. 14 (left). Plots of

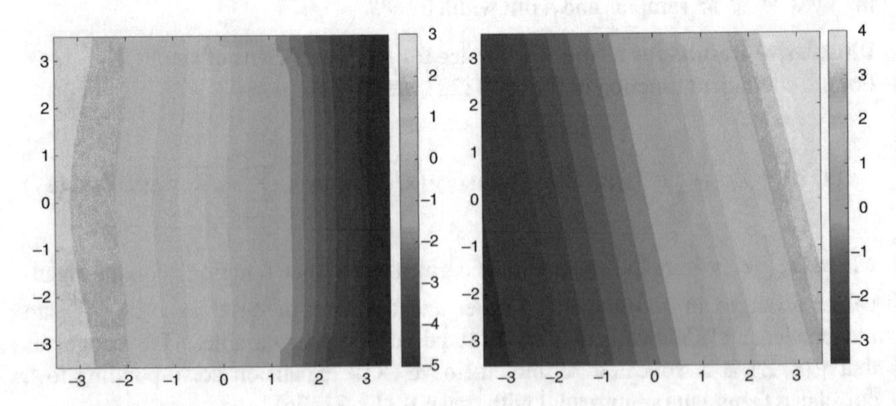

Fig. 14 For the coefficient (27) with $N_{input} = 2$, the outputs of interest (24) (left) and (25) (right) as a function of the input variable $\mathbf{Z} = (Z_1, Z_2)$

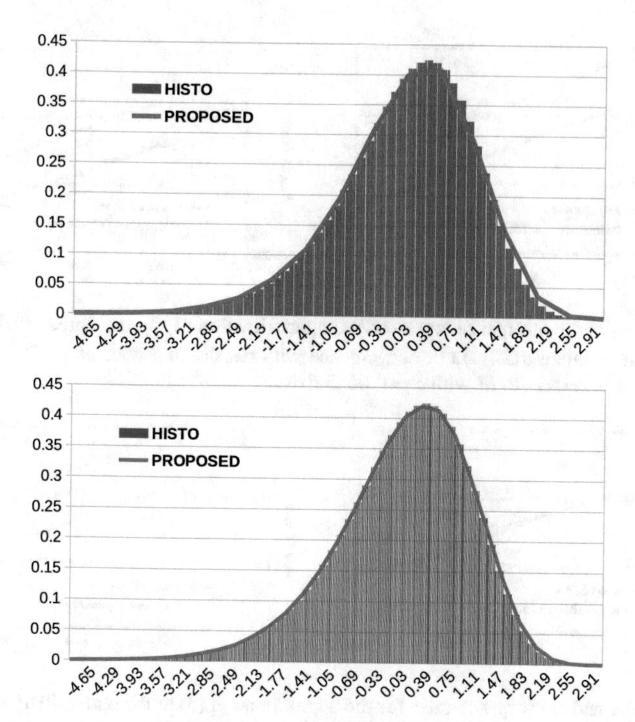

Fig. 15 For the coefficient (27) with $N_{input} = 2$ and the output of interest (24), a comparison between the histogram approximation (1) of the PDF $f_{output}(Y)$ with $\widehat{\delta} = 0.125$ and $\widehat{M} = 16^6$ and the approximation (5) with $h = 0.5$ and $M = 16^4$ (top) and with $h = 0.25$ and $M = 16^5$ (bottom)

the approximate output PDF $f_{\delta,M}(Y)$ determined using (5) is given in Fig. 15. For comparison purposes, a plot of the histogram approximation $f^{histo}_{\widehat{\delta},\widehat{M}}(Y)$ determined using (1) is also provided in that figure, but with larger sample size M and smaller bin size δ compared to those used for $f_{\delta,M}(Y)$.

We next examine the convergence behavior of the approximation (5) of the output PDF. Because the exact PDF is unknown, we measure the error using (26) with the histogram surrogate $f^{histo}_{\widehat{\delta},\widehat{M}}$ obtained with a bin size of $\widehat{\delta} = 8/2^8$ and $\widehat{M} = 10^8$ samples. To study the order of convergence with respect to δ, we choose

$$M = 10^7 \quad \text{and} \quad \delta = 8/2^k, \quad \text{for} \quad k = 2, 3, 4, 5, \tag{30}$$

whereas for the convergence with respect to M, we choose

$$\delta = 8/2^7 \quad \text{and} \quad M = 10^k \quad \text{for} \quad k = 3, 4, 5, 6. \tag{31}$$

Note that for these values, we have $M < \widehat{M}$ and $\delta > \widehat{\delta}$. In Fig. 16, we observe that the convergence rates with respect to δ and M are approximately 1.75, and

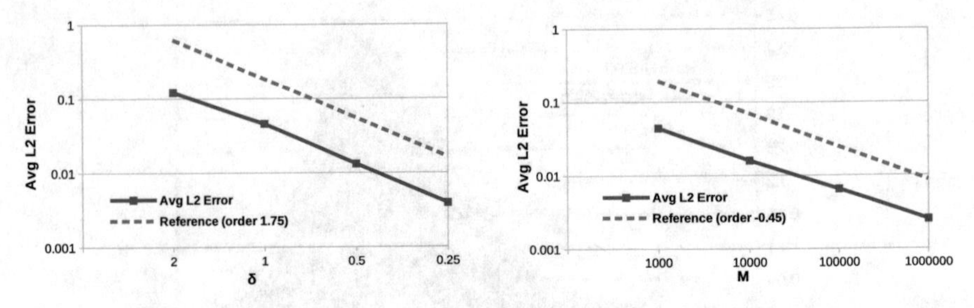

Fig. 16 Errors and convergence rates for the approximation (5) to the output PDF $f_{output}(Y)$ for the output of interest (24). Left: convergence with respect to δ with $M = 10^7$ fixed. Right: convergence with respect to M with $\delta = 0.0625$ fixed

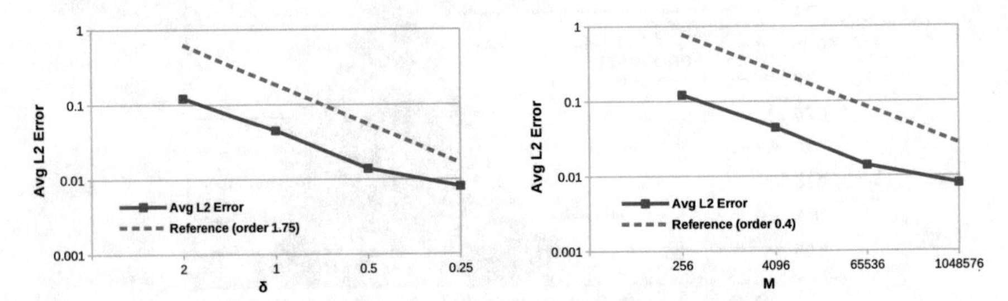

Fig. 17 Errors and convergence rates for the approximation (5) to the output PDF $f_{output}(Y)$ for the output of interest (24) with δ and M related through (16) with $r = 2$

0.45, respectively. Given that, from examining Fig. 15, the output PDF seems to be $C^2(\Gamma_{output})$, these rates are lower than the values 2 and 0.5, respectively, that one might expect. Likely causes of these lower rates are that errors are determined by comparing to a histogram approximation and not to an exact PDF and also because the values of $\widehat{M}$ and $\widehat{\delta}$ used for the histogram surrogate are "close" to the corresponding values used to determine the approximation (5).

Further results about errors and convergence rates are given in Fig. 17 for which (16) with $r = 2$ is used to relate δ and M. The histogram used for comparison to estimate errors is obtained with $\widehat{\delta} = 0.125$ and $\widehat{M} = 16^6$ samples. The results in this figure are consistent with those of Fig. 16.

Output of Interest (25) We next consider the approximation (5) of the PDF $f_{output}(Y)$ of the standardized output of interest Y given by (25). The various inputs are the same as those used for the output of interest (24) except that now $\Gamma_{output} = [-3.5, 4]$ and $\sigma_Y = 2$. A plot of this output of interest as a function of the input random variables $\mathbf{Z} \in [-3.5.3.5]^2$ is given in the right plot of Fig. 14. Plots of the approximate output PDF $f_{\delta,M}(Y)$ determined using (5) are given in Fig. 18. For comparison purposes, plots of the histogram approximation $f^{histo}_{\widehat{\delta},\widehat{M}}(Y)$ determined using (1) are also provided in that figure, but with larger sample size M and smaller bin size δ compared to those used for $f_{\delta,M}(Y)$. Figure (19) provides plots of the

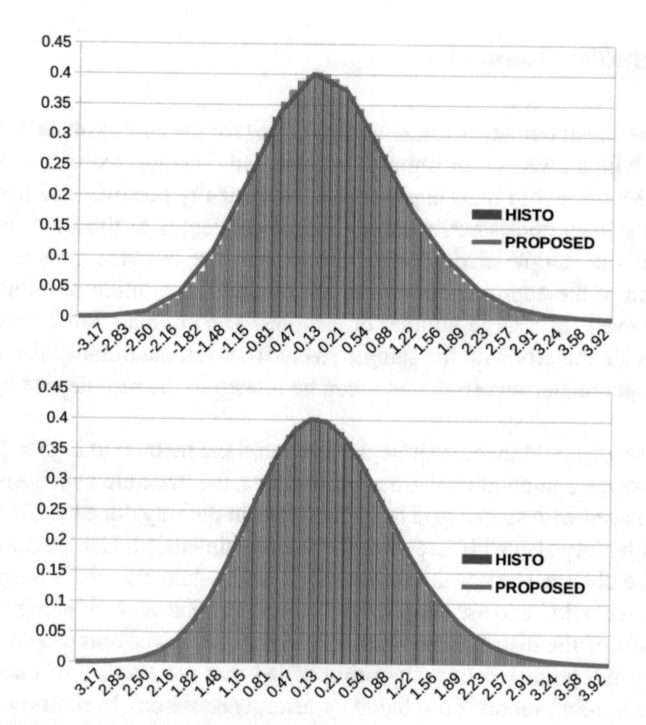

Fig. 18 For the coefficient (27) with $N_{input} = 2$ and the output of interest (25), a comparison between the histogram approximation (1) of the PDF $f_{output}(Y)$ with $\widehat{\delta} = 0.1171875$ and $\widehat{M} = 16^6$ and the approximation (5) with $\delta = 0.46875$ and $M = 16^4$ (top) and with $\delta = 0.234375$ and $M = 16^5$ (bottom)

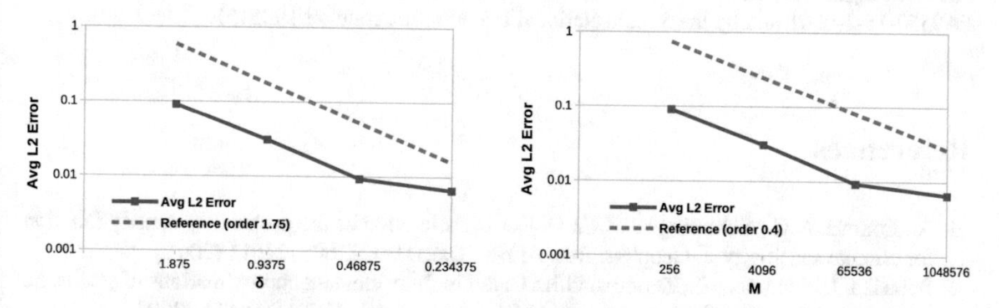

Fig. 19 Errors and convergence rates for the for the approximation (5) to the output PDF $f_{output}(Y)$ for the output of interest (25) with δ and M related through (16) with $r = 2$

errors in the approximation (5) determined through comparisons with histogram approximations determined with $\widehat{\delta} = 0.1171875$ and $\widehat{M} = 16^6$. Convergence rates of 1.75 and 0.4 are observed with respect to δ and M, respectively.

5 Concluding Remarks

A piecewise-linear density estimation method (5) for the approximation of the PDF associated with a given set of samples is presented. The approximation is naturally scalable with respect to the sample size, is intrinsically positive, and has a unitary integral. It is also consistent, meaning that it converges to the exact PDF in the L^2 norm if the sample size goes to infinity and the bin size goes to zero. The construction of the approximation does not require the solution of a linear system and is fast even for a large number of samples. The computational time has been shown to scale linearly with the sample size. Moreover, the binning size is the only smoothing parameter involved, and it can be related to the sample size by a simple rule.

Future work would involve strategies to extend the method to higher dimensions and decrease the computational time. For instance, the finite element basis functions may be replaced with sparse-grid basis functions in the way done in [25], in order to achieve scalability also with respect to the sample dimension. Computational speed-ups could be obtained by considering parallelization. Another means to speed up the computation would be to use adaptive refinement to coarsen the binning subdivision near the tails of the distribution. With this technique, fewer bins would have to be checked by the point locating algorithm of [8] that we employ to locate a given point in the binning subdivision shared by several processors. In addition, the Monte Carlo sampling used in our method can be replaced by, e.g., quasi-Monte Carlo or sparse-grid sampling, resulting in efficiency gains.

Acknowledgments This work was supported by US Air Force Office of Scientific Research grant FA9550-15-1-0001 and by the Sandia National Laboratories contract 1985151.

References

1. Andronova, N.G., Schlesinger, M.E.: Objective estimation of the probability density function for climate sensitivity. J. Geophys. Res. Atmos. **106**(D19), 22605–22611 (2001)
2. Babuška, I., Tempone, R., Zouraris, G.E.: Galerkin finite element approximations of stochastic elliptic partial differential equations. SIAM J. Numer. Anal. **42**(2), 800–825 (2004)
3. Babuška, I., Tempone, R., Zouraris, G.E.: Solving elliptic boundary value problems with uncertain coefficients by the finite element method: the stochastic formulation. Comput. Methods Appl. Mech. Eng. **194**(12–16), 1251–1294 (2005)
4. Babuška, I., Nobile, F., Tempone, R.: A stochastic collocation method for elliptic partial differential equations with random input data. SIAM J. Numer. Anal. **45**(3), 1005–1034 (2007)
5. Botev, Z.I., Kroese, D.P.: The generalized cross entropy method, with applications to probability density estimation. Methodol. Comput. Appl. Probab. **13**(1), 1–27 (2011)
6. Botev, Z.I., Grotowski, J.F., Kroese, D.P., et al.: Kernel density estimation via diffusion. Ann. Stat. **38**(5), 2916–2957 (2010)
7. Brenner, S., Scott, R.: The Mathematical Theory of Finite Element Methods, vol. 15. Springer Science & Business Media, Berlin (2007)

8. Capodaglio, G., Aulisa, E.: A particle tracking algorithm for parallel finite element applications. Comput. Fluids **159**, 338–355 (2017)
9. Capodaglio, G., Gunzburger, M., Wynn, H.P.: Approximation of probability density functions for SPDEs using truncated series expansions (2018). Preprint. arXiv:1810.01028
10. Ciarlet, P.: The Finite Element Method for Elliptic Problems. SIAM, Philadelphia (2002)
11. Criminisi, A., Shotton, J., Konukoglu, E., et al.: Decision forests: a unified framework for classification, regression, density estimation, manifold learning and semi-supervised learning. Found. Trends Comput. Graph. Vis. **7**(2–3), 81–227 (2012)
12. Fan, J., Marron, J.S.: Fast implementations of nonparametric curve estimators. J. Comput. Graph. Stat. **3**(1), 35–56 (1994)
13. Frauenfelder, P., Schwab, C., Todor, R.A.: Finite elements for elliptic problems with stochastic coefficients. Comput. Methods Appl. Mech. Eng. **194**(2–5), 205–228 (2005)
14. Gerber, M.S.: Predicting crime using twitter and kernel density estimation. Decis. Support. Syst. **61**, 115–125 (2014)
15. Ghanem, R.G., Spanos, P.D.: Stochastic finite element method: response statistics. In: Stochastic Finite Elements: A Spectral Approach, pp. 101–119. Springer, Berlin (1991)
16. Gunzburger, M.D., Webster, C.G., Zhang, G.: Stochastic finite element methods for partial differential equations with random input data. Acta Numerica **23**, 521–650 (2014)
17. Hall, P., Wand, M.P.: On the accuracy of binned kernel density estimators. J. Multivar. Anal. **56**(2), 165–184 (1996)
18. Hegland, M., Hooker, G., Roberts, S.: Finite element thin plate splines in density estimation. ANZIAM J. **42**, 712–734 (2009)
19. Heidenreich, N.-B., Schindler, A., Sperlich, S.: Bandwidth selection for kernel density estimation: a review of fully automatic selectors. AStA Adv. Stat. Anal. **97**(4), 403–433 (2013)
20. Izenman, A.J.: Review papers: recent developments in nonparametric density estimation. J. Am. Stat. Assoc. **86**(413), 205–224 (1991)
21. Li, C.F., Feng, Y.T., Owen, D.R.J., Li, D.F., Davis, I.M.: A Fourier–Karhunen–Loève discretization scheme for stationary random material properties in SFEM. Int. J. Numer. Methods Eng. **73**(13), 1942–1965 (2008)
22. Lopez-Novoa, U., Sáenz, J., Mendiburu, A., Miguel-Alonso, J., Errasti, I., Esnaola, G., Ezcurra, A., Ibarra-Berastegi, G.: Multi-objective environmental model evaluation by means of multidimensional kernel density estimators: Efficient and multi-core implementations. Environ. Model. Softw. **63**, 123–136 (2015)
23. Nobile, F., Tempone, R., Webster, C.G.: An anisotropic sparse grid stochastic collocation method for partial differential equations with random input data. SIAM J. Numer. Anal. **46**(5), 2411–2442 (2008)
24. Nobile, F., Tempone, R., Webster, C.G.: A sparse grid stochastic collocation method for partial differential equations with random input data. SIAM J. Numer. Anal. **46**(5), 2309–2345 (2008)
25. Peherstorfer, B., Pflüge, D., Bungartz, H.-J.: Density estimation with adaptive sparse grids for large data sets. In: Proceedings of the 2014 SIAM International Conference on Data Mining, pp. 443–451. SIAM, Philadelphia (2014)
26. Schevenels, M., Lombaert, G., Degrande, G.: Application of the stochastic finite element method for Gaussian and non-Gaussian systems. In: ISMA2004 International Conference on Noise and Vibration Engineering, pp. 3299–3314 (2004)
27. Silverman, B.W.: Density Estimation for Statistics and Data Analysis. Routledge, Abingdon (2018)
28. Turlach, B.A.: Bandwidth selection in kernel density estimation: a review. In: CORE and Institut de Statistique. Citeseer (1993)
29. Xie, Z., Yan, J.: Kernel density estimation of traffic accidents in a network space. Comput. Environ. Urban. Syst. **32**(5), 396–406 (2008)
30. Xu, X., Yan, Z., Xu, S.: Estimating wind speed probability distribution by diffusion-based kernel density method. Electr. Power Syst. Res. **121**, 28–37 (2015)
31. Zivkovic, Z., Van Der Heijden, F.: Efficient adaptive density estimation per image pixel for the task of background subtraction. Pattern Recogn. Lett. **27**(7), 773–780 (2006)

Analysis of Probabilistic and Parametric Reduced Order Models

Hermann G. Matthies

Abstract Stochastic models share many characteristics with generic parametric models. In some ways they can be regarded as a special case. But for stochastic models there is a notion of weak distribution or generalised random variable, and the same arguments can be used to analyse parametric models. Such models in vector spaces are connected to a linear map, and in infinite dimensional spaces are a true generalisation. Reproducing kernel Hilbert space and affine- / linear- representations in terms of tensor products are directly related to this linear operator. This linear map leads to a generalised correlation operator, and representations are connected with factorisations of the correlation operator. The fitting counterpart in the stochastic domain to make this point of view as simple as possible are algebras of random variables with a distinguished linear functional, the state, which is interpreted as expectation. The connections of factorisations of the generalised correlation to the spectral decomposition, as well as the associated Karhunen-Loève- or proper orthogonal decomposition will be sketched. The purpose of this short note is to show the common theoretical background and pull some lose ends together.

1 Introduction

Probabilistic and parametric models, used in many areas of science, engineering, and economics, share many similarities. Probabilistic models are used to describe uncertainties or random phenomena, whereas parametric models describe variations or changes of some system as some parameters are changed. Typically these are part of some larger mathematical model describing some system with such characteristics. A parameter can of course be a *random variable*, and this is the connection between these two kinds of models. Here the interest is mainly in system

---

H. G. Matthies (✉)
Institute of Scientific Computing, TU Braunschweig, Brunswick, Germany
e-mail: wire@tu-bs.de

M. D'Elia et al. (eds.), *Quantification of Uncertainty: Improving Efficiency and Technology*, Lecture Notes in Computational Science and Engineering 137,
https://doi.org/10.1007/978-3-030-48721-8_6

models with an infinite dimensional state space, e.g. systems described by ordinary or partial differential equations. This often also makes it necessary to theoretically consider infinitely many parameters. In an actual numerical computation this has of course to be reduced through some kind of discretisation to a finite number. And obviously one would like to have this number as small as possible while still retaining acceptable accuracy. This is the realm of *reduced order* models.

These reduced order models lessen the possibly high computational demand, and are hence probabilistic or parametrised reduced order models. The survey [1] and the recent collection [2], as well as the references therein, provide a good account of parametric reduced order models and some of the areas where they appear. The interested reader may find there further information on parametrised reduced order models and how to generate them.

Here we build on our recent work [17, 18] analysing parametrised reduced order systems, which itself is a continuation of [20]. In these publications the theoretical background of such parametrised models is treated in a functional analysis setting. The purpose of the present note is to use the same kind of techniques for stochastic or probabilistic models, where some generalisations are required due to the wish to cover infinite dimensional state spaces, and combine this with the description of parametric reduced order models.

As an example, assume that some physical system is investigated, which is modelled by an evolution equation for its state $v(t) \in \mathcal{V}$ at time $t \in [0, T]$, where $\mathcal{V}$ is assumed to be a Hilbert space for the sake of simplicity: $\dot{v}(t) = A(\varsigma, \mu; v(t)) + f(\varsigma, \mu; t); \quad v(0) = v_0$, where the superimposed dot signifies the time derivative, A is an operator modelling the physics of the system, and f is some external excitation. Here ς is a random variable (RV) defined on an event space Ω with values in some Hilbert space $\mathscr{S}$ (again for simplicity), and $\mu \in \mathcal{M}$ are parameters that can be controlled, and can be used to evaluate the *design* of the system, *control* its behaviour, or *optimise* the performance in some way. No specific structure is assumed for the set $\mathcal{M}$. We assume that for all possible values of ς and for all μ of interest the system is well-posed. This will make the system state $v(\varsigma, \mu; t)$ a random variable as well, depending on the value of the parameters μ.

One may be interested in the state of the system $v(\varsigma, \mu; t)$ and its statistics, or some functional of it, say $\Psi(\mu) = \mathbb{E}(\psi(v(\varsigma, \mu)))$, where $\mathbb{E}$ is an expectation operator. While evaluating $A(\varsigma, \mu)$ or $f(\varsigma, \mu)$ for a certain μ may be straightforward, evaluating $v(\varsigma, \mu; t)$ or $\Psi(\mu)$ may be very costly. This is why one wants representations of $v(\varsigma, \mu; t)$ or $\Psi(\mu)$ which allow a cheaper evaluation. This is achieved through reduced order models, which are often also called *proxy-* or *surrogate*-models. It turns out that such random and parametric objects can be analysed by associated linear maps [17, 18], which renders them much more accessible to the techniques of linear functional analysis, a well understood subject. This association with linear mappings has probably been known for a long time, see [15] for an exposition in the context of stochastic models. In Sect. 2 the association of parametric and stochastic models with linear maps will be explained, in passing touching on reproducing kernel Hilbert spaces. The classical probabilistic framework (cf. [26]), starting from measurable spaces and σ-algebras, can be used to define algebras of random variables (RVs) as measurable functions on these measure spaces, and the expectation operator as integral of these RVs w.r.t. the probability measure. These algebras of RVs can be used in the case of probabilistic models to build the range or image space for these linear maps as spaces of classical RVs. But alternatively one may also start by using as fundamental concepts algebras of objects that we want to call RVs together with the expectation operator (cf. [24]) as a linear functional, and if this algebra of RVs is Abelian or commutative one essentially recovers equivalence with classical probability. This approach allows for non-commuting algebras of RVs, which is important (cf. [21]) in order to deal with e.g. random matrices, random fields of tensors, quantum theory and quantum fields. More important for our immediate purposes here, this view greatly facilitates the specification of stochastic models on infinite dimensional spaces. Such an algebra of RVs, whether generated classically as derived concept as an algebra of measurable functions, or used as a primary model of possibly non-commuting RVs, seems to be a natural object to use in the case of stochastic models on infinite dimensional vector spaces, as it allows to generalise such stochastic models to so-called weak distributions or generalised processes (cf. [8, 9, 22, 23]), and thereby elegantly circumvent many problems which arise when one tries to define σ-additive set functions for example on Hilbert spaces. This algebraic and analytic view on probability will be explained in Sect. 3. Everything is tied together in Sect. 4 in the analysis of the generalised correlation operator, its factorisations, as well as its spectral decomposition, and the last Sect. 5 concludes by pointing out once more the connection between functions in high-dimensional spaces and the associated linear maps and correlation operators, where well-known methods can be used to analyse their structure.

2 Parametric and Stochastic Models

We start with a short recap of [17, 18], where the interested reader may find more detail. Let $r : \mathcal{M} \to \mathcal{U}$ be a generic substitute for any one of the parametric objects alluded to in the introduction, e.g. things like $\mu \mapsto v(\varsigma, \mu, t) \in \mathcal{V}$ or $\mu \mapsto \dot{v}(\varsigma, \mu, \cdot) \in \mathrm{L}_2([0, T]) \otimes \mathcal{V}$; $\omega \mapsto \varsigma(\omega) \in \mathscr{S}$—with Ω taking the rôle of $\mathcal{M}$; $(\mu, \omega) \mapsto v(\varsigma(\omega), \mu, t) \in \mathcal{V}$ —with $\mathcal{M} \times \Omega$ taking the rôle of $\mathcal{M}$; $\omega \mapsto f(\varsigma(\omega), \mu, t) \in \mathcal{V}$—with Ω taking the rôle of $\mathcal{M}$, or $\mu \mapsto A(\varsigma, \mu, \cdot) \in (\mathcal{V} \to \mathcal{V})$—the space of maps from $\mathcal{V}$ to $\mathcal{V}$, etc.

The space $\mathcal{U}$ is assumed for the sake of simplicity as a separable Hilbert space. The function r can thus be either a parametric input, or a random input—i.e. a random variable (RV), in which case $\mathcal{M}$ would be a measure space—to a model like that described in Sect. 1, or the operator of that model, or the state (solution) of that system. Assuming—without significant loss of generality—that the image $\operatorname{span} r(\mathcal{M}) = \operatorname{span} \operatorname{im} r \subseteq \mathcal{U}$ is dense in $\mathcal{U}$, one may to each such function r associate a linear map $R : \mathcal{U} \ni u \mapsto \langle r(\cdot) | u \rangle_{\mathcal{U}} \in \mathbb{R}^{\mathcal{M}}$ into the space $(\mathcal{M} \to \mathbb{R})$ of all real-valued functions on $\mathcal{M}$. By construction, R restricted to

$\operatorname{span}\operatorname{im} r = \operatorname{span} r(\mathcal{M})$ is injective. In Sect. 3 it will be explained how—in the case of a probabilistic or random model—the Hilbert space can be generated from an algebra of RVs.

As an aside, note that on its restricted range $\tilde{\mathcal{R}} := R(\operatorname{span}\operatorname{im} r) \subseteq \mathbb{R}^{\mathcal{M}}$ one may define an inner product as $\langle\phi|\psi\rangle_{\mathcal{R}} := \langle R^{-1}\phi|R^{-1}\psi\rangle_{\mathcal{U}}$ for all $\phi, \psi \in \tilde{\mathcal{R}}$. Denote the completion with this inner product by $\mathcal{R}$. This makes R and R^{-1} into bijective isometries, hence *unitary* maps between $\mathcal{U}$ and $\mathcal{R}$. It may easily be shown [17, 18] that $\mathcal{R}$ is a *reproducing kernel Hilbert space* (RKHS) [3, 12] with reproducing kernel $\varkappa(\mu_1, \mu_2) := \langle r(\mu_1)|r(\mu_2)\rangle_{\mathcal{U}}$, such that the reproducing property $\langle\varkappa(\mu, \cdot)|\phi\rangle_{\mathcal{R}} = \phi(\mu)$ holds for all $\phi \in \mathcal{R}$. In this note the RKHS $\mathcal{R}$ will not be used, but the important thing to keep in mind is that the map R and the space $\mathcal{R}$ of scalar functions on the set $\mathcal{M}$—one might view them as problem oriented co-ordinates—carry the same information as the parametric object $r(\mu)$.

Often some information of what is important in the set $\mathcal{M}$ is also available, here it is assumed to be given by a Hilbert subspace $\mathcal{Q} \subseteq \mathbb{R}^{\mathcal{M}}$, usually different from $\mathcal{R}$. From now on we shall by slight abuse of notation view the map R as mapping into $\mathcal{Q}$ and still assume that it is injective as well as closed, for the sake of simplicity. Details like the assumption that the subspace $R^{-1}(\mathcal{Q})$ is dense in $\mathcal{U}$ will not always be spelt out in detail for the sake of brevity. The idea is that with $u \in \mathcal{U}$ of unit length the vectors $Ru \in \mathcal{Q}$ with large norm are more important, and this will be considered in building reduced order models. As will be shown [17, 18] in Sect. 4, the map $C : \mathcal{U} \to \mathcal{U}$ defined by $C = R^* R$, where R^* is the adjoint of R, is central to the analysis. More precisely, with the above assumptions on R the adjoint R^* is surjective, and C is a densely defined self-adjoint positive definite operator, which we shall call the 'correlation' of the model $r(\mu)$.

A random variable or stochastic model as exemplified by the RV ς in Sect. 1 is usually formulated as a measurable map $\varsigma : \Omega \to \mathscr{S}$, where $(\Omega, \mathfrak{A}, \mathbb{P})$ is a probability space with σ-algebra $\mathfrak{A}$ and probability measure $\mathbb{P}$. One may view the set Ω as a parameter set like $\mathcal{M}$ above, and one can construct a linear map into the space $\mathbb{R}^{\Omega}$, i.e. the scalar random variables. Without loss of generality, we assume that $\operatorname{span}\varsigma(\Omega) = \operatorname{span}\operatorname{im}\varsigma \subseteq \mathscr{S}$ is dense in the separable Hilbert space $\mathscr{S}$, and define [15]

$$S : \mathscr{S} \ni \xi \mapsto \langle\varsigma(\cdot)|\xi\rangle_{\mathscr{S}} \in \mathbb{R}^{\Omega}. \tag{1}$$

It remains to define an inner product on $\mathbb{R}^{\Omega}$ and a subspace corresponding to $\mathcal{Q}$ for the parametric case above. This will be done in Sect. 3. For the time being assume that this has been defined, i.e. there is an inner product $\langle\cdot|\cdot\rangle_{\mathscr{V}}$ and a corresponding Hilbert space of (equivalence classes) of RVs $\mathscr{V} \subseteq \mathbb{R}^{\Omega}$, and we regard S as a map $S : \mathscr{S} \to \mathscr{V}$ with the same properties as assumed for R above. Obviously the densely defined self-adjoint positive definite operator $C_\varsigma = S^* S : \mathscr{S} \to \mathscr{S}$ corresponding to $C = R^* R$ above is indeed the correlation operator of the RV ς.

In case ς is an input to a dynamical system like the one alluded to in Sect. 1, the state of the system $v(\varsigma, \mu; t)$ also becomes a stochastic quantity, and inner product with a vector $w \in \mathcal{V}$ leads for fixed μ and t automatically to a linear mapping

$$P : \mathcal{V} \ni w \mapsto \langle v(\varsigma(\cdot), \mu; t) | w \rangle_{\mathcal{V}} \in \mathbb{R}^{\Omega}, \tag{2}$$

which we shall regard again as a map $P : \mathcal{V} \to \mathscr{V}$ into the just defined space $\mathscr{V}$. This defines a third correlation operator $C_v = P^* P : \mathcal{V} \to \mathcal{V}$.

It may be seen that with the correspondences

$$R : \mathcal{U} \to \mathcal{Q} \quad — \quad S : \mathscr{S} \to \mathscr{V} \quad — \quad P : \mathcal{V} \to \mathscr{V} \tag{3}$$

all three situations are completely analogous, and may in the simplest case be dealt with in the same formalism. The idea on how to obtain representations of $r(\mu)$ resp. $\varsigma(\omega)$ resp. $v(\omega)$ is the following [17, 18], which we shall mainly demonstrate for $r(\mu)$: choose a complete basis $\{q_j\}_j \subset \mathcal{Q}$, and represent $r(\mu)$ as

$$r(\mu) = \sum_j \alpha_j R^* q_j(\mu). \tag{4}$$

A good reduced order model is one where

$$r_{\text{ROM}}(\mu) = \sum_{j=1}^{J} \alpha_j R^* q_j(\mu) \tag{5}$$

is a good approximation to $r(\mu) \approx r_{\text{ROM}}(\mu)$ with a small J, i.e. with not too many terms. In Sect. 4 some other possibilities for the choice of basis $\{q_j\}_j$ will be discussed, where the μ-dependence is encoded in the scalar functions from $\mathcal{Q}$, but where a basis of μ-independent vectors is picked from $\mathcal{U}$, and where again for the sake of brevity and simplicity we shall confine ourselves to complete orthonormal systems (CONS). The important message here is that with R one has a factorisation of $C = R^* R$, and that the adjoint is the map which carries a representation on the function space to the space $\mathcal{U}$. Later we shall indicate [17, 18] how every representation leads to a factorisation of C, and that—with some additional assumptions on C—every factorisation leads to a representation. But the description and analysis via factorisations is more general [8, 9, 15, 22, 23], and this is needed in the formulation of probabilistic models where $\mathcal{U}$ resp. $\mathscr{S}$ is an infinite dimensional Hilbert space.

3 Algebras of Random Variables

Here we shall take a closer look at the stochastic or probabilistic model $\varsigma : \Omega \to \mathscr{S}$ and the associated linear map $S : \mathscr{S} \to \mathscr{V}$, as well as the space of RVs $\mathscr{V}$ and how it is generated. Although there are classical ways of specifying the space

$\mathscr{V}$, the most natural one seems to be the algebraic approach to probability. These ideas are certainly also used in the classical approach, but the algebraic probability approach distills the essential components in an abstract setting and allows at the same time generalisations. Historically, when looking back as how in the beginnings of probability theory the Bernoullis treated random variables (RVs), it is clear that they added them and took multiples—hence they form a vector space—and that they multiplied them with each other—so they form an algebra. Although the formalisation of probability as formulated by Kolmogorov used the concept of measure and this algebraic background was largely ignored, it was revived with the advent of quantum theory. It turns out that here this view is essential, as not all observables can be observed simultaneously, and this is reflected in the fact that they do not commute in the algebra. Another topic where this view is very advantageous are random matrices and more generally random fields of even-order tensors.

We are mainly interested in 'real' or self-adjoint RVs as they will later be called. But for analytical convenience we shall treat complex RVs, following Paul Painlevé's and Jacques Hadamard's adage that the shortest path between two truths in the real domain passes through the complex domain—"le plus court chemin entre deux vérités dans le domaine réel passe par le domaine complexe". Some algebraic language is needed, but most of the terms will be familiar from complex numbers and from matrices, which are indeed two simple but prime examples of algebras. Let us start right away with a simple and mostly familiar example from probability theory, which will at the same time serve as motivation, concrete example, and explanation of the abstract setting.

3.1 Specifying the Algebra

Consider a probability space $(\Omega, \mathfrak{A}, \mathbb{P})$ with a set of elementary events Ω, a σ-algebra $\mathfrak{A}$ of measurable subsets of Ω, and a probability measure $\mathbb{P}$. In the vector space $\mathrm{L}_0(\Omega, \mathfrak{A}, \mathbb{P}; \mathbb{C})$ of complex-valued measurable functions/classical random variables on Ω—which for the sake of brevity shall be denoted just by $\mathrm{L}_0(\Omega)$—let $\mathcal{A}_s := \mathrm{L}_{0s}(\Omega) \subset \mathrm{L}_0(\Omega)$ be the vector subspace of complex-valued simple measurable functions, i.e. complex linear combinations of functions $\mathbf{1}_{\mathcal{E}}$, which for $\mathcal{E} \in \mathfrak{A}$ are defined to be $\mathbf{1}_{\mathcal{E}}(\omega) = 1$ if $\omega \in \mathcal{E} \subseteq \Omega$, and zero otherwise. Hence $\mathcal{A}_s$ are the RVs where each one of them can only take finitely many different values.

On this vector space we may define a multiplication by just pointwise multiplication of two such RVs, and the product is obviously again a simple function; in fact for $\mathcal{E}, \mathcal{F} \in \mathfrak{A}$ one has $\mathbf{1}_{\mathcal{E}}\mathbf{1}_{\mathcal{F}} = \mathbf{1}_{\mathcal{E}\cap\mathcal{F}}$, i.e. the multiplication in $\mathcal{A}_s$ reflects the intersection in the σ-algebra $\mathfrak{A}$. This means that the space $\mathcal{A}_s$ is closed under multiplication and hence thanks to the properties of the multiplication on $\mathbb{C}$ is a complex, associative, and commutative or Abelian algebra, with the familiar distributive law from $\mathbb{C}$ coupling addition and multiplication also on $\mathcal{A}_s$. Another way of saying this is to state that the multiplication is a bilinear map from $\mathcal{A}_s \times \mathcal{A}_s$ to $\mathcal{A}_s$. Let us note in passing that with the same definition of pointwise multiplication also $\mathrm{L}_0(\Omega)$ is an associative and commutative algebra—with $\mathcal{A}_s$ a sub-algebra—as the pointwise product of two measurable functions is again measurable, but we shall see later that for our purposes $\mathrm{L}_0(\Omega)$ is in general too big. The element $\mathbf{1}_\Omega \in \mathcal{A}_s \subset \mathrm{L}_0(\Omega)$ which is constant equal to unity is obviously a neutral element or unit for the multiplication, and hence $\mathcal{A}_s$ and $\mathrm{L}_0(\Omega)$ are called unital algebras. For $\psi \in \mathcal{A}_s$ one can now compute *powers* $\psi^n = \psi\psi^{n-1}$ for any integer $n \geq 1$, and if we define $\psi^0 = \mathbf{1}_\Omega$ in a unital algebra even for any $n \geq 0$. Given a polynomial $Q(X) = \sum_{k=0}^n \alpha_k X^k \in \Pi_1$ in one unknown X with complex co-efficients $\alpha_k \in \mathbb{C}$, it is now possible to evaluate $Q(\psi) \in \mathcal{A}_s$ for any $\psi \in \mathcal{A}_s$. For some $\phi \in \mathcal{A}_s$ there is a $\psi \in \mathcal{A}_s$ such that $\phi\psi = \mathbf{1}_\Omega$. This is then called the (multiplicative) *inverse* $\psi = \phi^{-1}$, such that $\phi\phi^{-1} = \mathbf{1}_\Omega$.

For a complex number $\zeta \in \mathbb{C}$ its complex conjugate is denoted by $\zeta^* \in \mathbb{C}$, and this operation is an *involution*, as $(\zeta^*)^* = \zeta$. One may extend this involution from $\mathbb{C}$ to the algebra $\mathrm{L}_0(\Omega)$ through a pointwise definition of complex conjugation, and hence also to its sub-algebra $\mathcal{A}_s$. For $\phi, \psi \in \mathrm{L}_0(\Omega)$ and $\zeta \in \mathbb{C}$ this involution obviously satisfies $(\phi + \zeta\psi)^* = \phi^* + \zeta^*\psi^*$ and is thus *anti-linear*. As regards the product of two RVs, it satisfies $(\phi\psi)^* = \psi^*\phi^*$, and it is easy to verify that both $\mathcal{A}_s$ and $\mathrm{L}_0(\Omega)$ are closed under this involution. Associative algebras with such an anti-linear involution and the indicated behaviour on products are called *-algebras—the element ψ^* is usually called in algebraic terms the *adjoint* of ψ—and both $\mathrm{L}_0(\Omega)$ and its sub-algebra $\mathcal{A}_s = \mathrm{L}_{0s}(\Omega)$ are thus *-algebras.

Let Π_2^c denote the set of all polynomials $Q(X, Y)$ with complex co-efficients in two commuting variables X, Y. For $\phi \in \mathcal{A}_s$ the unital sub-*-algebra $\mathbb{C}[\phi, \phi^*] := \{Q(\phi, \phi^*) \mid Q \in \Pi_2^c\} \subset \mathcal{A}_s$ is called the sub-algebra *generated* by $\phi \in \mathcal{A}_s$. Observe that if $\psi \in \mathrm{L}_0(\Omega)$ is *self-adjoint*, i.e. $\psi = \psi^*$, then ψ has only real values, and if $\psi = \phi^*\phi$ for some $\phi \in \mathrm{L}_0(\Omega)$, then ψ is self-adjoint (real) and is called *positive* as it can not take negative values, i.e. $0 \leq \psi = \phi^*\phi$—in case $0 < \psi$ it is usually called *strictly positive*. One says that for self-adjoint $\phi, \psi \in \mathrm{L}_0(\Omega)$ one has $\psi \leq \phi$ iff $\phi - \psi$ is positive, and thus one can define a partial order on $\mathcal{A}_s$ and $\mathrm{L}_0(\Omega)$. Positive self-adjoint elements $\psi \in \mathrm{L}_0(\Omega)$ which are *idempotent*, i.e. satisfy $\psi^2 = \psi\psi = \psi$, are called *projections*. Observe that each $\mathbf{1}_\mathcal{E}$ is a projection, and that the unit $\mathbf{1}_\Omega$ is a maximal projection in the order mentioned. In fact all projections in $\mathrm{L}_0(\Omega)$ and $\mathcal{A}_s$ have the form $\mathbf{1}_\mathcal{E}$ for some $\mathcal{E} \in \mathfrak{A}$. Ultimately, one is only interested in the self-adjoint elements of the algebra $\mathcal{A}_s$, as they take real values; they are therefore often also called *observables*. The other elements of the algebra may be regarded as merely a kind of analytical completion to make the theory nice. It may be remarked that the self-adjoint elements of $\mathcal{A}_s$ form a *real* subspace of $\mathcal{A}_s$. Obviously an arbitrary $\phi \in \mathcal{A}_s$ may be decomposed into real and imaginary parts: $\phi = \Re\phi + \mathrm{i}\Im\phi$ with real resp. self-adjoint $\Re\phi = (\phi + \phi^*)/2$ and $\Im\phi = (\phi - \phi^*)/(2\mathrm{i})$, so that the whole algebra is the complex span of the self-adjoint elements or observables.

To extract the essential point from this example and generalise, we start with an associative algebra $\mathcal{A}$ of what we want to call random variables (RVs) $a, b, \dots \in \mathcal{A}$, i.e. a vector space [24] equipped with an associative and bi-linear multiplication which will be denoted just by juxtaposition: $\mathcal{A} \times \mathcal{A} \ni (a, b) \mapsto ab \in \mathcal{A}$.

As was noted before, it is advantageous to assume the algebra to be a complex algebra, which is no loss of generality as any real algebra may be embedded into a complex one. For $a \in \mathcal{A}$ the powers a^n are defined for any integer $n \geq 1$ in the natural recursive fashion. Additionally assume that the algebra is unital, i.e. has a multiplicative unit e such that $ae = ea = a$ for any $a \in \mathcal{A}$, and one defines the power a^n for $n = 0$ by $a^0 = e$. Hence for a polynomial $Q(X) \in \Pi_1$ it is now possible to evaluate $Q(a)$ for any $a \in \mathcal{A}$. Also assume that there is an anti-linear involution defined, called the 'adjoint', denoted as a^*, such that $(a^*)^* = a$ and $(ab)^* = b^* a^*$.

Let Π_2^n be a set of all polynomials $Q(X, Y)$ with complex co-efficients in two non-commuting variables X, Y, then for $a \in \mathcal{A}$ the unital sub-*-algebra $\mathbb{C}\{a, a^*\} := \{Q(a, a^*) \mid Q \in \Pi_2^n\} \subset \mathcal{A}$ is called the sub-algebra generated by $a \in \mathcal{A}$. Elements $a \in \mathcal{A}$ such that $a = a^*$ are called self-adjoint, and self-adjoint elements which may be factored as $a = b^* b$ are called positive. Positive elements form a salient pointed cone which defines an order relation on $\mathcal{A}$. Positive elements p which are idempotent $p = pp = p^2 = p^* p = p^*$ are called projections. Observe that e is a projection, and that it is maximal w.r.t. the order mentioned. Succinctly stated, we assume that $\mathcal{A}$ is a complex associative unital *-algebra, not *necessarily* commutative. As was shown, both $\mathrm{L}_0(\Omega)$ and $\mathcal{A}_s$ considered above are commutative examples of such algebras. Again, one is later ultimately interested in the self-adjoint elements of $\mathcal{A}$—the *observables*. Also in the general abstract case they form a *real* subspace of $\mathcal{A}$, and an arbitrary $a \in \mathcal{A}$ may be decomposed into two parts $a = a_s + \mathrm{i} a_w$ with self-adjoint $a_s = (a + a^*)/2$ and $a_w = (a - a^*)/(2\mathrm{i})$ —also called the *symmetric* and *skew* parts—so that the whole algebra is the complex span of the self-adjoint elements, the observables. And naturally, if for some $a \in \mathcal{A}$ there is a $c \in \mathcal{A}$ such that $ac = ca = e$, then $c = a^{-1}$ is the unique multiplicative inverse of a.

3.2 *States and the Expectation Functional*

To continue, we return to the example $\mathcal{A}_s$ above. Just as classical probability builds on the measurable space $(\Omega, \mathfrak{A})$ on one hand and the probability measure $\mathbb{P}$ on the other hand, in the algebraic framework the second entity needed is the linear expectation functional $\mathbb{E} : \mathcal{A}_s \to \mathbb{C}$. To define the expected value for a RV $\phi \in \mathcal{A}_s$ one only has to look at the generating elements $\mathbf{1}_{\mathcal{E}}$ with $\mathcal{E} \in \mathfrak{A}$. Here one defines $\mathbb{E}(\mathbf{1}_{\mathcal{E}}) := \int_\Omega \mathbf{1}_{\mathcal{E}}(\omega)\, \mathbb{P}(\mathrm{d}\omega) = \mathbb{P}(\mathcal{E})$ and extends this by linearity to all of $\mathcal{A}_s$. Thus the probability of an event $\mathcal{E} \in \mathfrak{A}$ is given in terms of the expected value of the associated projection $\mathbf{1}_{\mathcal{E}}$. For a typical $\phi(\omega) = \sum_k \alpha_k \mathbf{1}_{\mathcal{E}_k}(\omega) \in \mathcal{A}_s$ with $\alpha_k \in \mathbb{C}$ this gives $\mathbb{E}(\phi) = \int_\Omega \phi(\omega)\, \mathbb{P}(\mathrm{d}\omega) = \sum_k \alpha_k \mathbb{P}(\mathcal{E}_k) \in \mathbb{C}$. Obviously, as $\mathbb{P}(\Omega) = 1$, the expected value of the unit is $\mathbb{E}(\mathbf{1}_\Omega) = 1$, a kind of *normalisation* of the expectation functional.

This linear functional $\mathbb{E}$ additionally satisfies $\mathbb{E}(\phi^*) = (\mathbb{E}(\phi))^*$ and thus carries the adjoint to its complex conjugate and hence is real on self-adjoint elements. Such a linear functional is itself called *self-adjoint*. In addition, $\mathbb{E}(\phi^* \phi) =$

$\sum_k(\alpha_k^*\alpha_k)\mathbb{E}\left(\mathbf{1}_{\mathcal{E}_k}\right) = \sum_k |\alpha_k|^2\mathbb{P}(\mathcal{E}_k) \geq 0$, i.e. the functional is non-negative on positive $\psi = \phi^*\phi \in \mathcal{A}_s$. Such a self-adjoint linear functional is itself called *positive*. If $\rho \in \mathcal{A}_s$ is positive with unit expected value $\mathbb{E}(\rho) = 1$, one may define a new expectation functional—corresponding to a change of probability measure—via $\mathbb{E}_\rho(\phi) := \mathbb{E}(\rho\phi) = \int_\Omega \rho(\omega)\,\phi(\omega)\,\mathbb{P}(\mathrm{d}\omega)$. It is easily checked that $\mathbb{E}_\rho$ is linear, self-adjoint, positive, and normalised. Such linear functionals which can serve as expectation are called *states*, an element of the *dual* space $\mathcal{A}_s^*$.

The element $\bar{\phi} := \mathbb{E}(\phi)\,\mathbf{1}_\Omega \in \mathcal{A}_s$ is called the *mean* of $\phi \in \mathcal{A}_s$ and the additive rest $\tilde{\phi} = \phi - \bar{\phi} \in \mathcal{A}_s$ is its *zero-mean* or *centred* or *fluctuating* part. The one-dimensional unital *-algebra $\mathcal{A}_{sc} := \mathbb{C}[\mathbf{1}_\Omega] = \operatorname{span}\{\mathbf{1}_\Omega\} \subset \mathcal{A}_s$—isomorphic to $\mathbb{C}$—are the *constants*, whereas the subspace $\mathcal{A}_{s0} := \ker\mathbb{E}$ are the *zero-mean* or *centred* RVs, such that $\mathcal{A}_s = \mathcal{A}_{sc} \oplus \mathcal{A}_{s0} = \mathbb{C}[\mathbf{1}_\Omega] \oplus \ker\mathbb{E}$ as a direct sum.

One may observe that in general not every measurable $\phi \in \mathrm{L}_0(\Omega)$ has a finite integral. Thus the algebra of *all* classical RVs $\mathrm{L}_0(\Omega)$ is too big for our purpose as one would like $\mathbb{E}(\cdot)$ to be defined on the whole algebra. This is the reason to start with the 'smaller' algebra $\mathcal{A}_s = \mathrm{L}_{0s}(\Omega)$. It is a building block from which more complicated RVs can be built via limiting processes.

In the general abstract case one also wants a linear, self-adjoint, positive, and normalised functional—a *state*—$\mathbb{E} : \mathcal{A} \to \mathbb{C}$ with $\mathbb{E}(a^*) = \mathbb{E}(a)^*$. Such a state is called *faithful* if $\mathbb{E}(a^*a) = 0$ implies $a = 0$. If a state is not faithful, then one can start to work with an algebra of equivalence classes, where two elements $a, b \in \mathcal{A}$ are considered equivalent iff $\mathbb{E}((a-b)^*(a-b)) = 0$. It is therefore no loss of generality to assume that the state is faithful. The projections $p \in \mathcal{A}$ are also identified with events, and the *probability* of the event $p \in \mathcal{A}$ may be *defined* as $\mathbb{P}(p) := \mathbb{E}(p)$. As $\mathbb{E}$ is positive, one has $\mathbb{P}(p) \geq 0$, and as e is a maximal projection, $\mathbb{P}(p) \leq \mathbb{P}(e) = 1$. One defines the mean part of a RV as a multiple of the identity $\bar{a} := \mathbb{E}(a)\,e$ and the fluctuating zero-mean or centred part as $\tilde{a} := a - \bar{a}$ with $\mathbb{E}(\tilde{a}) = 0$. The one dimensional sub-*-algebra $\mathcal{A}_c = \mathbb{C}[e] = \operatorname{span}\{e\}$ of constants—isomorphic to $\mathbb{C}$—are multiples of the identity, and the subspace of zero-mean fluctuating parts $\mathcal{A}_0 = \ker\mathbb{E}$ is the kernel of the state, and the whole algebra is the direct sum of both parts $\mathcal{A} = \mathcal{A}_c \oplus \mathcal{A}_0 = \mathbb{C}[e] \oplus \ker\mathbb{E}$. An abstract algebra which satisfies all these requirements together with a distinguished faithful state as expectation is called a *probability algebra*. If $\varrho \in \mathcal{A}$ is positive with unit expectation $\mathbb{E}(\varrho) = 1$, then one may define a new weighted state by $\mathbb{E}_\varrho(a) := \mathbb{E}(\varrho\, a)$ for $a \in \mathcal{A}$.

A faithful state may be used to define an inner product on $\mathcal{A}$ [22–24] via a positive definite sesqui-linear form:

$$\mathcal{A}^2 \ni (a, b) \mapsto \langle a|b\rangle_2 := \mathbb{E}\left(b^*a\right) \in \mathbb{C}. \tag{6}$$

As usual, one may define the square of a norm via $\|a\|_2^2 := \langle a|a\rangle_2$. The completion of $\mathcal{A}$ in the uniform topology generated by this norm is a Hilbert space denoted by $\mathrm{L}_2(\mathcal{A})$, which is one candidate for $\mathscr{V} := \mathrm{L}_2(\mathcal{A})$. Later we shall see more possible ways of generating a Hilbert space of RVs. With this inner product the above direct sum $\mathcal{A} = \mathcal{A}_c \oplus \mathcal{A}_0$ is an *orthogonal* direct sum, i.e. $\mathcal{A}_c = \mathbb{C}[e] = \text{span}\{e\} = (\ker \mathbb{E})^{\perp} = \mathcal{A}_0^{\perp}$.

As the expectation or state is normally also continuous in the topology of the associated Hilbert space $\mathscr{V}$, it can be defined also on $\mathscr{V}$ giving an orthogonal decomposition $\mathscr{V} = \ker \mathbb{E} \oplus (\ker \mathbb{E})^{\perp} =: \mathscr{V}_0 \oplus \mathbb{C}[e]$. For the probabilistic model $S : \mathscr{S} \to \mathscr{V}$ this means that it can be extended to $\xi \in \mathscr{S}$ as $\mathbb{E}_{\mathscr{S}}(\xi) := \mathbb{E}(S\xi)$, and with it an orthogonal decomposition of $\mathscr{S} = \mathscr{S}_0 \oplus \mathscr{S}_0^{\perp} := \ker \mathbb{E}_{\mathscr{S}} \oplus (\ker \mathbb{E}_{\mathscr{S}})^{\perp}$, where $(\ker \mathbb{E}_{\mathscr{S}})^{\perp} = \text{span}\{S^* e\}$ are multiples of the *mean* $\bar{\varsigma} := S^* e \in \mathscr{S}$ of the RV ς. Instead of looking at the correlation operator $C_\varsigma = S^* S$, one is usually only interested in the correlation $\tilde{C}_\varsigma = \tilde{S}^* \tilde{S}$ of $\tilde{S}$, where $\tilde{S} : \mathscr{S} \ni \xi \mapsto S\xi - \mathbb{E}_{\mathscr{S}}(\xi) e \in \mathscr{V}_0$—$\tilde{C}_\varsigma$ is called the *covariance* operator. Completely analogous statements can be made for the map $P : \mathcal{V} \ni w \mapsto \langle v(\varsigma) | w \rangle_{\mathcal{U}} \in \mathscr{V}$, the associated expectation $\mathbb{E}_{\mathcal{V}}(w) := \mathbb{E}(Pw)$, the orthogonal split $\mathcal{V} = \mathcal{V}_0 \oplus \mathcal{V}_0^{\perp} := \ker \mathbb{E}_{\mathcal{V}} \oplus \text{span}\{P^* e\}$, and the associated covariance operator.

In the example algebra $\mathcal{A}_s = \mathrm{L}_{0s}(\Omega)$ from above, identifying $\mathbf{1}_{\mathcal{E}}$ and $\mathbf{1}_{\mathcal{F}}$ if $\mathcal{E}, \mathcal{F} \in \mathfrak{A}$ differ only by a null-set $\mathcal{N} \in \mathfrak{A}$ with $\mathbb{P}(\mathcal{N}) = 0$, the integral or expected value becomes a faithful state. As is well known [24], the construction in Eq. (6) defines the L_2 inner product $\langle \phi | \psi \rangle_2 = \mathbb{E}(\psi^* \phi) = \int_\Omega \psi(\omega)^* \phi(\omega)\, \mathbb{P}(\mathrm{d}\omega)$ for $\phi, \psi \in \mathcal{A}_s = \mathrm{L}_{0s}(\Omega)$, and the completion is the familiar Hilbert space $\mathrm{L}_2(\Omega) = \mathrm{L}_2(\mathcal{A}_s)$. The inner product $\langle \phi | \psi \rangle_2$ of two RVs $\phi, \psi \in \mathcal{A}_s$ is also called their *correlation*, and one may continue and define the *covariance* in the usual way by $\text{cov}(\phi, \psi) := \mathbb{E}\left(\tilde{\psi}^* \tilde{\phi}\right) = \langle \tilde{\phi} | \tilde{\psi} \rangle_2$, i.e. the inner product or correlation of the fluctuating parts. The variance of a RV $\phi \in \mathcal{A}_s$ is then $\text{var}(\phi) := \text{cov}(\phi, \phi)$, and one has from Pythagoras's theorem $\|\phi\|_2^2 = \|\bar{\phi}\|_2^2 + \|\tilde{\phi}\|_2^2 = \mathbb{E}(\phi)^2 + \text{var}(\phi)$. Two RVs $\phi, \psi \in \mathcal{A}_s$ are *uncorrelated* iff their covariance vanishes: $\text{cov}(\phi, \psi) = 0$, i.e. their fluctuating parts are *orthogonal*. Two such RVs are *independent* iff $\text{cov}(Q_1(\phi, \phi^*), Q_2(\psi, \psi^*)) = 0$ for all $Q_1, Q_2 \in \Pi_2^c$ with $\mathbb{E}(Q_1(\phi, \phi^*)) = \mathbb{E}(Q_2(\phi, \phi^*)) = 0$, i.e. if the centred subspaces of the algebras generated by them are orthogonal, i.e. $(\mathbb{C}[\phi, \phi^*] \cap \ker \mathbb{E}) \perp (\mathbb{C}[\psi, \psi^*] \cap \ker \mathbb{E})$.

Completely analogous in the general case, for two RVs $a, b \in \mathcal{A}$ one defines the *correlation* as the inner product $\langle a | b \rangle_2$, the *covariance* as the inner product of the fluctuating parts $\text{cov}(a, b) := \langle \tilde{a} | \tilde{b} \rangle_2$, and the *variance* as $\text{var}(a) := \text{cov}(a, a)$. Pythagoras's theorem can be applied here as well to give $\|a\|_2^2 = \|\bar{a}\|_2^2 + \|\tilde{a}\|_2^2 = \mathbb{E}(a)^2 + \text{var}(a)$. Two RVs $a, b \in \mathcal{A}$ are *uncorrelated* iff their covariance vanishes: $\text{cov}(a, b) = 0$, i.e. if their fluctuating parts are orthogonal $\langle \tilde{a} | \tilde{a} \rangle_2 = 0$. The two RVs $a, b \in \mathcal{A}$ are *independent* iff $\text{cov}(Q_1(a, a^*), Q_2(b, b^*)) = 0$ for all $Q_1, Q_2 \in \Pi_2^n$ with $\mathbb{E}(Q_1(a, a^*)) = \mathbb{E}(Q_2(b, b^*)) = 0$, i.e. if the centred subspaces of the algebras generated by them are orthogonal, i.e. $(\mathbb{C}\{a, a^*\} \cap \ker \mathbb{E}) \perp (\mathbb{C}\{b, b^*\} \cap \ker \mathbb{E})$. In the non-commutative case, the concept of *freeness* and *free independence* becomes more important, cf. [10, 21, 25, 27], but we shall not further pursue this topic here.

We have seen that the example algebra $\mathcal{A}_s = \mathrm{L}_{0s}(\Omega)$ satisfies all the requirements and is thus a concrete example of a probability algebra, and generates the

Hilbert space $L_2(\Omega)$, which is one concrete example of the abstract Hilbert space $\mathscr{V} := L_2(\mathcal{A})$ for a general probability algebra $\mathcal{A}$.

3.3 More Examples

For the example algebra $\mathcal{A}_s = L_{0s}(\Omega)$ it is also well known that one may define the L_p-norms for any $1 \leq p < \infty$ via $\|\phi\|_p^p := \mathbb{E}\left((\phi^*\phi)^{p/2}\right) = \int_\Omega |\phi(\omega)|^p \, \mathbb{P}(\mathrm{d}\omega)$. For $p = \infty$ one sets $\|\phi\|_\infty := \operatorname{ess\,sup}_\Omega |\phi|$. The completion of $\mathcal{A}_s = L_{0s}(\Omega)$ in any of the norms $\|\cdot\|_p$ for $1 \leq p \leq \infty$ gives the familiar Banach spaces $L_p(\Omega)$. This gives two more concrete examples of probability algebras, namely $L_\infty(\Omega)$ and $L_{\infty-}(\Omega) := \bigcap_{1 \leq p < \infty} L_p(\Omega)$. The last example contains *unbounded* RVs, e.g. all the Gaussian RVs. Obviously one has $\mathcal{A}_s = L_{0s}(\Omega) \subset L_\infty(\Omega) \subset L_{\infty-}(\Omega) \subset L_0(\Omega)$, i.e. the classical simple RVs in $\mathcal{A}_s$ are a probability sub-algebra of the classical bounded RVs $L_\infty(\Omega)$, which is a probability sub-algebra of the algebra $L_{\infty-}(\Omega)$ of unbounded RVs which have finite moments of any order, which in turn is a sub-*-algebra of the *-algebra of all RVs, which is not a probability algebra as not every element has a finite expected value.

One more classical example which should be mentioned is the case when Ω is in addition a compact Hausdorff topological space, the σ-algebra $\mathfrak{A}$ is the Borel algebra $\mathfrak{B}(\Omega)$, and the probability measure a Radon measure. Then the RVs given by the continuous complex-valued functions $C(\Omega; \mathbb{C})$—for brevity only $C(\Omega)$—are a sub-probability algebra of $L_\infty(\Omega)$, in fact a C^*-algebra—a Banach space in the $\|\cdot\|_\infty$ norm such that $\|\phi\psi\|_\infty \leq \|\phi\|_\infty \|\psi\|_\infty$ and $\|\phi\phi^*\|_\infty = \|\phi\|_\infty \|\phi^*\|_\infty = \|\phi\|_\infty^2$, so that the product and adjoint are continuous—called the *uniform* algebra on Ω.

These are all examples of classical commutative resp. Abelian algebras of RVs with the state the usual Lebesgue integral (i.e. the usual expected value) w.r.t the measure $\mathbb{P}$. The bounded RVs $L_\infty(\Omega)$ are a *maximal Abelian W^*-algebra* [24] —a W^*-algebra is in simplest terms defined as a C^*-algebra which as Banach space is the dual of another Banach space. It may be shown conversely that any complex maximal Abelian W^*-probability algebra $\mathcal{A}$ is isomorphic to an L_∞-algebra on a probability space, a result that will be used in the sequel—this is the Segal representation. Thus the algebraic approach to probability can completely recover the classical approach due to Kolmogorov which starts from measure spaces and defines RVs as measurable functions. Similarly it can be shown that unital Abelian C^*-algebras are isomorphic to the uniform algebra on a compact space—the Gel'fand representation. Abelian algebras of this kind are therefore often called 'function algebras'.

Let us now consider some non-commutative examples. A simple one is $\mathbb{M}(\mathbb{C}, n) = \mathbb{C}^{n \times n}$, the algebra of complex $n \times n$ matrices with complex conjugate transposition as involution. The language of the algebra is completely the same, except that projections in the abstract setting—which are self-adjoint—are called *orthogonal projections* here. This kind of algebra corresponds to RVs which can take no more than n different values. Let $\boldsymbol{\varrho} \in \mathbb{M}(\mathbb{C}, n)$ be a self-adjoint positive definite matrix with $\operatorname{tr} \boldsymbol{\varrho} = 1$, called a *density matrix*. Then $\mathbb{E}_{\boldsymbol{\varrho}}(\boldsymbol{A}) := \operatorname{tr}(\boldsymbol{\varrho} \boldsymbol{A})$ is a faithful state. Of course any sub-algebra of $\mathbb{M}(\mathbb{C}, n)$ which contains the identity matrix is another example, and the diagonal matrices are an example of a commutative sub-algebra. More powerful is the algebra $\mathbb{M}(\mathrm{L}_\infty(\Omega), n)$ of $n \times n$ random matrices with entries from $\mathrm{L}_\infty(\Omega)$, and the expectation is the expected value of a matrix state, i.e. for $\boldsymbol{A} \in \mathbb{M}(\mathrm{L}_\infty(\Omega), n)$ one may set $\mathbb{E}(\boldsymbol{A}) := \int_\Omega \mathbb{E}_{\boldsymbol{\varrho}}(\boldsymbol{A}(\omega))\, \mathbb{P}(\mathrm{d}\omega)$.

An example generalising the previous case is $\mathscr{L}(\mathcal{H})$, the algebra of bounded linear maps on a complex Hilbert space $\mathcal{H}$ with the adjoint taking the rôle of the involution, or any unital sub-algebra thereof. $\mathscr{L}(\mathcal{H})$ is a W^*-algebra, non-commutative if $\dim \mathcal{H} > 1$. If $\varrho \in \mathscr{L}(\mathcal{H})$ is a nuclear resp. trace-class positive definite operator with unit trace $\operatorname{tr} \varrho = 1$—called again a density matrix—then a state may be defined for $A \in \mathscr{L}(\mathcal{H})$ as $\mathbb{E}_\varrho(A) := \operatorname{tr}(\varrho A)$. The example is in some way universal, as with the *Gel'fand-Naimark-Segal* (GNS) construction any algebra with faithful state may be embedded (faithfully represented) into an algebra of operators on a complex Hilbert space [22–24, 27]; namely $a \in \mathcal{A}$ is represented as $L_a : \mathcal{A} \ni b \mapsto ab \in \mathcal{A}$ in $\mathscr{L}(\mathrm{L}_2(\mathcal{A}))$.

When the Hilbert space $\mathcal{H}$ in question is a Lebesgue space $\mathrm{L}_2(\Omega)$, then any $\kappa \in \mathrm{L}_\infty(\Omega)$ can be represented as a linear map $M_\kappa : \mathrm{L}_2(\Omega) \ni \varphi \mapsto M_\kappa \varphi = \kappa\varphi \in \mathrm{L}_2(\Omega)$. Thus the Abelian algebra $\mathrm{L}_\infty(\Omega)$ is represented as a maximal Abelian W^*-sub-algebra of $\mathscr{L}(\mathrm{L}_2(\Omega))$, it is called the *multiplication algebra* of $\mathrm{L}_2(\Omega)$.

3.4 Weights, Spectrum, and Spectral Functional Calculus

In this abstract setting we have now seen RVs and their expectation and what can be deduced from these concepts. The question arises now as to what an actual observation or sample of such an RV really is. To this end a bit more theory is needed. First it turns out that with non-commuting observables, in an experiment or other observation, only commuting observables (self-adjoint elements) can be observed simultaneously [28]. This is implied by the *uncertainty relation*. Let $a, b \in \mathcal{A}$ be two self-adjoint elements resp. observables, and $[a, b] = ab - ba$ be their commutator. The Cauchy-Bunyakovsky-Schwarz inequality for non-commutative variables easily gives the *uncertainty relation* $\operatorname{var}(a)\operatorname{var}(b) \geq \mathbb{E}(\mathrm{i}[a, b])^2 / 4 \geq 0$; where the expected value on the right hand side is real, as it is easy to see that $\mathrm{i}[a, b]$ is self-adjoint. Once say a has been observed, it is known and its variance vanishes. This shows that it is not possible to observe a and b simultaneously, unless they commute.

Therefore the way to approach this is to consider for some observation or experiment all relevant commuting RVs which can be observed simultaneously, say $a_1, \ldots, a_k \in \mathcal{A}$. They, and hence any powers or polynomials in commuting variables of them can be observed simultaneously, in fact any element of the Abelian sub-probability algebra $\mathcal{A}_x := \mathbb{C}[a_1, \ldots, a_k] \subseteq \mathcal{A}$ generated by them. We shall shortly add more functions beyond polynomials to this list.

As $a_1, \ldots, a_k$ commute, so do the linear operators $L_{a_1}, \ldots, L_{a_k}$ in the GNS-representation, and the algebra $\mathcal{L}_x := \mathbb{C}[L_{a_1}, \ldots, L_{a_k}] \subseteq \mathscr{L}(\mathrm{L}_2(\mathcal{A}))$ generated by them is an Abelian algebra isomorphic to $\mathcal{A}_x$. It is worthwhile at this point to remember that for linear operators the fact that they commute means that they have the same spectral resolution, and the Gel'fand representation of Abelian C^*-algebras and the Segal representation of maximal Abelian W^*-algebras can now be used [7, 8, 24]. This can in fact be employed to obtain a version of the spectral theorem for linear operators. We defer this for a moment in order to point out the importance of spectral theory to the subject.

The concept of a state as a self-adjoint positive normalised linear functional was already introduced. The set of all possible states $S(\mathcal{A}_x)$ is clearly a subset of the dual $\mathcal{A}_x^*$, and due to the normalisation they are actually on the unit ball of $\mathcal{A}_x^*$. One can easily show that $S(\mathcal{A}_x)$ is a closed, convex, and hence weak-* compact subset of the unit ball of the dual. The extreme points of $S(\mathcal{A}_x)$ are called *pure states*, and their convex combinations are weak-* dense in $S(\mathcal{A}_x)$. In the case of classical RVs, the states are naturally represented by probability measures, which are known to form a convex weak-* compact subset of the unit ball in the space of all measures of bounded total variation. The extreme points in that case are well known to be Dirac-δ-measures.

A *weight*, or more specifically a representational weight, also called a multiplicative character, $\alpha \in S(\mathcal{A}_x)$ is a special kind of state, namely one that is also an algebra *-homomorphism $\mathcal{A}_x \to \mathbb{C}$. This means that for $b, c \in \mathcal{A}_x$ and $\eta, \zeta \in \mathbb{C}$ it holds not only that $\alpha(\eta b + \zeta c) = \langle \alpha, \eta b + \zeta c \rangle = \eta\alpha(b) + \zeta\alpha(c)$ (linearity), but also that $\alpha(b^*) = (\alpha(a))^*$ and $\alpha(bc) = \alpha(b)\alpha(c)$. The set of all weights —one-dimensional representations of $\mathcal{A}_x$—is denoted by $\hat{\mathcal{A}}_x$ and is called the *spectrum* of $\mathcal{A}_x$; it is also a weak-* compact subset $\hat{\mathcal{A}}_x \subset S(\mathcal{A}_x) \subset B_1(0) \subset \mathcal{A}_x^*$ of the unit ball of the dual. In the case of classical algebras of RVs the Dirac-δ-measures are a good example of weights.

The best known meaning of the term spectrum is certainly when used with regard to a linear map or an element $c \in \mathcal{A}$ as the set $\sigma(c) = \{\lambda \in \mathbb{C} \mid c - \lambda e \text{ is not invertible}\}$. Now let $\alpha \in \hat{\mathcal{A}}_x$ be any weight, and $b \in \mathcal{A}_x$. If b is invertible with inverse b^{-1}, then $e = bb^{-1}$ implies $1 = \alpha(e) = \alpha(bb^{-1}) = \alpha(b)\alpha(b^{-1})$, and hence $\alpha(b) \neq 0$. Invertible elements can thus not be mapped to 0 by any weight, i.e. any element in the spectrum $\hat{\mathcal{A}}_x$. Looking at $b = c - \alpha(c)e$, one sees that $\alpha(b) = \alpha(c - \alpha(c)e) = \alpha(c) - \alpha(c)\alpha(e) = 0$, hence $b = c - \alpha(c)e$ can not be invertible and therefore $\alpha(c) \in \sigma(c)$ for any weight $\alpha \in \hat{\mathcal{A}}_x$. This explains the name spectrum for the set of weights $\hat{\mathcal{A}}_x$, i.e. each $\alpha(c)$ is in the spectrum of c. In fact, for any $\lambda \in \sigma(c)$ there is an $\alpha \in \hat{\mathcal{A}}_x$ such that $\alpha(c) = \lambda$.

The interpretation now is that when one observes a RV, i.e. sees a sample, then one sees the action of some weight on the RV. Hence the possible values (sample observations) of an abstract RV $a \in \mathcal{A}$ are given by the action of all weights on the RV, $\{\alpha(a) = \langle \alpha, a \rangle \mid \alpha \in \hat{\mathcal{A}}_x\}$. Therefore one concludes that all possible observations of a RV a are given by its spectrum $\sigma(a)$; and as the observables are self-adjoint the spectrum is real, $\sigma(a) \subseteq \mathbb{R}$.

Considering general non-commutative probability algebras, the spectrum of the algebra is often empty as there are no non-zero one-dimensional representations—another sign that these observables cannot be observed simultaneously—but in the case of Abelian algebras like $\mathcal{A}_x$ or $\mathcal{L}_x$, the ones we are considering when examining a concrete experiment or observation, the Gel'fand and Segal representations tell us that the spectrum is rich enough. One may hence use spectral theory of linear operators to determine the set of possible values, as $a \in \mathcal{A}_x$ and $L_a \in \mathcal{L}_x$ in the GNS-construction have the same spectrum.

The representation theorems state [24] that an Abelian probability algebra is isomorphic to a sub-algebra of $\mathrm{L}_\infty(\mathcal{X})$ on a compact Hausdorff space $\mathcal{X}$. In fact, the compact space may be chosen as $\mathcal{X} := \widehat{\mathcal{A}}_x$. The version of the spectral theorem for linear operators which is most useful here—and will be used again for a different purpose in Sect. 4—is that an Abelian algebra of operators like $\mathcal{L}_x$ is not only isomorphic but unitarily equivalent to a sub-algebra of the multiplication algebra on some measure space $\mathcal{Y}$ [7, 24] with total measure equal to unity, i.e. a classical probability space. The spectrum of such a multiplication operator M_κ with the function or RV $\kappa \in \mathrm{L}_\infty(\mathcal{Y})$ [24] is the *essential range* of the function κ. Hence any of the commuting RVs a_ℓ resp. L_{a_ℓ} is represented by a multiplication operator M_{κ_ℓ}, and hence as algebra by an RV $\kappa_\ell \in \mathrm{L}_\infty(\mathcal{Y})$. We may thus say that $\sigma(a_\ell) = \sigma(L_{a_\ell}) = \sigma(M_{\kappa_\ell}) = \sigma(\kappa_\ell) = \text{ess range}\, \kappa_\ell$.

In the classical framework where RVs are measurable maps on a probability space, one important and relevant fact is that the composition of measurable functions is again a measurable function, and one can form new RVs by applying a measurable function to an existing RV. In the algebraic framework presented so far only polynomials—which are kind of natural when dealing with algebras—have appeared. Now if $f : \mathbb{R} \to \mathbb{R}$—or more generally $f : \sigma(a_\ell) \subseteq \mathbb{R} \to \mathbb{R}$—is an essentially bounded measurable function, so is $\gamma = f \circ \kappa_\ell \in \mathrm{L}_\infty(\mathcal{Y})$. Hence there is a corresponding $M_\gamma := f(M_{\kappa_\ell})$ in the multiplication algebra, and a $L_g := f(L_{a_\ell}) \in \mathscr{L}(\mathrm{L}_2(\mathcal{A}))$, and a $g := f(a_\ell)$ in the weak-* closure of $\mathcal{A}_x$. This defines the function f now on the algebra $\mathcal{L}_x$ or $\mathcal{A}_x$, and is the essence of spectral functional calculus, used here to obtain new RVs by applying a measurable function f.

3.5 Extensions

With the spectral functional calculus one may define non-commutative analogues of the classical L_p-spaces for all $1 \leq p \leq \infty$ by extending any probability algebra $\mathcal{A}$ through completion in a certain uniform topology, and not just for $p = 2$ as above. First note that for a positive element $a = b^* b \in \mathcal{A}$ one can always find a unique positive $c \in \mathcal{A}$ such that $a = cc = c^2$ via spectral functional calculus, as this $c = a^{1/2} \in \mathcal{A}$ is the square root. This allows one to define for any $a \in \mathcal{A}$ the absolute value as the positive element $|a| := (a^* a)^{1/2} \in \mathcal{A}$. Similarly one may compute the p-th power for real $p > 0$. For $1 \leq p < \infty$ the expression $\|a\|_p^p := \mathbb{E}(|a|^p)$ defines the p-th power of a norm. Completion of $\mathcal{A}$ w.r.t. any of those norms gives non-commutative Banach spaces $\mathrm{L}_p(\mathcal{A})$, and this agrees for $p = 2$ with the previous definition. It also immediately gives a new algebra $\mathrm{L}_{\infty-}(\mathcal{A}) := \bigcap_{1 \leq p < \infty} \mathrm{L}_p(\mathcal{A})$.

Recalling the functional calculus from the end of the previous Sect. 3.4, one may now state that $\mathrm{L}_p(\mathcal{A})$ contains elements $f(a)$ for $a \in \mathcal{A}$ and certain measurable functions $f \in \mathrm{L}_0(\sigma(a))$. These measurable functions have to be such that in the representation of the Abelian probability sub-algebra $\mathbb{C}[a]$, where a is represented by the multiplication operator M_κ on $\mathrm{L}_2(\mathcal{Y})$ with $\kappa \in \mathrm{L}_\infty(\mathcal{Y})$, and where $\sigma(a) = \sigma(\kappa) = \text{ess range}\,\kappa$, the composite function satisfies $f \circ \kappa \in \mathrm{L}_p(\mathcal{Y})$.

For $p = \infty$ one has to look at the representation of $a \in \mathcal{A}$ through the linear map L_a in the GNS-construction above and define the $\|a\|_\infty := \|L_a\|_{op}$ as the operator norm of L_a, effectively $\|a\|_\infty := \sup_{b \neq 0} \|ab\|_2/\|b\|_2$. One may also define a topology corresponding to the weak operator topology through the semi-norms $q_{b,c}(a) := |\langle L_a b|c\rangle_2| = |\mathbb{E}(c^* ab)|$. Completion of the sub-algebra $\mathcal{A}_\infty := \{a \mid \|a\|_\infty < \infty\} \subseteq \mathcal{A}$ with finite ∞-norm w.r.t. the uniform locally convex topology generated by the semi-norms $q_{b,c}(\cdot)$ gives the probability W^*-algebra $\mathrm{L}_\infty(\mathcal{A})$. This shows that the L_p-spaces of non-commutative RVs can be generated just as in the classical Abelian case.

As already mentioned, the space $\mathrm{L}_2(\mathcal{A})$ is a possible candidate for the space $\mathscr{V}$ appearing in the probabilistic model $S : \mathscr{S} \to \mathscr{V}$. Other candidates may be generated by the following very general construction: if $\mathcal{H}$ is a Hilbert space with inner product $\langle\cdot|\cdot\rangle_0$, and A a possibly unbounded self-adjoint positive operator in $\mathcal{H}$ with dense domain $\operatorname{dom} A$, one may via spectral calculus define A^s for any $s > 0$ with dense domain $\operatorname{dom} A^s$. The positive definite sesqui-linear form given by $\langle f|g\rangle_s := \langle f|g\rangle_0 + \langle A^s f|g\rangle_0$ for $f, g \in \operatorname{dom} A^s$ defines an inner product on $\operatorname{dom} A^s$, the completion of which in the associated topology defines the densely embedded Hilbert space $\mathcal{H}_s \hookrightarrow \mathcal{H}$. Obviously one also has dense embeddings $\mathcal{H}_s \hookrightarrow \mathcal{H}_t$ for $s > t > 0$. Identifying $\mathcal{H}$ with its dual and denoting the dual of $\mathcal{H}_s$ by $\mathcal{H}_{-s}$, one obtains Gel'fand triplets [7, 8] or 'sandwiched' dense embeddings $\mathcal{H}_s \hookrightarrow \mathcal{H} \hookrightarrow \mathcal{H}_{-s}$ of Hilbert spaces. One may even go a step further and introduce the projective limit $\mathcal{S} = \varprojlim_{s>0} \mathcal{H}_s$, depending on A often a nuclear space, which in our case usually will be a new probability algebra. The dual construction of inductive limit $\mathcal{S}^* = \varinjlim_{s>0} \mathcal{H}_{-s}$ then generates the dual space of generalised objects, like the distributions in the sense of Sobolev and Schwartz.

It is worthwhile to recall that the familiar Sobolev-Hilbert spaces $\mathrm{H}^s(\mathbb{R}^n)$ are generated in this way by taking $\mathcal{H} = \mathrm{H}^0(\mathbb{R}^n) = \mathrm{L}_2(\mathbb{R}^n)$ and $A = -\Delta + M_{|x|^2}$, essentially the negative Laplacian added to a multiplication operator. Then the Schwartz space of rapidly decaying smooth functions $\mathscr{S}(\mathbb{R}^n)$ is the projective limit and additionally an Abelian algebra, and its dual $\mathscr{S}'(\mathbb{R}^n)$, the inductive limit, is the Schwartz space of tempered distributions.

The same device can be used here by choosing $\mathcal{H} = \mathrm{L}_2(\mathcal{A})$—a space which is naturally given by the expectation state—and an appropriate operator A; then all the spaces $\mathcal{H}_t, t \in \mathbb{R}$, are possible candidates for $\mathscr{V}$, and the 'regularity' of the RVs in $\mathscr{V} := \mathcal{H}_t$ can be controlled by the parameter $t \in \mathbb{R}$. For $t < 0$ these are spaces of 'generalised' RVs, only defined via the duality, similar to the Sobolev-Hilbert spaces with negative exponent.

One possible classical choice for the linear operator A for $\mathcal{H} = \mathrm{L}_2(\Omega) = \mathrm{L}_2(\mathrm{L}_{\infty-}(\Omega))$ is the following: denote by $H^{:n:}$, $n \in \mathbb{N}_0$, the n-th homogeneous chaos [11, 12] in Wiener's polynomial chaos decomposition $\mathcal{H} = \overline{\bigoplus_{n=0}^{\infty} H^{:n:}}$, and define A by $A\,h := n\,h$ for any $h \in H^{:n:}$; a self-adjoint operator with spectrum $\sigma(A) = \mathbb{N}_0$, called the number operator. More examples of Hilbert spaces of RVs which can be generated in this way may be found in [11, 12], they are all practically defined with the help of the Wiener-Itô polynomial chaos expansion and are all possible candidates for the space $\mathscr{V}$.

3.6 Weak or Generalised Distributions

In any case, this construction of a unital algebra with involution and faithful state leads to an inner product and Hilbert space $\mathscr{V}$, and the state $\mathbb{E}$ may be extended as continuous functional onto the whole space $\mathscr{V}$. This may be used in the mapping $S : \mathscr{S} \to \mathscr{V}$ in Sect. 2. With the possibility of also using non-commutative algebras, this approach also allows to deal with objects such as random matrices, or more generally random fields of tensors of even order [17, 18], which is much more cumbersome in the traditional measure space approach. Our first example $\mathcal{A}_s = \mathrm{L}_{0s}(\Omega)$ also indicates that the algebraic approach is more general and can completely recover the measure space approach [10, 21, 24, 25, 27]. The state takes the place of the usual expectation operator, and it has all its usual properties.

Nevertheless, even in the general abstract setting of a probability algebra, it is possible to define a distribution probability measure or 'law' on $\mathbb{R}$ for any non-commutative self-adjoint RV, i.e. an observable. Classically, for a real-valued or self-adjoint RV $\phi \in \mathrm{L}_0(\Omega)$ the law of ϕ is the *push-forward* $\phi_*\mathbb{P}$ of the probability measure $\mathbb{P}$, given for an element $\mathcal{B}$ of the Borel-σ-algebra $\mathfrak{B}(\mathbb{R})$ by $\phi_*\mathbb{P}(\mathcal{B}) := \mathbb{P}(\phi^{-1}(\mathcal{B}))$.

In the abstract setting, for any $a \in \mathcal{A}$ one may define the *law* of a as a map $\tau_a : \Pi_1 \to \mathbb{C}$ which assigns to any polynomial $Q \in \Pi_1$ the number $\tau_a(Q) := \mathbb{E}\,(Q(a))$. With $a \in \mathcal{A}$ self-adjoint, we know that the spectrum is real: $\sigma(a) \subseteq \mathbb{R}$. Let $\mathcal{J} \subset \mathbb{R}$ be a compact interval which contains the spectrum $\sigma(a)$. The polynomials Π_1^r with real co-efficients are known to be dense in $\mathrm{C}(\mathcal{J}; \mathbb{R})$ due to the Stone-Weierstrass theorem, and τ_a can be shown to be a continuous map, hence may be extended to all of $\mathrm{C}(\mathcal{J}; \mathbb{R})$. From the Riesz-Markov representation theorem it now follows that there is a Radon probability measure $\mathbb{P}_a$ such that $\int_{\mathcal{J}} Q(t)\,\mathbb{P}_a(\mathrm{d}t) = \tau_a(Q)$ for any $Q \in \Pi_1^r$, called the *distribution measure* or *law* of the self-adjoint RV $a \in \mathcal{A}$.

This more general approach via a mapping like $S : \mathscr{S} \to \mathscr{V}$ and abstract probability algebras $\mathcal{A}$ related to $\mathscr{V}$ is also needed in many concrete analytic situations. As a simple example, consider, as in Sects. 1 and 2, a RV ς with values in an infinite dimensional Hilbert space $\mathscr{S}$. For this to be an 'honest' RV, the push-forward distribution $\varsigma_*\mathbb{P} = \mathbb{P} \circ \varsigma^{-1}$ of the probability measure $\mathbb{P}$ should be a

σ-additive measure on the Borel sets $\mathfrak{B}(\mathscr{S})$ of $\mathscr{S}$. It is well known that on a Hilbert space this is only possible (Sazonov's theorem, cf. e.g. [4, 26]) if the correlation C_ς already mentioned in Sect. 2 is a nuclear or trace-class operator. In particular, there is no iso-Gaussian measure—i.e. where $C_\varsigma = I$ is the identity, invariant under unitaries—on an infinite-dimensional Hilbert space; one has to resort to so-called cylindrical pro-measures (which are not σ-additive) or enlargements of the Hilbert space.

The formulations such as with the mapping S or P from above or Sect. 2 circumvent all the difficulties mentioned in the previous paragraph with non-nuclear correlation or covariance operators, and such an assignment is called a *weak distribution* or *generalised* RV [4, 9, 22, 23] resp. a *generalised process* [8]. For example the aforementioned iso-Gaussian weak distribution resp. generalised process—this is also called *white noise* on the Hilbert space $\mathscr{S}$—is very simply defined: Pick any complete orthonormal system $\{\varsigma_n\}_n$ in $\mathscr{S}$ and an infinite sequence of independent identically distributed (iid) standard Gaussian RVs $\{\zeta_n\}$ (zero mean, unit variance) as CONS, and let $\mathscr{H}$ be the Hilbert space generated by them. Define a linear map $W : \mathscr{S} \ni \varsigma_n \mapsto \zeta_n \in \mathscr{H}$, and it is clear that its covariance is $C_W = W^*W = I$, as W is by construction unitary. Hence W defines a weak white noise distribution on $\mathscr{S}$. Other extensions covered by this use of weak distributions are the cases when the covariance has continuous spectrum, as often happens for translation invariant covariance kernels [16] which are diagonalised by the Fourier transform [5].

From all this we conclude that one may define a stochastic model as a weak distribution on $\mathscr{S}$ via a linear map $S : \mathscr{S} \to \mathscr{V}$, where $\mathscr{V}$ was generated by a probability algebra $\mathcal{A}$ as described above, and similarly for $P : \mathcal{V} \to \mathscr{V}$. For a conventional probability model we assume that the algebra is Abelian, but the non-commutative case is useful to model e.g. random matrices or tensor fields [17, 18]. For a dynamical system like the one mentioned in Sect. 1, the equality in the equation is to be understood in a probabilistically weak sense as just described: both sides of the equation are mapped into the space $\mathscr{V}$, and have to be equal as elements of that space, i.e. in a $\mathscr{V}$-weak sense. First we spell out the meaning of the map P:

$$P(\dot{v}(t)) = P(A(\varsigma, \mu; v(t))) + P(f(\varsigma, \mu; t)) \quad \Leftrightarrow$$
$$\forall w \in \mathcal{V} : \quad \langle \dot{v}(t)|w\rangle_{\mathcal{V}} = \langle A(\varsigma, \mu; v(t))|w\rangle_{\mathcal{V}} + \langle f(\varsigma, \mu; t)|w\rangle_{\mathcal{V}}, \tag{7}$$

as an element of $\mathscr{V}$, which in detail in $\mathscr{V}$ means

$$\forall \varphi \in \mathscr{V} : \quad \langle P(\dot{v}(t))|\varphi\rangle_{\mathscr{V}} = \langle P(A(\varsigma, \mu; v(t)))|\varphi\rangle_{\mathscr{V}} + \langle P(f(\varsigma, \mu; t))|\varphi\rangle_{\mathscr{V}}. \tag{8}$$

This allows one to deal with a much wider range of probabilistic situations, including white noise as already alluded to, as well as white noise or a Wiener process in time, as the Itô-integral can be understood as a weak stochastic distribution [11]. The way Eq. (7) and Eq. (8) are formulated also immediately suggests numerical approximations by Galerkin's method—called the *stochastic Galerkin* method [19]—using finite dimensional subspaces $\mathcal{V}_n \subseteq \mathcal{V}$ and $\mathscr{V}_m \subseteq \mathscr{V}$.

It may be noted that this whole development is analogous on how generalised functions or distributions are introduced in the Sobolev-Schwartz framework. There they are linear maps from a 'nice' space—in fact an algebra—such as $\mathscr{S}(\mathbb{R}^n)$ into the *algebra* $\mathbb{C}$ with the distinguished state given as the identity. Here the generalised probabilistic models on a Hilbert space $\mathscr{S}$ are linear maps into an *algebra* $\mathcal{A}$ of random variables with distinguished state $\mathbb{E}$, which again maps the *algebra* $\mathcal{A}$ into the algebra $\mathbb{C}$.

4 Correlation Factorisations

The correlation operators $C = R^*R$, $C_\varsigma = S^*S$, and $C_\upsilon = P^*P$ have already been mentioned in Sect. 2. We shall show the development in terms of the map R defining the parametric variable $r(\mu)$, for the maps S and P which define the stochastic content, everything has to be just repeated with different symbols, which we leave for the reader. In general, one may specify [15, 17, 18] a densely defined map C in $\mathcal{U}$ through the bilinear form

$$\forall u, v \in \mathcal{U}: \quad \langle Cu|v\rangle_{\mathcal{U}} := \langle Ru|Rv\rangle_{\mathcal{Q}}. \tag{9}$$

The map $C = R^*R$ may be called the *'correlation'* operator and is by construction self-adjoint and positive, and if R is continuous so is C. In case the inner product $\langle\cdot|\cdot\rangle_{\mathcal{Q}}$ comes from a measure ϖ on $\mathcal{M}$, so that for two functions ϕ and ψ on $\mathcal{M}$, one has

$$\langle\phi|\psi\rangle_{\mathcal{Q}} := \int_{\mathcal{M}} \phi(\mu)\psi(\mu)\,\varpi(\mathrm{d}\mu), \text{ so that } C = R^*R = \int_{\mathcal{M}} r(\mu)\otimes r(\mu)\,\varpi(\mathrm{d}\mu),$$

the usual formula for the correlation. The space $\mathcal{Q}$ may then be taken as $\mathcal{Q} := \mathrm{L}_2(\mathcal{M}, \varpi)$. A special case is when ϖ is a probability measure, $\varpi(\mathcal{M}) = 1$, the situation we have for $\mathcal{M} \leftarrow \Omega$ and $\varpi \leftarrow \mathbb{P}$, this inspired the term 'correlation' operator. In terms of the developments in Sect. 3 the Hilbert space $\mathcal{Q}$ would be replaced by any of the candidates for $\mathscr{V}$ and instead of $C = R^*R$ we would be investigating $C_\varsigma = S^*S$ or $C_\upsilon = P^*P$.

The spectral theorem for operators in a Hilbert space was already used in Sect. 3, but here we start in a gentler way. To make everything as simple as possible to explain the main underlying idea, assume first that C is a non-singular *trace class* or *nuclear* operator. This means that it is compact, the spectrum $\sigma(C)$ is a point spectrum, has a CONS $\{v_m\}_m \subset \mathcal{U}$ consisting of eigenvectors, with each eigenvalue $\lambda_m \geq \lambda_{m+1} \cdots \geq 0$ positive and counted decreasingly according to their finite multiplicity, and has finite trace $\operatorname{tr} C = \sum_m \lambda_m < \infty$. Then a version of the spectral decomposition of C is

$$C = \sum_m \lambda_m (v_m \otimes v_m). \tag{10}$$

Use this CONS to define a new CONS $\{s_m\}_m$ in $\mathcal{Q}$: $\lambda_m^{1/2} s_m := R v_m$, to obtain the corresponding *singular value decomposition* (SVD) of R and R^*:

$$R = \sum_m \sqrt{\lambda_m} (s_m \otimes v_m); \quad R^* = \sum_m \sqrt{\lambda_m} (v_m \otimes s_m);$$

$$r(\mu) = \sum_m \sqrt{\lambda_m}\, s_m(\mu) v_m = \sum_m (R^* s_m)(\mu), \tag{11}$$

The set $\varsigma(R) = \{\sqrt{\lambda_m}\}_m = \sqrt{\sigma(C)} \subset \mathbb{R}_+$ are the *singular values* of R and R^*. The last relation in Eq. (11) is the so-called *Karhunen-Loève expansion* or *proper orthogonal decomposition* (POD). The finite trace condition of C translates into the fact that r is in $\mathcal{U} \otimes \mathcal{Q}$. If in that relation the sum is *truncated* at $n \in \mathbb{N}$, i.e.

$$r(\mu) \approx r_{\text{ROM}}(\mu) = \sum_{m=1}^{n} \sqrt{\lambda_m}\, s_m(\mu) v_m = \sum_{m=1}^{n} (R^* s_m)(\mu), \tag{12}$$

we obtain the *best n-term approximation* to $r(\mu)$ in the norm of $\mathcal{U}$. Observe that r is linear in the s_m. This means that by choosing the 'co-ordinate system' $\mathcal{M} \ni \mu \mapsto (s_1(\mu), \ldots, s_m(\mu), \ldots) \in \mathbb{R}^{\mathbb{N}}$, one obtains a *linear / affine* representation where the first co-ordinates are the most important ones. For the stochastic cases $C_\varsigma = S^* S$ and $C_v = P^* P$ we point out again as in Sect. 3 that the nuclearity of C_ς resp. C_v is necessary for the existence of a measurable map $\varsigma : \Omega \to \mathscr{S}$ resp. $v(\varsigma(\cdot), \mu; t) : \Omega \to \mathcal{V}$.

Equivalently this means that S resp. P has to be a *Hilbert-Schmidt* operator, e.g. [7], a condition which severely restricts stochastic models. There is a practical need to consider more general classes of correlation operators, as already evidenced in the seminal paper by Karhunen [13, 14], where integral transforms for representations as in Eq. (12) were investigated. This more general view is for example necessary to consider homogeneous or stationary random fields or stochastic processes, cf. e.g. [16].

One formulation of the spectral decomposition extending Eq. (10), already used implicitly in Sect. 3, which does not require C to be nuclear [7, 24], nor do C or R have to be continuous, which was used already in Sect. 3 and has to be applied here to the Abelian algebra $\mathbb{C}[C]$, is as follows. The densely defined self-adjoint and positive operator $C : \mathcal{U} \to \mathcal{U}$ is unitarily equivalent with a multiplication operator M_γ on an appropriate measure space $\mathcal{T}$,

$$C = V M_\gamma V^*, \tag{13}$$

where the unitary map is $V : \mathrm{L}_2(\mathcal{T}) \to \mathcal{U}$, and M_γ multiplies a $\psi \in \mathrm{L}_2(\mathcal{T})$ with a real-valued function γ; $M_\gamma : \psi \mapsto \gamma\psi$. In case C is bounded, so is $\gamma \in \mathrm{L}_\infty(\mathcal{T})$. As C is positive, $\gamma(t) \geq 0$ for $t \in \mathcal{T}$, and the essential range of γ is the spectrum of C. In Sect. 3 this was already used for the Abelian algebra $\mathbb{C}[a]$ resp. $\mathbb{C}[L_a]$, which says then that any member of that algebra is unitarily equivalent to a multiplication operator.

As already indicated, via spectral calculus one may define the square root $M_\gamma^{1/2} := M_{\sqrt{\gamma}}$, and a factorisation similar to $C = R^*R$ is obtained via $C = (VM_{\sqrt{\gamma}})(VM_{\sqrt{\gamma}})^* =: G^*G$. From this factorisation and the spectral decomposition Eq. (13) follows another singular value decomposition (SVD) of R and R^*, which is

$$R = UM_{\sqrt{\mu}}V^*, \quad R^* = VM_{\sqrt{\mu}}U^*, \tag{14}$$

where $U : \mathrm{L}_2(\mathcal{T}) \to \mathcal{Q}$ is a unitary operator. Having $M_\gamma^{1/2}$ allows us to compute the square root of C: $C^{1/2} = VM_\gamma^{1/2}V^*$, and from it the self-adjoint positive definite factorisation $C = C^{1/2}C^{1/2}$.

Consider now an arbitrary factorisation $C = B^*B$, where $B : \mathcal{U} \to \mathcal{H}$ is a map to a Hilbert space $\mathcal{H}$. Any two such factorisations $B_1 : \mathcal{U} \to \mathcal{H}_1$ and $B_2 : \mathcal{U} \to \mathcal{H}_2$ with $C = B_1^*B_1 = B_2^*B_2$ are [18] *unitarily equivalent* in that there is a unitary map $X_{21} : \mathcal{H}_1 \to \mathcal{H}_2$ such that $B_2 = X_{21}B_1$. Each such factorisation is also unitarily equivalent to R, i.e. there is a unitary $X : \mathcal{H} \to \mathcal{Q}$ such that $R = XB$. For finite dimensional spaces, a favourite choice for such a decomposition of C is the Cholesky factorisation $C = LL^*$, where $B = L^*$ is represented by an upper triangular matrix.

Let us go back to the situation of Eq. (10) and how the SVD of the factors R in Eq. (11) in the factorisation $C = R^*R$ was generated. In the same way a SVD of any of the factorisations just considered may be generated with left-singular vectors $h_m := BC^{-1}R^*s_m = BC^{-1/2}v_m$, plus the analogue of Eq. (14), i.e.

$$B = \sum_m \sqrt{\lambda_m}(h_m \otimes v_m); \quad B^* = \sum_m \sqrt{\lambda_m}(v_m \otimes h_m);$$

$$r = \sum_m \sqrt{\lambda_m}\, h_m v_m = \sum_m B^* h_m,$$

and with $W = X^*U$:

$$B = WM_{\sqrt{\mu}}V^*, \quad R^* = VM_{\sqrt{\mu}}W^*.$$

The left-singular vectors h_m can now be thought of living on any of the spaces which appeared in the factorisation, i.e. generically $\mathcal{H}$, for which we have just seen the examples $\mathcal{H} = \mathrm{L}_2(\mathcal{T})$ and $\mathcal{H} = \mathcal{U}$ (not necessarily very useful) [18].

Instead of $C = B^*B$, one may of course consider

$$C_{\mathcal{H}} = BB^* = WM_\gamma W^* \tag{15}$$

on $\mathcal{H}$, which has the same spectrum as C—with C nuclear, $C_{\mathcal{H}}$ is also nuclear—and the whole game can be repeated by looking at the spectral decompositions of $C_{\mathcal{H}}$.

When one takes the special case $\mathcal{H} = \mathcal{Q}$ with $C_{\mathcal{Q}} = RR^*$, we see that $C_{\mathcal{Q}} s_m = \lambda_m s_m$, and $s_m = UV^* v_m$, as well as $C_{\mathcal{Q}} = UV^*CVU^*$. This abstract equation can be spelt out in more analytical detail for the special case when the inner product on $\mathcal{Q}$ is given by a measure ϖ on $\mathcal{P}$, as it then becomes

$$\langle C_{\mathcal{Q}}\phi|\psi\rangle_{\mathcal{Q}} = \langle R^*\varphi|R^*\psi\rangle_{\mathcal{U}} = \iint_{\mathcal{M}\times\mathcal{M}} \varphi(\mu_1)\varkappa(\mu_1,\mu_2)\psi(\mu_2)\ \varpi(\mathrm{d}\mu_1)\varpi(\mathrm{d}\mu_2),$$

i.e. $C_{\mathcal{Q}}$ is a Fredholm integral operator with kernel $\varkappa$—on $\mathcal{Q}$ the kernel is in general not reproducing—and its spectral decomposition $C_{\mathcal{Q}} = \sum_m \lambda_m s_m \otimes s_m$ is nothing but the familiar theorem of Mercer [6]. Factorisations of $C_{\mathcal{Q}}$ are then factorisations of the kernel $\varkappa(\mu_1, \mu_2)$ and the corresponding representations of $r(\mu)$ are obtained by integral transforms [17, 18], as already indicated by Karhunen in [13, 14]. The abstract setting outlined in this section can now be applied to the analysis of a great number of different situations, see [18] for more detail.

As already indicated, the spectral decomposition Eq. (13) allows one to go beyond the requirement that C be nuclear, but in the case of a probability assignment the push-forward is not a measure any more on $\mathcal{U}$, though it can still be useful in the computation considering weak distributions. Another formulation of the spectral decomposition in the same vein as Eq. (10) allows also to cover the general case [7, 8]. The space $\mathcal{U} = \overline{\bigoplus}_j \mathcal{U}_j$ can be decomposed into a orthogonal direct sum of invariant subspaces $\mathcal{U}_j$ on each of which the operator has a simple spectrum. So we may assume for this that the operator has a simple spectrum, otherwise consider each subspace $\mathcal{U}_j$ in turn. It turns out that one can find a so-called *rigged Hilbert space* or *Gel'fand triplet*: $\mathcal{N} \hookrightarrow \mathcal{U} \hookrightarrow \mathcal{N}^*$ with $\mathcal{N}$ nuclear and densely embedded in $\mathcal{U}$. The eigenvalue equation for a self-adjoint operator C can be written in weak form: for $\lambda \in \sigma(C)$ find $v_\lambda \in \mathcal{U}$ s.t. for all $w \in \mathcal{U}$ $\langle w|Cv_\lambda\rangle = \lambda\langle w|v_\lambda\rangle$, but there may be no $v_\lambda \in \mathcal{U}$ if λ is merely in the spectrum and not also an eigenvalue. Using duality, this is now weakened to: for $\lambda \in \sigma(C)$ find $v_\lambda \in \mathcal{N}^*$ s.t. for all $w \in \mathcal{N}$ $\langle Cw, v_\lambda\rangle = \lambda\langle w, v_\lambda\rangle$, and it turns out that one can find such $v_\lambda \in \mathcal{N}^*$, in the larger space $\mathcal{N}^*$. With this the Eq. (10) may be generalised, where, as the spectrum $\sigma(C)$ may be continuous, the sum in general has to be replaced by an integral w.r.t. a measure ρ on $\sigma(C) \subseteq \mathbb{R}$. As $C = R^*R$, the operator $C_{\mathcal{Q}} = RR^*$ has the same spectrum, and can be decomposed in a Gel'fand triplet or rigged Hilbert space $\mathcal{P} \hookrightarrow \mathcal{Q} \hookrightarrow \mathcal{P}^*$ with $s_\lambda \in \mathcal{P}^*$:

$$C = \int_{\sigma(C)} \lambda\ v_\lambda \otimes v_\lambda\ \rho(\mathrm{d}\lambda); \qquad C_{\mathcal{Q}} = \int_{\sigma(C)} \lambda\ s_\lambda \otimes s_\lambda\ \rho(\mathrm{d}\lambda). \tag{16}$$

The $s_\lambda \in \mathcal{P}^*$ may be seen as generalised functions, and both decompositions together in Eq. (16) allow to write a SVD-like decomposition of R and R^*, corresponding to Eq. (11), and have a representation of $r(\mu)$ in a weak sense as a Karhunen-Loève integral over $\mathcal{P}^*$-generalised functions:

$$R = \int_{\sigma(C)} \sqrt{\lambda}\,(s_\lambda \otimes v_\lambda)\,\rho(\mathrm{d}\lambda); \quad R^* = \int_{\sigma(C)} \sqrt{\lambda}\,(v_\lambda \otimes s_\lambda)\,\rho(\mathrm{d}\lambda);$$
$$r(\mu) = \int_{\sigma(C)} \sqrt{\lambda}\, s_\lambda(\mu) v_\lambda\,\rho(\mathrm{d}\lambda) = \int_{\sigma(C)} (R^* s_\lambda)(\mu)\,\rho(\mathrm{d}\lambda). \tag{17}$$

One familiar and frequent place where this occurs (e.g. [16]) is the classical spectral representation of a stationary stochastic process

$$q(t) = \int_{\mathbb{R}} \sqrt{S(\omega)}\,\exp(\mathrm{i}\omega t)\,Z(\mathrm{d}\omega),$$

where $\sqrt{S(\omega)}$ is the square root of the spectral density—corresponding to $\sqrt{\lambda}$—and $Z(\mathrm{d}\omega)$ is a random measure with orthogonal increments and unit variance. This random measure corresponds to $v_\lambda\,\rho(\mathrm{d}\lambda)$ in Eq. (17), the space $\mathcal{Q}$ corresponds to $\mathrm{L}_2(\mathbb{R})$, the space of generalised functions $\mathcal{P}^*$ corresponds to the Schwartz space of tempered distributions $\mathcal{S}'(\mathbb{R})$, and the generalised eigenfunction $s_\lambda(\mu)$ corresponds to $\exp(\mathrm{i}\omega t)$, a generalised eigenfunction of a stationary covariance kernel which is in $\mathcal{S}'(\mathbb{R})$ but not in $\mathrm{L}_2(\mathbb{R})$ [5].

5 Conclusion

Parametric mappings have been analysed together with random variables with values in infinite dimensional spaces and their generalisations via an associated linear map, enabling the analysis by using well known techniques for the analysis of linear mappings. In the case of stochastic elements this leads to what is called weak distributions, a generalisation of the usual concept of a random variable.

In this connection algebras of random variables, the so-called algebraic approach to probability, leads to a concise description of the generation of appropriate spaces of random variables, and can naturally be used to specify randomness on infinite dimensional spaces via weak distributions. This has as a fundamental building block, next to the algebra of random variables, a distinguished self-adjoint, positive, and normalised linear functional called the state, which may be interpreted as an expectation operator. It is this setting that turns out to be conceptually much simpler than the measure-theoretic point of view, especially in the infinite dimensional setting. In particular this allows a natural approach to random matrices and tensor fields, where the random variables do not necessarily have to commute, and the interesting object is the behaviour of their spectra, a distinctly analytic and algebraic concept which is much more complicated to treat with the usual measure-theoretic background.

The associated linear map leads to the self-adjoint and positive definite so-called 'correlation operator', as well as its different factorisations. Different representations generate different factorisations and thus allow a uniform analysis of their behaviour via an analysis of linear maps. It is in particular the different factorisations, and especially the spectral decomposition, which lead to suggestions for reduced order models and their analysis.

Not only does each separated representation define an associated linear map, but conversely under the restrictive conditions of a nuclear or trace-class correlation operator each factorisation induces a Karhunen-Loève- or proper orthogonal decomposition (POD)-like separated representation. The extension of this idea to arbitrary non-nuclear correlations operators is indicated through integral transforms, exemplified through the use of appropriate spectral decompositions, either via multiplication operators or as spectral integrals with rigged Hilbert spaces. These representations must be classed as generalised maps or generalised random variables, they can only be considered in a duality framework in a weak sense. This can be seen as an analogy to how normal generalised functions or distributions in the Sobolev-Schwartz sense are treated as a dual space of very smooth functions, and in fact the theoretical treatment follows along similar lines.

As this is a very short note touching on many diverse subjects to show their interconnection, it can naturally only be brief and in many cases just provides hints which have to be followed further with the references indicated. The analytic techniques used are 'classical' and have been developed along with the growth of quantum theory in the 1940s. It is their combination and uniform view from the point of linear functional analysis which is novel here.

Acknowledgments Partly supported by the Deutsche Forschungsgemeinschaft (DFG) through SPP 1886 and SFB 880.

References

1. Benner, P., Gugercin, S., Willcox, K.: A survey of projection-based model reduction methods for parametric dynamical systems. SIAM Rev. **57**, 483–531 (2015). https://doi.org/10.1007/10.1137/130932715
2. Benner, P., Ohlberger, M., Patera, A.T., Rozza, G., Urban, K. (eds.): Model reduction of parametrized systems. In: MS&A – Modeling, Simulation & Applications, vol. 17. Springer, Berlin (2017). https://doi.org/10.1007/978-3-319-58786-8
3. Berlinet, A., Thomas-Agnan, C.: Reproducing Kernel Hilbert Spaces in Probability and Statistics. Kluwer, Dordrecht (2004)
4. Bogachev, V.I., Smolyanov, O.G.: Topological Vector Spaces and Their Applications. Springer, Berlin (2017). https://doi.org/10.1007/978-3-319-57117-1
5. Bracewell, R.N.: The Fourier Transform and Its Applications. McGraw-Hill, New York (1978)
6. Courant, R., Hilbert, D.: Methods of Mathematical Physics. John Wiley & Sons, Chichester (1989)

7. Dautray, R., Lions, J.L.: Spectral theory and applications. In: Mathematical Analysis and Numerical Methods for Science and Technology, vol. 3. Springer, Berlin (1990)
8. Gel'fand, I.M., Vilenkin, N.Y.: Applications of harmonic analysis. In: Generalized Functions, vol. 4. Academic Press, New York (1964)
9. Gross, L.: Measurable functions on Hilbert space. Trans. Am. Math. Soc. **105**(3), 372–390 (1962). https://doi.org/10.2307/1993726
10. Hiai, F., Petz, D.: The semicircle law, free random variables and entropy. In: Mathematical Surveys and Monographs, vol. 77. American Mathematical Society, Providence (2000)
11. Holden, H., Øksendal, B., Ubøe, J., Zhang, T.S.: Stochastic Partial Differential Equations. Birkhäuser, Basel (1996)
12. Janson, S.: Gaussian Hilbert spaces. Cambridge Tracts in Mathematics, 129. Cambridge University Press, Cambridge (1997)
13. Karhunen, K.: Über lineare Methoden in der Wahrscheinlichkeitsrechnung. Ann. Acad. Sci. Fennicae. Ser. A. I. Math. Phys. **37**, 1–79 (1947)
14. Karhunen, K., Selin (transl.), I.: On linear methods in probability theory – Über lineare Methoden in der Wahrscheinlichkeitsrechnung – 1947. U.S. Air Force – Project RAND T-131, The RAND Corporation, St Monica, CA (1960). https://www.rand.org/pubs/translations/T131.html. English Translation
15. Krée, P., Soize, C.: Mathematics of Random Phenomena—Random Vibrations of Mechanical Structures. D. Reidel, Dordrecht (1986)
16. Matthies, H.G.: Uncertainty quantification with stochastic finite elements. In: Stein, E., de Borst, R., Hughes, T.J.R. (eds.) Encyclopaedia of Computational Mechanics, vol. 1. John Wiley & Sons, Chichester (2007). https://doi.org/10.1002/0470091355.ecm071. Part 1. Fundamentals. Encyclopaedia of Computational Mechanics
17. Matthies, H.G., Ohayon, R.: Analysis of parametric models for coupled systems (2018). arXiv: 1806.07255 [math.NA]. http://arxiv.org/1806.07255
18. Matthies, H.G., Ohayon, R.: Analysis of parametric models – linear methods and approximations (2018). arXiv: 1806.01101 [math.NA]. http://arxiv.org/1806.01101
19. Matthies, H.G., Keese, A.: Galerkin methods for linear and nonlinear elliptic stochastic partial differential equations. Comput. Methods Appl. Mech. Eng. **194**(12–16), 1295–1331 (2005). https://doi.org/10.1016/j.cma.2004.05.027
20. Matthies, H.G., Litvinenko, A., Pajonk, O., Rosić, B.V., Zander, E.: Parametric and uncertainty computations with tensor product representations. In: Dienstfrey, A., Boisvert, R. (eds.) Uncertainty Quantification in Scientific Computing. IFIP Advances in Information and Communication Technology, vol. 377, pp. 139–150. Springer, Boulder (2012). https://doi.org/10.1007/978-3-642-32677-6
21. Mingo, J.A., Speicher, R.: Free probability and random matrices. In: Fields Institute Monographs, vol. 35. Springer, Berlin (2017). https://doi.org/10.1007/978-1-4939-6942-5
22. Segal, I.E.: Distributions in Hilbert space and canonical systems of operators. Trans. Am. Math. Soc. **88**(1), 12–41 (1958). https://doi.org/10.2307/1993234
23. Segal, I.E.: Nonlinear functions of weak processes. I. J. Funct. Anal. **4**(3), 404–456 (1969). https://doi.org/10.1016/0022-1236(69)90007-X
24. Segal, I.E., Kunze, R.A.: Integrals and Operators. Springer, Berlin (1978)
25. Speicher, R.: Free probability theory. Jahresber Dtsch Math-Ver **119**, 3–30 (2017). https://doi.org/10.1365/s13291-016-0150-5
26. Sullivan, T.J.: Introduction to uncertainty quantification. In: Texts in Applied Mathematics, vol. 63. Springer, Berlin (2015). https://doi.org/10.1007/978-3-319-23395-6
27. Voiculescu, D.V., Dykema, K.J., Nica, A.: Free random variables. In: CRM Monograph Series, vol. 1. American Mathematical Society, Providence (1992)
28. Whittle, P.: Probability via Expectation, 4th edn. Springer Texts in Statistics. Springer, Berlin (2000)

Reduced Order Isogeometric Analysis Approach for PDEs in Parametrized Domains

Fabrizio Garotta, Nicola Demo, Marco Tezzele, Massimo Carraturo, Alessandro Reali, and Gianluigi Rozza

Abstract In this contribution, we coupled the isogeometric analysis to a reduced order modelling technique in order to provide a computationally efficient solution in parametric domains. In details, we adopt the free-form deformation method to obtain the parametric formulation of the domain and proper orthogonal decomposition with interpolation for the computational reduction of the model. This technique provides a real-time solution for any parameter by combining several solutions, in this case computed using isogeometric analysis on different geometrical configurations of the domain, properly mapped into a reference configuration. We underline that this reduced order model requires only the full-order solutions, making this approach non-intrusive. We present in this work the results of the application of this methodology to a heat conduction problem inside a deformable collector pipe.

F. Garotta
Department of Civil Engineering and Architecture, Università degli Studi di Pavia, Pavia, Italy
e-mail: fabrizio.garotta01@universitadipavia.it

N. Demo · M. Tezzele
mathLab, Mathematics Area, SISSA, Trieste, Italy
e-mail: ndemo@sissa.it; mtezzele@sissa.it

M. Carraturo · A. Reali
Department of Civil Engineering and Architecture, Università degli Studi di Pavia, Pavia, Italy
e-mail: massimo.carraturo01@universitadipavia.it; alereali@unipv.it

G. Rozza (✉)
mathLab, Mathematics Area, SISSA, Trieste, Italy
e-mail: grozza@sissa.it

M. D'Elia et al. (eds.), *Quantification of Uncertainty: Improving Efficiency and Technology*, Lecture Notes in Computational Science and Engineering 137,
https://doi.org/10.1007/978-3-030-48721-8_7

1 Introduction

Nowadays, in the industrial and engineering fields, as well as in the biomedical sciences, fast and accurate simulations are crucial in several applications, such as, for example, shape design optimization and real-time patient specific diagnosis and control. To this end many reduced order modelling (ROM) techniques have been developed in the last decade [9, 32, 37, 38, 41]. We cite among others reduced basis methods [22, 35], proper orthogonal decomposition (POD) [10], proper generalized decomposition [8], and hierarchical model reduction [4, 28, 29]. Reduced order modelling can be integrated to various high fidelity methods such as finite element [33], spectral element, or finite volume methods [21, 46, 47]. We do mention also recent features of the reduced methods able to provide useful algorithms for uncertainty quantification as well as data science and better exploitation of high performance computing [7, 50].

Reduced order methods allow a fast and reliable approximation of parameterized PDEs by constructing small-sized approximation spaces. Using these spaces for the discretization of the original problem, it is possible to build a reduced order model that is a sufficiently accurate approximation of the original full order problem. The fundamental characteristic that makes the method functional from an engineering and industrial point of view is that the offline phase (more expensive), where the actual analysis is carried out, is performed only once in high performance computing (HPC) structures and then remains. The online phase exploits the calculations already performed and therefore a small computational power, like the one of laptops or portable devices, is sufficient. This ensures real-time processing of the problem without having to access HPC facilities for the analysis of new parameters.

ROM is crucial in industrial simulation-based design optimization problems in naval and nautical engineering [16, 50], but also in biomedical applications for coronary bypass [1, 2] and carotid occlusions [48] for example.

The focus of this work is to embed in a ROM framework the isogeometric analysis (IGA) [11, 12, 23] for the simulation of heat diffusion inside a collector pipe. The proposed approach is integrated in a numerical pipeline with efficient geometrical parameterization of the domain through free form deformation (FFD) [24, 42], an IGA solver as high fidelity discretization, and POD with interpolation (PODI) [6, 34, 40] for a fast evaluation of the solution field at untested parameters. Figure 1 depicts the schema of the complete computational pipeline.

We chose FFD instead of other general purpose geometrical parameterization techniques such as radial basis functions (RBF) interpolation [5, 25, 27], or inverse distance weighting (IDW) interpolation [3, 20, 43, 54], because of the possibility to use only few parameters to deform the entire domain of interest.

The IGA approach allows to integrate classical finite element analysis (FEA) into conventional industrial CAD tools. To this end IGA directly employs standard CAD representation bases, e.g., B-splines or Non-uniform rational B-splines (NURBS), as basis for the analysis. In this way we can avoid the classical mesh generation and the consequent geometrical approximation error, obtaining a direct design-to-

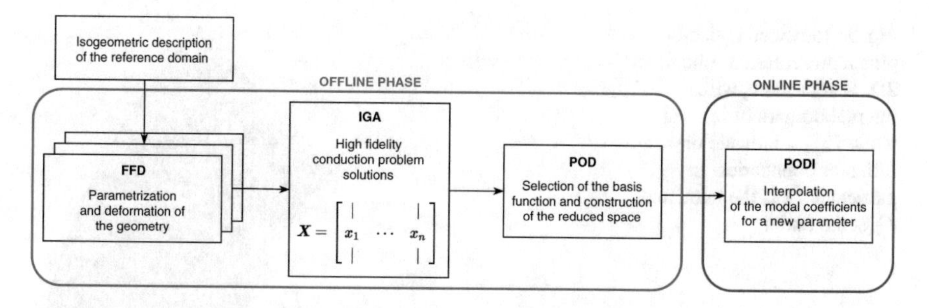

Fig. 1 Offline-online numerical pipeline. We consider the heat conduction problem described by PDEs defined on parametrized geometry. The parametrization of the geometry is managed through the FFD which allows the realization of different deformation settings. In the offline stage we first solve a full order model using IGA to derive the solutions and then, we create a ROM applying the POD as space reduction technique. Finally, through the PODI, in the online stage we look for the real-time solution of the reduced problem for a new parameter

analysis simulation, since we are employing the same class of functions for both the geometry parameterization and the solution fields approximation. In this context, the IGA is ideal for solving elliptic and parabolic PDEs on domains of very general shape. However, when the objective is to solve the same problem repeatedly on different domains, the cost of setting up the problem (meshing, matrix assembly) every time from scratch can be too high. An optimal solution to this problem is a reduction of the model.

Previous IGA-ROMs works were developed in the last years [26, 39, 55], but we underline that the novelty of this work is related to the POD with interpolation integration into the numerical pipeline, for a non-intrusive approach. Even if in this work we present a proof of concept we stress the fact that it can play an important role for the integration with industrial CAD files being independent from the IGA full order solver used.

2 The Parametrized Heat Conduction Problem Inside a Collector Pipe

The problem of interest we are going to solve throughout this work is a parametrized heat diffusion problem inside a collector pipe.

Let $\Omega \subset \mathbb{R}^2$ be a domain that describes an idealized collector pipe in 2D, as shown in Fig. 2. We will refer to Ω as the reference domain, and for practical reasons it represents the undeformed geometry.

We also introduce $\mathbb{D} \subset \mathbb{R}^m$ which is our parameter space, and for convenience it will be an hypercube. For every $\mu \in \mathbb{D}$, which is a vector of geometrical parameters describing a particular deformation of the domain, we can define a shape morphing map $\mathscr{M}(x; \mu) : \mathbb{R}^2 \to \mathbb{R}^2$. We will indicate the deformed domain as $\Omega(\mu) = \mathscr{M}(\Omega; \mu)$. We refer to Sect. 4 for the specific characterization of such mapping.

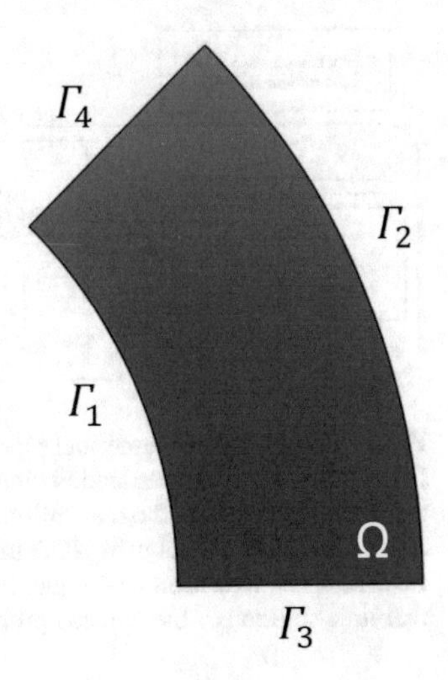

Fig. 2 Idealized collector pipe representation scheme in 2D. Ω represents with internal domain of interest, while $\Gamma_{1,\dots,4}$ indicate the different boundaries. In particular Γ_3 is the inlet, and Γ_4 is the outlet

The parametrized heat diffusion problem reads: find $u(\mu)$ such that

$$\begin{cases} \Delta u(\mu) = 0 & \text{in } \Omega(\mu) \\ u(\mu) = 0 & \text{in } \Gamma_{1,2,4} \\ \nabla u(\mu) \cdot n = g & \text{in } \Gamma_3, \end{cases} \tag{1}$$

where u is the temperature distribution inside the domain, and g represents the prescribed heat flux at the inlet. The Dirichlet boundary conditions describe a perfect insulator with no flux. For sake of simplicity from now on $g = 1$.

We can introduce the weak formulation of the problem (1). We denote with

$$V = H^1_{0_{124}}(\Omega) := \left\{ v \in H^1(\Omega) \text{ such that } v|_{\Gamma_1,\Gamma_2,\Gamma_4} = 0 \right\}$$

the Sobolev space for the temperature. Multiplying the first equation of the system by a test function and integrating by parts we obtain the following problem: given $\mu \in \mathbb{D}$, find $u \in V$ such that

$$a(u, v; \mu) = L(v; \mu) \qquad \forall v \in V, \tag{2}$$

where the bilinear form $a(u, v; \mu)$, and the linear form $L(v; \mu)$, are defined as follows

$$a(u, v; \mu) = \int_\Omega \nabla u(\mu)\, \nabla v \, dV \qquad \forall u, v \in V, \tag{3}$$

$$L(v; \mu) = \int_{\Gamma_3} g\, v \, dS \qquad \forall v \in V. \tag{4}$$

3 Isogeometric Paradigm for Both the Geometry and the Solution Field

Usually a CAD representation of the domain is obtained through B-splines or NURBS, which are able to exactly describe all conic sections. Here we are going to briefly present both.

It is possible to derive the B-spline basis functions of order p using Cox-de Boor's recursion formula [13, 14]

$$N_{i,p}(\xi) = \frac{\xi - \xi_i}{\xi_{i+p} - \xi_i} N_{i,p-1}(\xi) + \frac{\xi_{i+p+1} - \xi}{\xi_{i+p+1} - \xi_{i+1}} N_{i+1,p-1}(\xi), \tag{5}$$

where

$$N_{i,0}(\xi) = \begin{cases} 1 & \text{if} \quad \xi_i \le \xi < \xi_{i+1}, \\ 0 & \text{otherwise}, \end{cases} \tag{6}$$

and $\Xi = \{\xi_1, \xi_2, \ldots, \xi_{n+p+1}\}$ is the knot vector, a non-decreasing set of coordinates with $\xi_i \in \mathbb{R}$, and n is the number of basis functions which comprise the B-spline. B-spline curves in $\mathbb{R}^d$ are constructed by taking a linear combination of B-spline basis functions. The vector valued coefficients of the basis functions are referred to as *IGA control points*. Given n basis functions $\{N_{i,p}\}_{i=1}^n$ of order p, and the corresponding control points P_i, a piecewise polynomial B-spline curve is given by

$$C(\xi) = \sum_{i=1}^{n} P_i N_{i,p}(\xi). \tag{7}$$

IGA control points are points that define the so called control mesh, which is a mesh made up by the multilinear elements that define and control the geometry of the problem. It is important to emphasize that the control mesh does not coincide with the actual geometry of the physical domain. The control points can be considered as the analog of the nodal coordinates of the finite element method, with the difference that, in IGA contest they represent the coefficients of the basis functions of a B-spline having non-interpolatory nature. A generalization of B-splines are the NURBS, which are a rational version of them and can thus represent exactly any kind of geometry. This feature of NURBS allows to bypass altogether the computationally expensive mesh generation and refinement cycle and at the same time to preserve the exact geometry of the CAD model. The key insight of IGA is to use the geometrical map of the NURBS representation as a basis for the push forward used in the analysis. NURBS basis functions of order p are defined through B-spline basis functions as

$$R_{i,p}(\xi) = \frac{N_{i,p}(\xi) w_i}{W(\xi)} = \frac{N_{i,p}(\xi) w_i}{\sum_{j=1}^{n} N_{j,p}(\xi) w_j}, \tag{8}$$

where w_i are the associated weights. Taking a linear combination of basis functions and control points, we express a NURBS curve as

$$C(\xi) = \sum_{i=1}^{n} P_i R_{i,p}(\xi). \tag{9}$$

The isoparametric concept is utilized for both FEA and IGA. However, the difference between FEA and IGA lies in the bases employed for the analysis. In IGA, the inputs for the calculations come from a CAD model defined by NURBS curves, which can be used directly for analysis, while in FEA the finite element mesh is generated starting from an approximation of the original geometry. The mapping from the parametric domain to the physical domain is then given by

$$x = \sum_{k=1}^{n} R_k(\xi) P_k, \tag{10}$$

where $R_k(\xi)$ are the NURBS basis functions, n is the number of control points, ξ the parametric coordinate and P_k is the k-th control point. In an isoparametric formulations the displacement field is approximated by the same shape functions formally:

$$u = \sum_{k=1}^{n} R_k(\xi) u_k, \tag{11}$$

where u_k is the value of the displacement field at the control point P_k. It is therefore referred to as a control variable or more generally a degree of freedom.

In Fig. 3 we present the IGA representation of the domain Ω we described in Sect. 2. In red the six IGA control points defining the NURBS curves. In particular the knot vectors Ξ and H are defined as follows:

$$\Xi = \{0, 0, 1, 1\} \qquad k = 2 \quad p = 1,$$
$$H = \{0, 0, 0, 1, 1, 1\} \qquad k = 3 \quad p = 2,$$

where k and p respectively indicate the multiplicity and the degree of the polynomial $(k = p + 1)$.

4 Shape Parameterization and Deformation Through Free Form Deformation

The FFD method has been proposed in [42]. It was initially used as a tool for computer-assisted geometric design and animation, nowadays instead it is mostly adopted in academia, industry and several engineering application fields as

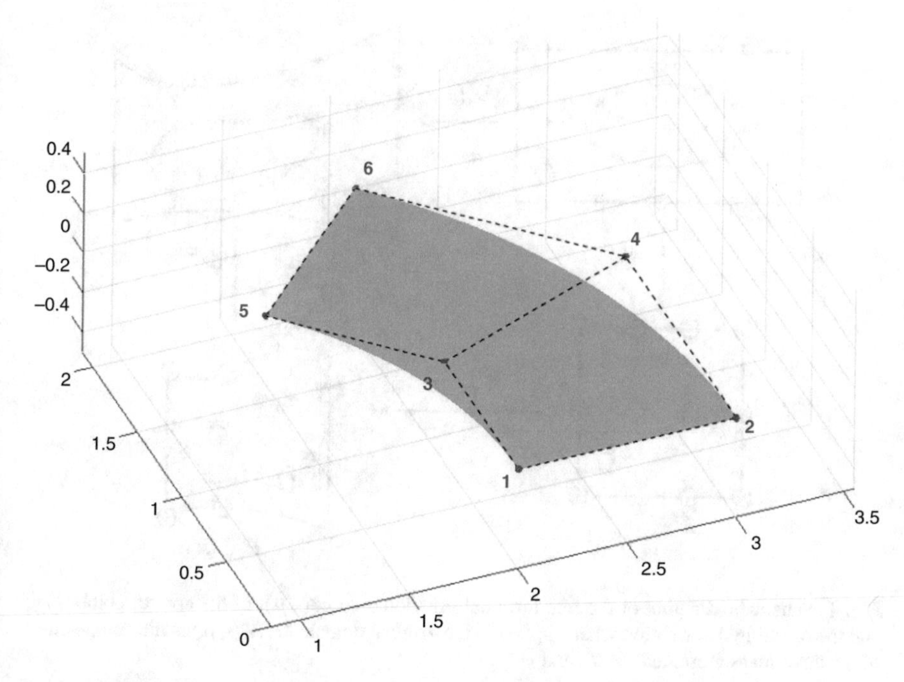

Fig. 3 Idealized 2D collector pipe and its IGA control points mesh. The six control points are indicated with red dots

morphing technique for complex geometries thanks to its features [17, 44, 49, 51]. In the FFD procedure, the object to be deformed is embedded into a rectangular lattice of points, then some of these points are moved to deform the whole embedded domain. This technique has three main benefits: (1) with few parameters—the displacement of the lattice points—it is possible to perform global deformations, (2) it allows to preserve continuity also in the surface derivatives and (3) it is completely independent with respect to the object, so it results applicable also to computational grids [24].

Initially, FFD maps the original domain Ω to the reference one using the affine map ψ defined as $\psi : D \to [0, 1]^n$, where $D \supset \Omega$ is the parallelepiped containing the domain and n is the number of dimensions. We select a regular grid of control points P in the unitary hypercube and we perturb the space by moving these points. The displacements, the so called FFD weights, control the basis functions whose tensor product constitute the deformation map $\hat{T}$. We underline that it is also possible to move only some points: typically we fix several rows/columns of control points to obtain desired levels of continuity and to fix certain parts of the domain.

Finally, we need the back mapping to the physical domain, that is the map ψ^{-1}. Formally, we obtain the FFD map as the composition of the three maps, i.e. $\mathscr{M}(\cdot, \mu) := (\psi \circ \hat{T} \circ \psi^{-1})(\cdot, \mu)$, where μ refers to the parametric displacement of the control points (see Fig. 4 for a schematic summary).

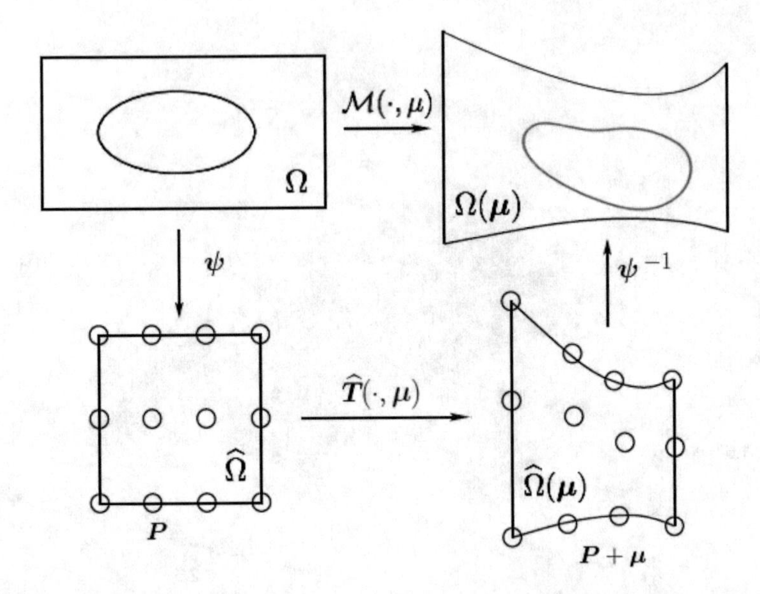

Fig. 4 Schematic diagram of the free form deformation map $\mathscr{M}(\cdot, \mu)$, of the control points $P_{i,k}$, and the resulting deformation when applied to the original domain Ω. $\mathscr{M}(\cdot, \mu)$ is the composition of the three maps presented: ψ, $\hat{T}$, and ψ^{-1}

It must be remarked that, although FFD is characterized by high flexibility and easiness of handling, it suffers from some limitations. The first lies in the fact that the design variables may have no physical significance: they are defined in a parametric domain that can not be expressed into a particular unit of measurement by definition. Moreover all the control points are restricted to lie on a regular lattice and, in that way, local refinements could not be performed.

In this work, we apply the FFD to parametrize the initial 2D domain. We embed the domain with a square lattice of length 3, using 2×2 control points. The lattice origin coincides with the axes origin. We use two different parameters that are the displacement along the x direction of the FFD control points P_{11} and P_{12} depicted in Fig. 5. In particular we define $\mathbb{D} := [-0.3, 0.3]^2$. We use Bernstein polynomials as basis function to deform the geometry in the reference domain. In Fig. 5 we present on the left the undeformed configuration of the idealized collector pipe, where in red we highlight the IGA control points, while the white big dots are the FFD control points. On the right there is just an example of deformation corresponding to a displacement of 0.4 for P_{11} and -0.5 for P_{12}, for now on express as $\mu = (0.4, -0.5)$.

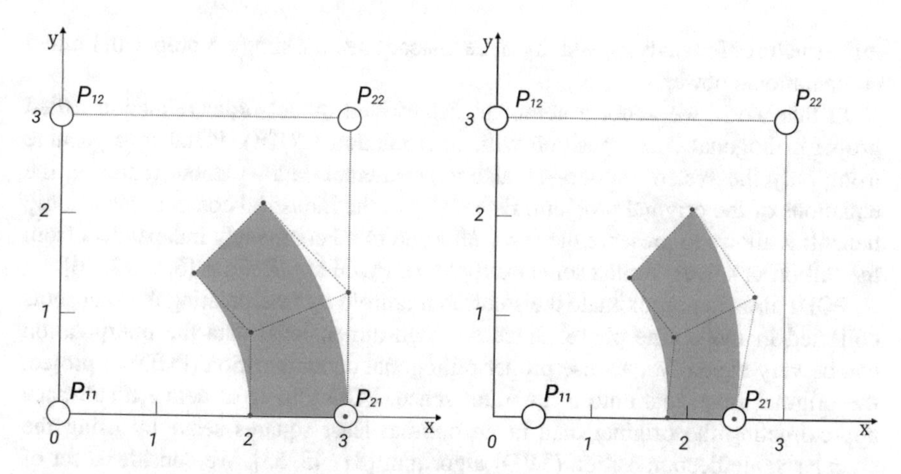

Fig. 5 The initial unperturbed domain (left) and an example of deformed domain (right) using the FFD technique with $\mu = (0.4, -0.5)$. The red dots are the NURBS control points, the white big dots are the FFD control points

5 Data-Driven Reduced Order Modelling by Proper Orthogonal Decomposition with Interpolation

The reduced basis (RB) method is a computational reduction technique allowing to quickly and accurately obtain the solution of parametric PDEs. The need to solve parameterized differential problems, possibly in a very rapid calculation time, emerges in various contexts, particularly when we are interested in characterizing the response of a system in numerous scenarios or operating conditions [30, 32, 35, 36]. The goal of an RB approximation is the representation of the full-order problem as combination of the (few) essential characteristics of the problem itself. In this way the dimensions are considerably lower than those of a problem discretized with a classic Galerkin method. Any discretization leading to a large system to be solved to achieve a certain accuracy is referred to as *high fidelity* (or *full order*) *approximation*. The basic idea of an RB approximation is a computationally efficient solution of the parametric problem keeping the approximation error lower than a given tolerance. In particular, the aim is to approximate the solution of a parametric PDE using a very small number of degrees of freedom instead of the large number required by an high fidelity approximation.

To do this, the full order problem is solved only for a few instances of the input parameters during the computationally expensive *offline* phase. The so stored snapshots are used in the *online* phase for the approximation of the solution for any new parameter. The generation of the snapshots database can be done only once and it is completely decoupled from any new input-output calculation related to a new parameter. The online phase exploits the calculations already performed and therefore not necessary of a large computational power. This ensures real-time processing of the problem without having to use high performance computing infrastructures for analysis, which can be instead run on a simple laptop with limited computational power.

In this work, we adopt a complete data-driven model order reduction called proper orthogonal decomposition with interpolation (PODI). PODI is applicable using only the system output—so also experimental data—without requiring the equations of the original problem. Especially in the industrial context, this is a big benefit: it allows to preserve the *know how* and to be completely independent from the full-order solver. We list some examples of PODI applications [6, 9, 22, 40].

PODI aims to approximate the solution manifold by interpolating the snapshots collected in the offline phase. Since for high-dimensional data the interpolation can be very expensive, we use proper orthogonal decomposition (POD) to project the original snapshots onto a low-rank space. POD allows to define a subspace approximating the original data in an optimal least squares sense by using the singular value decomposition (SVD) algorithm [31, 45, 53]. We consider a set of n_{train} snapshots $s_1, \dots, s_{n_{\text{train}}} = s(\mu_1), \dots, s(\mu_{n_{\text{train}}}) \in V^{\mathcal{N}}$, where $V^{\mathcal{N}}$ is the high-dimensional space and $\mathcal{N}$ refers to its dimension. We define the snapshots matrix $\mathbf{S}$ as the matrix that contains the snapshots in the columns $\mathbf{S} = \left[s_1 \dots s_{n_{\text{train}}}\right]$. We apply the SVD to $\mathbf{S}$:

$$\mathbf{S} = \mathbf{V}\boldsymbol{\Sigma}\mathbf{W}^*, \tag{12}$$

where

$$\mathbf{V} = \left[\zeta_1 \dots \zeta_{n_{\text{train}}}\right] \in \mathbb{C}^{\mathcal{N} \times n_{\text{train}}}, \tag{13}$$

$$\mathbf{W} = \left[\Psi_1 \dots \Psi_{n_{\text{train}}}\right] \in \mathbb{C}^{n_{\text{train}} \times n_{\text{train}}}, \tag{14}$$

are orthogonal matrices whose columns are the left and right singular vectors of $\mathbf{S}$ respectively, and

$$\boldsymbol{\Sigma} = diag(\sigma_1 \dots \sigma_{n_{\text{train}}}) \in \mathbb{C}^{n_{\text{train}} \times n_{\text{train}}}, \tag{15}$$

is a diagonal matrix such that $\sigma_1 \geq \sigma_2 \geq \dots \geq \sigma_{\text{train}} \geq 0$ are the computed singular values of $\mathbf{S}$. The POD modes of dimension N are defined as the first N left singular vectors of $\mathbf{S}$, that correspond to the N largest singular values

$$\mathbf{Z} = \left[\zeta_1 \dots \zeta_N\right]. \tag{16}$$

Now we project the original snapshots onto the space spanned by the modes: the snapshots are so described as linear combination of the modes such that

$$s_i = \sum_{j=1}^{N} \mathbf{C}_{j,i}\zeta_j \qquad \text{for } i = 1, \dots, n_{\text{train}}, \tag{17}$$

where the columns of matrix $\mathbf{C}$ are called *modal coefficients*. We can compute these coefficients as $\mathbf{C} = \mathbf{Z}^T\mathbf{S}$, where $\mathbf{C} = \begin{bmatrix} c_1 \dots c_{n_{\text{train}}} \end{bmatrix} \in \mathbb{R}^{N \times n_{\text{train}}}$. We remark the relation between these coefficients and the parameters; hence we can interpolate the modal coefficients to compute the coefficient for any new point belonging to the parameter space. Finally, using the modes, we are able to approximate the new high-dimensional solution.

Since the PODI technique relies on interpolation, the accuracy of the approximated solution depends mostly by the chosen interpolation method.

6 Numerical Results

In order to construct the reduced order model, we firstly need to sample the solution manifold using several high-fidelity snapshots. We select 100 different configurations applying the FFD technique to the initial domain. The parameters μ are equispaced in the parameter space $[-0.3, 0.3] \times [-0.3, 0.3]$. This strategy allows us to cover the entire parametric space with a linear interpolation.

For each configuration, IGA is performed testing *GeoPDEs* [15, 52], an open source and free package introduced in 2010 by Rafael Vázquez, written in Octave and fully compatible with Matlab. In GeoPDEs the IGA is efficiently implemented in its classic Galerkin version. For the resolution of the full-order problem we created a mesh with 400 degrees of freedom. Figure 6 shows the graphical representation of the numerical solution.

Once the snapshots are collected, we create the reduced order model using the PODI method. The modes are so computed by applying the SVD algorithm to the snapshots matrix. We show in Fig. 7 the obtained singular values: we note that the first one retains ~96% of the total energy, while the 10th singular value is below 10^{-6}. We expect that even with only few modes we can generate a reduced order model introducing only a negligible error.

Using the modes, we can calculate the modal coefficients by projecting the original snapshots. Hence, we can approximate any new solution in the parametric space trough the interpolation of the modal coefficients. Among the various interpolation techniques we choose *linear interpolation*. We report an example where the reduced solution is calculated for the undeformed object by setting the parameter to zero. In Fig. 8 a visual comparison between the high-fidelity solution and the reduced one is presented: it is very intuitive to note that the two solutions are almost identical.

In Fig. 9 instead we can see the error between the reduced solution and the IGA solution. We calculate the error $\mathrm{e}(\mu)$ as follows

$$\mathrm{e}(\mu) = |u_{\mathcal{N}}(\mu) - u_N(\mu)|. \tag{18}$$

The maximum error is around 6×10^{-4} so it is possible to state that it is an acceptable error.

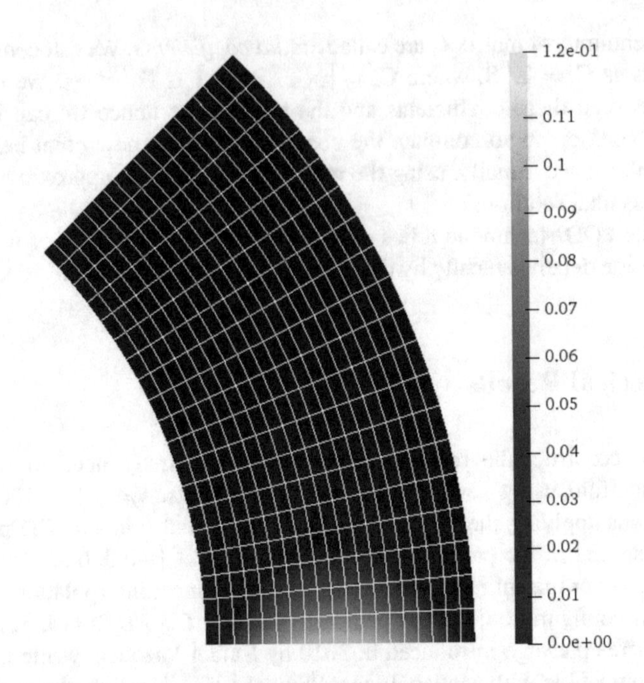

Fig. 6 The adopted computational grid and the graphical representation of the numerical solution of the Laplace problem for the undeformed configuration

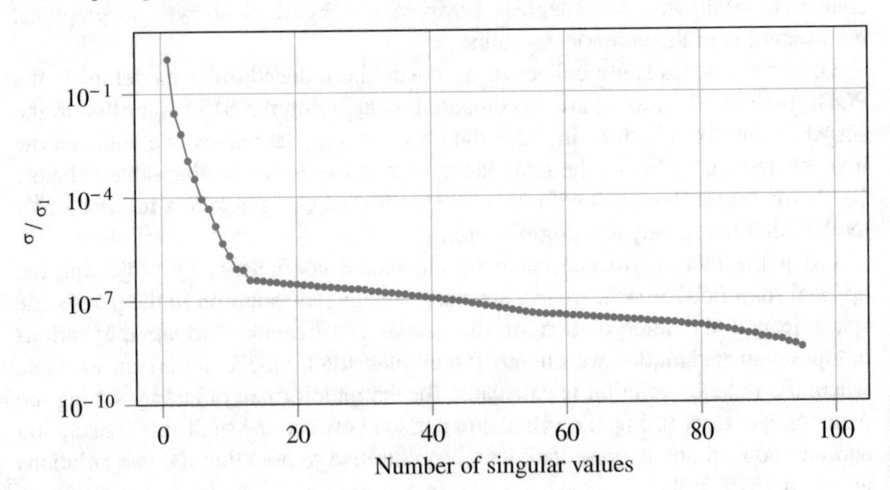

Fig. 7 The singular values obtained by the snapshots matrix using the SVD technique

We can also evaluate the a posteriori error committed on a test dataset. A posteriori error estimation allows to minimize the dimension N of the snapshot database used to generate the reduced space and to quantify the error of the approximation with respect to the number of modes selected. The error is calculated

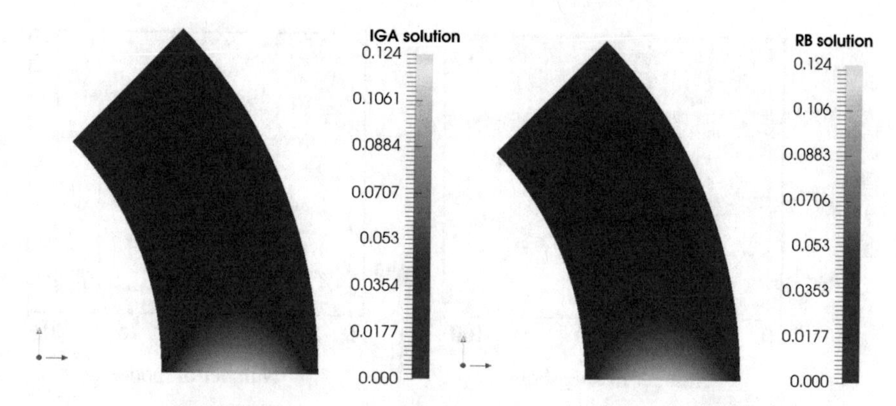

Fig. 8 Comparison between the full-order solution (left) and the reduced order model solution (right) for the undeformed configuration

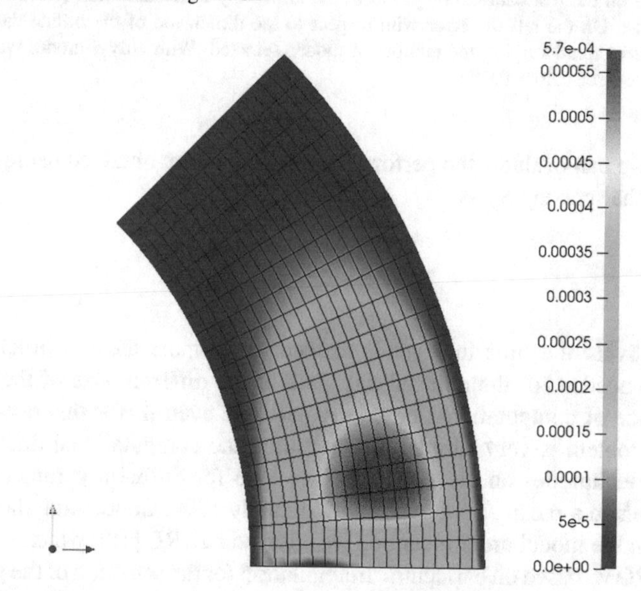

Fig. 9 Error between the full-order solution and the reduced order solution for the unperturbed configuration. The different color corresponds to ascending values of the error. The error gradually increases from the blue zone to the red zone where there is the greatest diffusive heat effect. The results show how, even in the red zone, the error assumes acceptable values

computing the relative L^2 norm of the difference between the approximated solution obtained using PODI approach and the IGA truth solution, on a test dataset composed by high fidelity solutions corresponding to 20 uniformly distributed random samples in the parameter space. The plots in Fig. 10 show the relative error against the dimension of the database and the number of modes. We see that 100 samples and only 4 modes are enough for an average error below 10^{-3}. We refer to [19] for a posteriori error bounds in an RB-IGA setting.

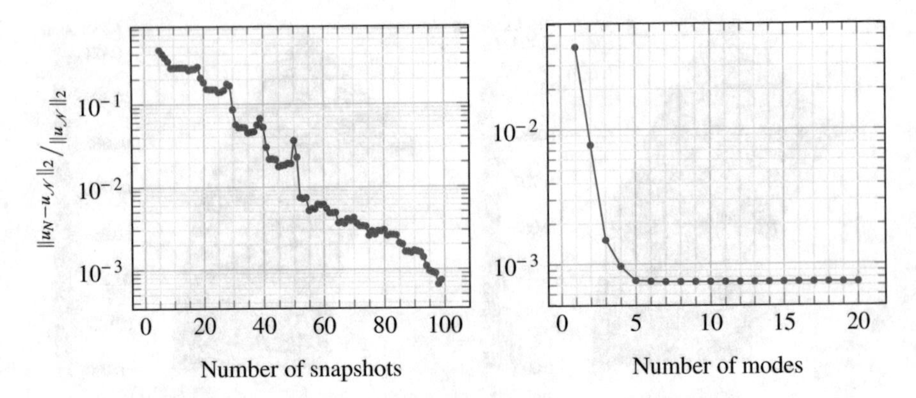

Fig. 10 A posteriori L^2 relative error between the reduced order solution and the high fidelity one, computed on the test dataset composed by 20 uniformly distributed random samples in the parameter space. On the left the error with respect to the dimension of the offline database. On the right the error trend varying the number of modes selected. With only 4 modes we obtain an average relative error below 10^{-3}

Finally, we can evaluate the performance improvement obtained using ROM by calculating the speedup S_p as

$$S_p = \frac{u_{\mathcal{N}}(s)}{u_N(s)}, \tag{19}$$

where we divide the time in seconds needed to compute the full order solution by the time needed for the reduced one. Due to the different size of the systems, the difference of computational time is remarkable even if, for this test-case, the full-order problem is very simple. We measured the computational time required by the two techniques on the same machine, and for different parameter values, and we obtained a mean speedup of approximately 1000. Concerning the software involved, for the model order reduction we adopted EZyRB [18], which is a Python library for ROM, based on baricentric triangulation for the selection of the parameter points and on POD for the selection of the modes. The software uses a non-intrusive approach in which the solutions are projected on the low dimensional space then interpolated for the approximation of the solution.

7 Conclusions and Future Developments

In this work we presented a complete non-intrusive computational pipeline involving geometrical parameterization through free form deformation, isogeometric analysis, and reduced order model, for fast and reliable field evaluation. We applied this pipeline to a diffusion problem in an idealized 2D collector pipe. We used a data-driven non-intrusive approach for the model order reduction, that is the proper orthogonal decomposition with interpolation. This setting, even if tested on a simple problem, will allow us to deal with more complex industrial CAD files, since we used a geometrical parameterization technique independent from the object of interest, and the ROM chosen uses only the snapshots of the IGA high fidelity simulations.

Results and speedup achieved look promising to continue with the implementation of more complex problems on 3D geometries. The effectiveness of an RB approach would be exploited even better increasing the complexity of the simulation in cases where a large number of analysis has to be computed, e.g. in parameter optimization studies. The developed RB-IGA method is thus interesting from both academic and industrial points of view. As a matter of fact, since IGA is directly interfaced with CAD, an undergoing development of the work is the implementation of a dedicated software based on the RB-IGA method, allowing real-time evaluations of outputs of interest for different NURBS parameterizations.

Acknowledgments This work was partially funded by the project SOPHYA, "Seakeeping Of Planing Hull YAchts", supported by Regione FVG, POR-FESR 2014–2020, Piano Operativo Regionale Fondo Europeo per lo Sviluppo Regionale, and partially supported by European Union Funding for Research and Innovation—Horizon 2020 Program—in the framework of European Research Council Executive Agency: H2020 ERC CoG 2015 AROMA-CFD project 681447 "Advanced Reduced Order Methods with Applications in Computational Fluid Dynamics" P.I. Gianluigi Rozza.

This work was also partially supported by Fondazione Cariplo–Regione Lombardia through the project "Verso nuovi strumenti di simulazione super veloci ed accurati basati sull'analisi isogeometrica", within the program RST—rafforzamento.

References

1. Ballarin, F., Faggiano, E., Ippolito, S., Manzoni, A., Quarteroni, A., Rozza, G., Scrofani, R.: Fast simulations of patient-specific haemodynamics of coronary artery bypass grafts based on a POD–Galerkin method and a vascular shape parametrization. J. Comput. Phys. **315**, 609–628 (2016)
2. Ballarin, F., Faggiano, E., Manzoni, A., Quarteroni, A., Rozza, G., Ippolito, S., Antona, C., Scrofani, R.: Numerical modeling of hemodynamics scenarios of patient-specific coronary artery bypass grafts. Biomech. Model. Mechanobiol. **16**(4), 1373–1399 (2017)
3. Ballarin, F., D'Amario, A., Perotto, S., Rozza, G.: A POD-selective inverse distance weighting method for fast parametrized shape morphing. Int. J. Num. Meth. Eng. **117**, 860–884 (2019)
4. Baroli, D., Cova, C.M., Perotto, S., Sala, L., Veneziani, A.: Hi-POD solution of parametrized fluid dynamics problems: Preliminary results. In: Model Reduction of Parametrized Systems, pp. 235–254. Springer, Berlin (2017)
5. Buhmann, M.D.: Radial Basis Functions: Theory and Implementations, Vol. 12. Cambridge University Press, Cambridge (2003)
6. Bui-Thanh, T., Damodaran, M., Willcox, K.: Proper orthogonal decomposition extensions for parametric applications in compressible aerodynamics. In: 21st AIAA Applied Aerodynamics Conference, p. 4213 (2003). https://doi.org/10.2514/6.2003-4213
7. Chen, P., Quarteroni, A., Rozza, G.: Reduced basis methods for uncertainty quantification. SIAM/ASA J. Uncertain. Quantif. **5**, 813–869 (2017). https://doi.org/10.1137/151004550

8. Chinesta, F., Keunings, R., Leygue, A.: The Proper Generalized Decomposition for Advanced Numerical Simulations: A Primer. Springer Science & Business Media, Berlin (2013)
9. Chinesta, F., Huerta, A., Rozza, G., Willcox, K.: Model order reduction: A survey. In: Wiley Encyclopedia of Computational Mechanics. Wiley, Hoboken (2016). http://eu.wiley.com/WileyCDA/WileyTitle/productCd-1119003792.html
10. Christensen, E.A., Brøns, M., Sørensen, J.N.: Evaluation of proper orthogonal decomposition–based decomposition techniques applied to parameter-dependent nonturbulent flows. SIAM J. Sci. Comput. **21**(4), 1419–1434 (1999)
11. Cottrell, J.A., Hughes, T.J., Reali, A.: Studies of refinement and continuity in isogeometric structural analysis. Comput. Method. Appl. Mech. Eng. **196**(41–44), 4160–4183 (2007)
12. Cottrell, J.A., Hughes, T.J., Bazilevs, Y.: Isogeometric Analysis: Toward Integration of CAD and FEA. Wiley, Hoboken (2009)
13. Cox Maurice, G.: The numerical evaluation of B-splines. IMA J. Appl. Math. **10**(2), 134–149 (1972)
14. De Boor, C.: On calculating with B-splines. J. Approx. Theory **6**(1), 50–62 (1972)
15. De Falco, C., Reali, A., Vázquez, R.: GeoPDEs: a research tool for isogeometric analysis of PDEs. Adv. Eng. Softw. **42**(12), 1020–1034 (2011)
16. Demo, N., Tezzele, M., Gustin, G., Lavini, G., Rozza, G.: Shape optimization by means of proper orthogonal decomposition and dynamic mode decomposition. In: Technology and Science for the Ships of the Future: Proceedings of NAV 2018: 19th International Conference on Ship & Maritime Research, pp. 212–219. IOS Press, Amsterdam (2018). https://doi.org/10.3233/978-1-61499-870-9-212
17. Demo, N., Tezzele, M., Mola, A., Rozza, G.: An efficient shape parametrisation by free-form deformation enhanced by active subspace for hull hydrodynamic ship design problems in open source environment. In: The 28th International Ocean and Polar Engineering Conference, ISOPE (2018)
18. Demo, N., Tezzele, M., Rozza, G.: EZyRB: easy reduced basis method. J. Open Source Softw. **3**(24), 661 (2018). https://doi.org/10.21105/joss.00661
19. Devaud, D., Rozza, G.: Certified Reduced Basis Method for Affinely Parametric Isogeometric Analysis NURBS Approximation, vol. 119. Springer, Berlin (2017). https://doi.org/10.1007/978-3-319-65870-4_3
20. Forti, D., Rozza, G.: Efficient geometrical parametrisation techniques of interfaces for reduced-order modelling: application to fluid–structure interaction coupling problems. Int. J. Comput. Fluid Dyn. **28**(3–4), 158–169 (2014)
21. Haasdonk, B., Ohlberger, M.: Reduced basis method for finite volume approximations of parametrized linear evolution equations. ESAIM: Math. Model. Numer. Anal. **42**(2), 277–302 (2008)
22. Hesthaven, J.S., Rozza, G., Stamm, B.: Certified Reduced Basis Methods for Parametrized Partial Differential Equations. Springer Briefs in Mathematics, 1st edn. Springer, Berlin (2015). https://doi.org/10.1007/978-3-319-22470-1
23. Hughes, T.J., Cottrell, J.A., Bazilevs, Y.: Isogeometric analysis: CAD, finite elements, NURBS, exact geometry and mesh refinement. Comput. Method. Appl. Mech. Eng. **194**(39–41), 4135–4195 (2005)
24. Lassila, T., Rozza, G.: Parametric free-form shape design with PDE models and reduced basis method. Comput. Method. Appl. Mech. Eng. **199**(23–24), 1583–1592 (2010)
25. Manzoni, A., Quarteroni, A., Rozza, G.: Model reduction techniques for fast blood flow simulation in parametrized geometries. Int. J. Numer. Meth. Bio. Eng. **28**(6–7), 604–625 (2012)
26. Manzoni, A., Salmoiraghi, F., Heltai, L.: Reduced basis isogeometric methods (RB-IGA) for the real-time simulation of potential flows about parametrized NACA airfoils. Comput. Method. Appl. Mech. Eng. **284**, 1147–1180 (2015)
27. Morris, A., Allen, C., Rendall, T.: CFD-based optimization of aerofoils using radial basis functions for domain element parameterization and mesh deformation. Int. J. Numer. Method. Fluids **58**(8), 827–860 (2008)

28. Perotto, S., Ern, A., Veneziani, A.: Hierarchical local model reduction for elliptic problems: a domain decomposition approach. Multiscale Model. Simul. **8**(4), 1102–1127 (2010)
29. Perotto, S., Reali, A., Rusconi, P., Veneziani, A.: Higamod: a hierarchical isogeometric approach for model reduction in curved pipes. Comput. Fluids **142**, 21–29 (2017)
30. Peterson, J.S.: The reduced basis method for incompressible viscous flow calculations. SIAM J. Sci. Stat. Comput. **10**(4), 777–786 (1989)
31. Quarteroni, A.: Numerical Models for Differential Problems, vol. 2. Springer, Berlin (2009)
32. Quarteroni, A., Rozza, G.: Reduced Order Methods for Modeling and Computational Reduction. MS&A – Modeling, Simulation and Applications, vol. 9. Springer, Berlin (2014)
33. Quarteroni, A., Rozza, G., Manzoni, A.: Certified reduced basis approximation for parametrized partial differential equations and applications. J. Math. Ind. **1**(1), 3 (2011)
34. Ripepi, M., Verveld, M., Karcher, N., Franz, T., Abu-Zurayk, M., Görtz, S., Kier, T.: Reduced-order models for aerodynamic applications, loads and MDO. CEAS Aeronaut. J. **9**(1), 171–193 (2018)
35. Rozza, G., Huynh, D.B.P., Patera, A.T.: Reduced basis approximation and a posteriori error estimation for affinely parametrized elliptic coercive partial differential equations. Arch. Comput. Meth. Eng.**15**(3), 1 (2007)
36. Rozza, G., Lassila, T., Manzoni, A.: Reduced basis approximation for shape optimization in thermal flows with a parametrized polynomial geometric map. In: Spectral and High Order Methods for Partial Differential Equations, pp. 307–315. Springer, Berlin (2011)
37. Rozza, G., Malik, M.H., Demo, N., Tezzele, M., Girfoglio, M., Stabile, G., Mola, A.: Advances in reduced order methods for parametric industrial problems in computational fluid dynamics. In: Owen, R., de Borst, R., Reese, J., Chris, P. (eds.) ECCOMAS ECFD 7 - Proceedings of 6th European Conference on Computational Mechanics (ECCM 6) and 7th European Conference on Computational Fluid Dynamics (ECFD 7), Glasgow, pp. 59–76 (2018)
38. Salmoiraghi, F., Ballarin, F., Corsi, G., Mola, A., Tezzele, M., Rozza, G.: Advances in Geometrical Parametrization and Reduced Order Models and Methods for Computational Fluid Dynamics Problems in Applied Sciences and Engineering: Overview and Perspectives. ECCOMAS, Crete (2016). https://doi.org/10.7712/100016.1867.8680
39. Salmoiraghi, F., Ballarin, F., Heltai, L., Rozza, G.: Isogeometric analysis-based reduced order modelling for incompressible linear viscous flows in parametrized shapes. Adv. Model. Simul. Eng. Sci. **3**(1), 21 (2016)
40. Salmoiraghi, F., Scardigli, A., Telib, H., Rozza, G.: Free-form deformation, mesh morphing and reduced-order methods: enablers for efficient aerodynamic shape optimisation. Int. J. Comput. Fluid Dyn. **32**(4–5), 233–247 (2018). https://doi.org/10.1080/10618562.2018.1514115
41. Schilders, W.H., Van der Vorst, H.A., Rommes, J.: Model Order Reduction: Theory, Research Aspects and Applications, vol. 13. Springer, Berlin (2008). https://doi.org/10.1007/978-3-540-78841-6
42. Sederberg, T., Parry, S.: Free-form deformation of solid geometric models. In: Proceedings of SIGGRAPH - Special Interest Group on Graphics and Interactive Techniques. SIGGRAPH, pp. 151–159. (1986)
43. Shepard, D.: A two-dimensional interpolation function for irregularly-spaced data. In: Proceedings-1968 ACM National Conference, pp. 517–524. ACM, New York (1968)
44. Sieger, D., Menzel, S., Botsch, M.: On shape deformation techniques for simulation-based design optimization. In: Perotto, S., Formaggia, L. (eds.) New Challenges in Grid Generation and Adaptivity for Scientific Computing, pp. 281–303. Springer, Berlin (2015)
45. Sirovich, L.: Turbulence and the dynamics of coherent structures. I. coherent structures. Q. Appl. Math. **45**(3), 561–571 (1987)
46. Stabile, G., Rozza, G.: Finite volume POD-Galerkin stabilised reduced order methods for the parametrised incompressible Navier–Stokes equations. Comput. Fluid. **173**, 273–284 (2018). https://doi.org/10.1016/j.compfluid.2018.01.035

47. Stabile, G., Hijazi, S., Mola, A., Lorenzi, S., Rozza, G.: POD-Galerkin reduced order methods for CFD using finite volume discretisation: vortex shedding around a circular cylinder. Commun. Appl. Ind. Math. **8**(1), 210–236 (2017). https://doi.org/10.1515/caim-2017-0011
48. Tezzele, M., Ballarin, F., Rozza, G.: Combined parameter and model reduction of cardiovascular problems by means of active subspaces and POD-Galerkin methods. In: Mathematical and Numerical Modeling of the Cardiovascular System and Applications. SEMA SIMAI Springer Series, vol. 16. Springer, Berlin (2018). https://doi.org/10.1007/978-3-319-96649-6_8
49. Tezzele, M., Demo, N., Gadalla, M., Mola, A., Rozza, G.: Model order reduction by means of active subspaces and dynamic mode decomposition for parametric hull shape design hydrodynamics. In: Technology and Science for the Ships of the Future: Proceedings of NAV 2018: 19th International Conference on Ship & Maritime Research, pp. 569–576. IOS Press, Amsterdam (2018). https://doi.org/10.3233/978-1-61499-870-9-569
50. Tezzele, M., Demo, N., Mola, A., Rozza, G.: An integrated data-driven computational pipeline with model order reduction for industrial and applied mathematics. Special Volume ECMI (2020). https://arxiv.org/abs/1810.12364
51. Tezzele, M., Salmoiraghi, F., Mola, A., Rozza, G.: Dimension reduction in heterogeneous parametric spaces with application to naval engineering shape design problems. Adv. Model. Simul. Eng. Sci. **5**(1), 25 (2018). https://doi.org/10.1186/s40323-018-0118-3
52. Vázquez, R.: A new design for the implementation of isogeometric analysis in Octave and Matlab: GeoPDEs 3.0. Comput. Math. Appl. **72**, 523–554 (2016). http://dx.doi.org/10.1016/j.camwa.2016.05.010
53. Volkwein, S.: Proper orthogonal decomposition: theory and reduced-order modelling. Lect. Notes Univ. Konstanz **4**(4), 1–29 (2013)
54. Witteveen, J., Bijl, H.: Explicit mesh deformation using inverse distance weighting interpolation. In: 19th AIAA Computational Fluid Dynamics. AIAA (2009)
55. Zhu, S., Dedè, L., Quarteroni, A.: Isogeometric analysis and proper orthogonal decomposition for parabolic problems. Numer. Math. **135**(2), 333–370 (2017)

Uncertainty Quantification Applied to Hemodynamic Simulations of Thoracic Aorta Aneurysms: Sensitivity to Inlet Conditions

Alessandro Boccadifuoco, Alessandro Mariotti, Katia Capellini, Simona Celi, and Maria Vittoria Salvetti

Abstract In this work, the numerical simulation of the blood flow inside a patient specific aorta in presence of an aneurysm is considered. A systematic sensitivity analysis of numerical predictions to the shape of the inlet flow rate waveform is carried out. In particular, two parameters are selected to describe the inlet waveform: the stroke volume and the period of the cardiac cycle. In order to limit the number of hemodynamic simulations required, we used a stochastic method based on the generalized polynomial chaos (gPC) approach, in which the selected parameters are considered as random variables with a given probability distribution. The uncertainty is propagated through the numerical model and a continuous response surface of the output quantities of interest in the parameter space can be recovered through a "surrogate" model. For both selected uncertain parameters, we first assumed uniform Probability Density Functions (PDFs) on a given variation range, and then we used clinical data to construct more accurate beta PDFs. In all cases, the two input parameters appeared to have a significant influence on wall shear stresses, confirming the need of using patient-specific inlet conditions.

A. Boccadifuoco
Institute of Life Sciences, Scuola Superiore Sant'Anna, Pisa, Italy
e-mail: alessandro.boccadifuoco@sssup.it

A. Mariotti (✉) · M. V. Salvetti
Dipartimento di Ingegneria Civile e Industriale, Università di Pisa, Pisa, Italy
e-mail: alessandro.mariotti@for.unipi.it; mv.salvetti@ing.unipi.it

K. Capellini · S. Celi
BiocardioLab, Bioengineering Unit, Fondazione Toscana "G. Monasterio", Heart Hospital, Massa, Italy

Dipartimento di Ingegneria dell'Informazione, Università di Pisa, Pisa, Italy
e-mail: katia.capellini@ftgm.it; simona.celi@ftgm.it

M. D'Elia et al. (eds.), *Quantification of Uncertainty: Improving Efficiency and Technology*, Lecture Notes in Computational Science and Engineering 137,
https://doi.org/10.1007/978-3-030-48721-8_8

1 Introduction

An Ascending Thoracic Aortic Aneurysm (ATAA) is a permanent dilatation occurring in the ascending part of the thoracic aorta. This disease represents a clinical challenge due to the significant mortality risk carried by both complications [11, 18] and surgical repair [7]. Currently, decisions about clinical management are based on the maximum aneurysm diameter: surgery is recommended whether the aneurysm diameter reaches 5.5 cm. However, [15] reported that 60% of ascending aortic dissections occur with smaller diameters. In this context, several studies [5, 6] have showed that simple morphological features like maximum diameter are not able to accurately predict the potential critical evolutions of aneurysms.

Among the hemodynamic descriptors, the wall shear stresses can have an important effect on vessel wall mechanical properties, and, therefore, on the aneurysm rupture risk. For instance, a recent result by [16], related to Type B dissections (i.e., dissections involving the descending aorta), indicated a significantly higher time-averaged wall shear stress in dissections complicated by rapid aneurysm progression than in patients with stable aortic diameters.

In this context, Computational Fluid Dynamics (CFD) permits the investigation of pressure and flow field at a temporal and spatial resolution unachievable by any clinical methodology. A variety of variables and indicators difficult to be obtained from *in vivo* measurements can be easily quantify through CFD and, even better, the combination of medical imaging with CFD permits to investigate hemodynamics on a patient-specific basis (see e.g. [4, 10, 12]).

Nonetheless, the accuracy of CFD predictions strongly depends on modeling assumptions and computational set-up. An important critical aspect is represented by boundary conditions, which must be correctly prescribed to reproduce the effect of organs and vessels outside the portion of aorta that is actually simulated. Both inflow and outflow boundary conditions should be patient specific and can only seldom be obtained from *in vivo* measurements (and, when available, experimental data are often characterized by a space and time resolution not adequate for numerical simulation).

In two our previous works we investigated on the impact of outflow boundary conditions based on the three-element Windkessel model, by performing a stochastic analysis of the effect of the uncertainties in the Windkessel parameters [1, 2]. The results highlighted that although the uncertainties in the outflow parameters may give significant variability of the instantaneous shear stresses in regions characterized by flow recirculation or large streamline curvature, the impact on cycle-averaged shear stresses is moderate. Therefore, the usual procedure of estimating these parameters in order to obtain a physiological wave pressure form in a 0D model seems to be adequate for a correct prediction of wall shear stresses, clearly within all the other sources of error possibly present in the simulations, and a fine tuning of RCR parameters is not needed (see [1, 2, 9]). Moreover, the effect of wall compliance has been investigated though a stochastic approach in [3], where hemodynamic simulations and MRI data for a patient-specific simulation have been integrated. Again, as previously found for the outlet boundary conditions, a significant sensitivity to wall compliance of instantaneous shear stresses was observed during large part of the cardiac cycle period; however, the variability of the time-averaged wall shear stresses remains very low.

We focus herein on inlet boundary conditions. A common practice is the imposition of a flow rate waveform [8, 9]. We consider a real aneurysm geometry acquired by Magnetic Resonance Imaging (MRI) and perform the simulations with the open-source software *SimVascular* [21]. We carried out a systematic sensitivity analysis to the shape of the inlet flow rate waveform, and, in particular, to the flow stroke volume and to the period of the cardiac cycle. A deterministic analysis of the influence of these parameters is difficult because of the significant cost of each simulation. An efficient alternative is to use a stochastic approach, in which the selected parameters are considered as random variables with a given probability distribution. The uncertainty can, thus, be propagated through the CFD model and a continuous response surface of the output quantities of interest in the parameter space can be recovered through a "surrogate" model, which requires a limited number of deterministic simulations. In the present work, we use the generalized Polynomial Chaos (gPC) approach [19]. For both selected uncertain parameters, viz. the flow stroke volume and to the period of the cardiac cycle, we first assumed uniform Probability Density Functions (PDFs) on a given variation range, and then we used clinical data to construct more accurate beta PDFs. In all analyses, the two input parameters showed a deep influence on wall shear stresses, and in particular the stroke volume in the case of beta PDFs.

2 Modeling and Computational Set-Up

The patient-specific geometry of thoracic aorta subjected to aneurysm in its ascending region is the same as the one object of the study in [2] (see Fig. 1). It was acquired by MRI before the surgical procedure. The patient object of the study is male with tricuspid aortic valve, age 65 years, weight 74 kg. The maximum diameter ratio is $D_r = D_{\text{max}}/D_{\text{healthy}} = 1.5$, where D_{max} is the diameter measured in correspondence of the widest section of the aneurysm and D_{healthy} a diameter characteristic of the healthy segment. The maximum diameter is $D_{\text{max}} = 3.8$ cm.

At the inlet section of the computational domain we specified a Dirichlet boundary condition on the velocity, by imposing idealized physiological profile of Fig. 2 with different values of the stroke volume and of the period of the cardiac cycle. Since the stroke volume is the amount of blood pumped by the left ventricle of the heart in one cardiac cycle, the stroke volume can be evaluated as the integral in time of the flow rate during the systolic part of the cardiac cycle. For a given cardiac cycle period, the stroke volume is changed by multiplying the flow rate in Fig. 2 by a constant factor only in the systolic part of the cycle. The ranges of variation of

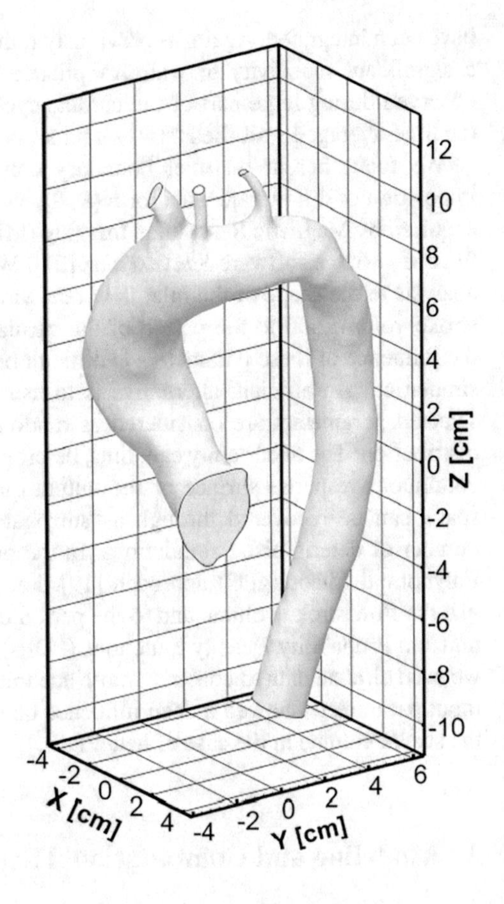

Fig. 1 Sketch of the considered geometry ($D_r = 1.5$). Aortic geometry from MRI images before surgical procedure

the stroke volume and of the cardiac period will be in-depth discussed in Sect. 4. The resulting Reynolds number, based on the diameter of the inlet section and on the bulk inlet velocity averaged during the cardiac cycle, was all below $Re \simeq 2200$. The spatial distribution was assumed to be uniform in the whole inlet section. As a further assumption, in-plane velocity components were not considered, as often made in this type of simulations (see e.g. [13, 20]).

At each outlet, boundary conditions based on the three-element Windkessel model were imposed, as sketched in Fig. 3. The values of the RCR parameters were set in order to obtain a desired physiological behavior of the pressure profile (120/80 mmHg) in the simplified fully lumped model in which the contribution of the computational domain is neglected (see [9]). In addition, stochastic sensitivity analyses on the effect of outlet conditions in ATAA simulations [1, 2] highlighted that a fine tuning of RCR parameters is not needed for these simulations. The so-

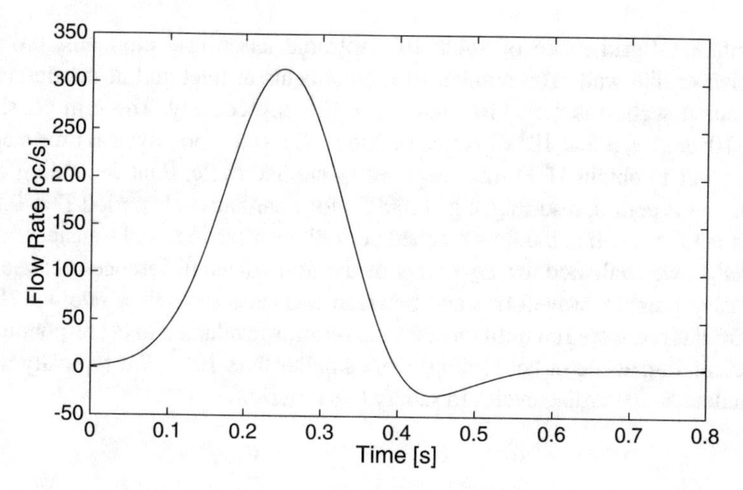

Fig. 2 Flow rate waveform imposed at the inlet section (plug flow)

Fig. 3 Sketch of the three-element Windkessel model

obtained values of R_p, R_d and C were then distributed to the various outlets as follows:

$$
\begin{aligned}
R_{p_i} &= R_p \frac{A_{\text{tot}}}{A_i} \\
R_{d_i} &= R_d \frac{A_{\text{tot}}}{A_i} \qquad \text{for } i = 1, \ldots, n_{\text{outlets}} \\
C_i &= C \frac{A_i}{A_{\text{tot}}}
\end{aligned}
\tag{1}
$$

where n_{outlets} is the number of the outlets (i.e., four in the considered case), A_i is the area of the outlet i and A_{tot} is the sum of the areas of all the outlets.

Finally, on the arterial wall we imposed a no-slip condition between the fluid and the wall and the effect of wall compliance is not taken into account in the considered simulations.

An unstructured computational grid consisting of tetrahedral elements was used. Based on the preliminary grid sensitivity analyses performed in [2], we used a computational grid made of 1.4×10^6 isotropic tetrahedral elements, 6.6×10^4 of which on the wall. The number of the elements at inlet and at the descending aorta outlet section is 1.5×10^3 and 6.3×10^2, respectively. The grid consists of 2.4×10^5 nodes, 3.3×10^4 of which on the aortic wall. The physical time step was set in order to obtain 1600 time steps every cardiac cycle, thus depending on the cardiac cycle period, resulting, e.g., 0.0005 s for a cardiac cycle period $T = 0.8$ s. In order to make sure that the flow reached periodicity after an initial transient, in each simulation we evaluated the L_2-norms of the normalized differences between two successive pressure waveforms and between two successive flow rate waveforms. The simulations were run until the above L_2-norms (evaluated in correspondence of the descending aortic outlet section) were smaller than 10^{-3}. We typically needed to simulate 5–10 cardiac cycles to satisfy this criterion.

3 Uncertainty Quantification Methodology

The generalized Polynomial Chaos (gPC) is the strategy used in the present work to obtain a continuous response surface in the parameter space starting from a few deterministic numerical simulations. We briefly recall here its main features. The gPC approach is based on the projection of a given stochastic response in terms of an orthogonal polynomial basis [19]. The gPC expansion for a given quantity of interest, say X, may be expressed as follows (term-based indexing):

$$X(\omega) = \sum_{k=0}^{\infty} a_k \Phi_k(\boldsymbol{\xi}(\omega)) \tag{2}$$

where ω is an elementary event, $\boldsymbol{\xi}(\omega)$ is the vector consisting of the independent random variables (i.e., the set of considered uncertain parameters), $\Phi_k(\boldsymbol{\xi})$ is the gPC polynomial of index k and a_k is the corresponding coefficient.

The response surface is obtained by a truncation of the expansion (2) to a finite limit Q. Using the maximum polynomial order for all one-dimensional polynomials (i.e., full tensor-product polynomial expansion), Q is obtained as follows:

$$Q = \prod_{i=1}^{M} (P_i + 1) - 1 \tag{3}$$

where M is the number of the uncertain parameters and P_i is the maximum polynomial order for the ith parameter. The coefficient a_k can be computed as follows:

$$a_k = \frac{\langle X, \Phi_k \rangle}{\langle \Phi_k, \Phi_k \rangle} = \frac{1}{\langle \Phi_k, \Phi_k \rangle} \int_{\omega \in \Xi} X \, \Phi_k \rho(\boldsymbol{\xi}) \, d\boldsymbol{\xi} \tag{4}$$

where $\langle \cdot, \cdot \rangle$ denotes the usual L_2 scalar product involving a weight function depending on the polynomial family chosen. The integrals in the scalar products were computed numerically by using Gaussian quadrature. The polynomial family, Φ_k, must be *a priori* specified and its choice affects the speed of the convergence of the gPC expansion: a suitable polynomial family is able to approximate the stochastic response by means of fewer degrees of freedom. When dealing with Gaussian quadrature, an optimal family has a weight function similar to the probability measure of the random variables. The choice of the polynomial family thus depends on the Probability Density Function (PDF) shape of the uncertain parameters.

Finally, in case of stochastic analysis with 2 or more input variables, it is interesting to estimate their influence (individual or due to their interactions) on the uncertainty of the output quantities. To this purpose, we use the Sobol indices, also referred to in the following as partial sensitivities, which are global sensitivity indices defined in [17].

4 Choice of the Distributions of the Uncertain Parameters

We aim at investigating the sensitivity of the output hemodynamic quantities of interest to the shape of the flow rate waveform imposed at the inlet section. For this purpose, we considered the idealized curve presented in Fig. 2 and scaled it in order to vary the stroke volume and the cardiac cycle period T. Two different distributions of the uncertain parameters are considered in the present work. For both selected uncertain parameters, we first assumed uniform Probability Density Functions (PDFs) on a given variation range, and then we used clinical data to construct more accurate beta PDFs.

The choice of a uniform PDF is motivated by the fact that, among the classically-used distributions, it is the least informative distribution with the highest variance in given intervals. It is expected that, for a given variation interval of the input parameters, the uniform PDF distribution should give the largest variability of the output quantities, thus providing a "conservative" estimation of this sensitivity to the considered input parameters. Therefore, when no information is available on the output parameter PDF, a uniform distribution is usually chosen. Since in a first phase of this work we had no clinical information, an assumption should be made also on the parameter variation range. The chosen variation intervals were determined by imposing a standard deviation of 10% of the reference value obtained from the reference case, i.e. SV $= 69.5$ ml and $T = 0.8$ s. The resulting variation ranges are: SV $\in [61.5, 85.6]$ml, $T \in [0.71, 0.99]$s. The optimal polynomial family for the gPC basis in case of uniform PDF distribution of the input uncertain parameters is Legendre polynomials. The polynomial expansion was truncated to the third order for each dimension and thus four quadrature points for each random variable were needed to compute the coefficients of the expansion. These values are shown in Table 1 for all the parameters.

Table 1 Quadrature points for the input parameter, case of uniform PDF distributions

Random variable	1st	2nd	3rd	4th
SV [ml]	63.2	69.5	77.7	83.9
T [s]	0.73	0.8	0.89	0.97

Table 2 Quadrature points for the input parameters, case of beta PDF distributions

Random variable	1st	2nd	3rd	4th
SV [ml]	41.4	69.1	101.5	133.8
T [s]	0.67	0.79	0.95	1.14

For the second stochastic analysis, we used clinical data regarding 23 patients, shown in Fig. 4 transposed in the normalized support range. In order to extract a PDF distribution suitably representing the experimental data, we chose beta PDF distributions, which have two free parameters α and β to be tuned:

$$f(x; \alpha, \beta) = K(1 - x)^{\alpha}(1 + x)^{\beta} \tag{5}$$

where K is a constant such that the integral of $f(x)$ in its support range $[-1, 1]$ is one. With $\alpha = 4.3$ and $\beta = 2.6$ for the PDF of the stroke volume and $\alpha = 5.5$ and $\beta = 1.3$ for the PDF of the cardiac period, we obtained the distributions in Fig. 4. The variation ranges are: SV $\in [20, 170]$ml, $T \in [0.6, 1.4]$s. It is worth noting that the resulting standard deviation of the two obtained distributions is about 33% and 14% of the reference case values, for SV and T respectively. In this case of beta PDF distributions, Jacobi polynomials represent the optimal polynomial family. Again, the polynomial expansion was truncated to the third order for each dimension; hence, four quadrature points for each random variable were sufficient to compute the coefficients of the expansion. The values of the quadrature points are shown in Table 2 for the two parameters.

Since a full tensor grid is used, each stochastic analysis introduced above requires 16 evaluations of the output quantities of interest and thus 16 hemodynamic simulations. It is worth noting that also sparse tensor grids have been proposed in the literature to mitigate the so-called *curse of dimensionality*, i.e. the exponential growth of quadrature points with increasing number of input random variables [14]. However, we did not use such an approach in the present work because in each performed stochastic analysis the number of random variables was low, in particular never larger than 2.

5 Definition of the Output Quantities of Interest and Indicators

When considering the effects of the blood flow on the arterial wall, particular attention is given to the Wall Shear Stress (WSS), namely the tangential stress $\boldsymbol{\tau}$

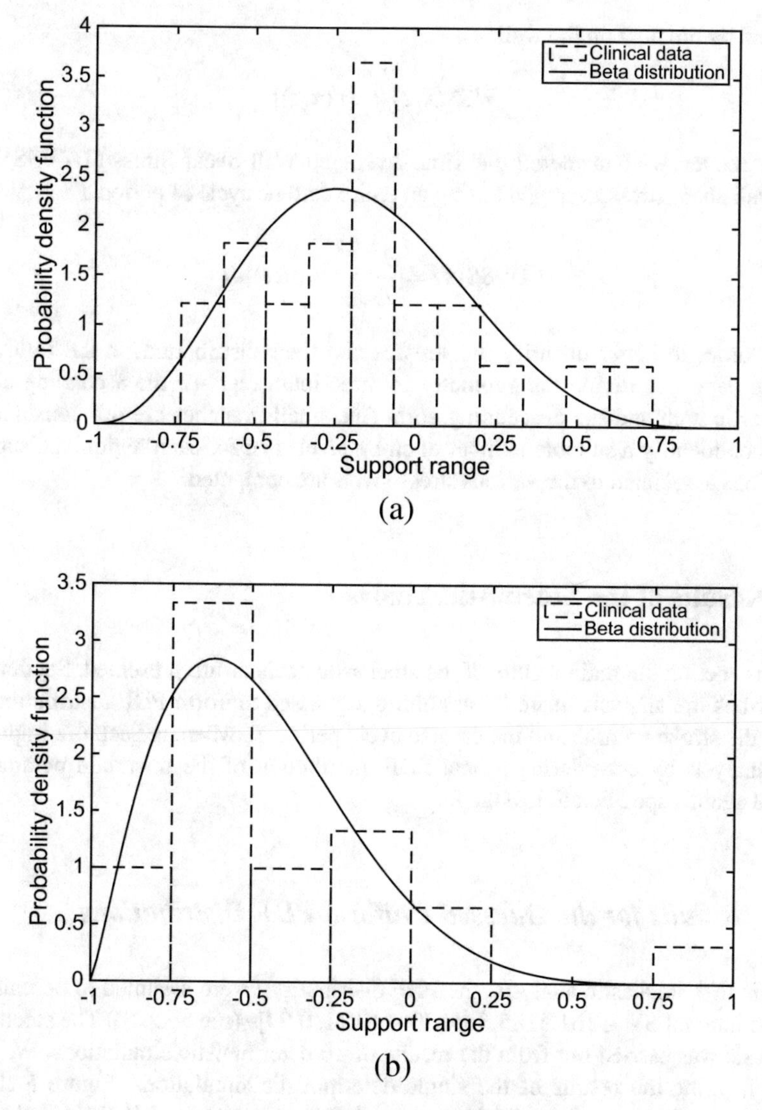

Fig. 4 Comparison between the clinical data and the beta PDF distributions obtained by tuning the free parameters. (**a**) Stroke volume SV. (**b**) Cardiac cycle period T

exerted by the flow on the wall:

$$\text{WSS}(\mathbf{x}, t) = |\boldsymbol{\tau}(\mathbf{x}, t)| \tag{6}$$

Moreover, we considered the Time-Averaged Wall Shear Stress (TAWSS), i.e. the wall shear stress averaged during an entire cardiac cycle of period T:

$$\text{TAWSS}(\mathbf{x}) = \frac{1}{T}\int_0^T |\boldsymbol{\tau}(\mathbf{x}, t)|dt \tag{7}$$

In order to better quantify the stresses and their distribution, in the following of the paper we divide the geometry in three main regions: the ascending aorta, the aortic arch and the descending aorta (the small branches are not considered). We then identify a suitable number of stress levels and for each region, the surface fractions associated to the various stress levels are computed.

6 Results of the Stochastic Analysis

In this section the main results of the stochastic analysis are presented. Section 6.1 describes the analysis made by assuming a guessed uniform PDF distribution for both the stroke volume and the cardiac cycle period T, whereas Sect. 6.2 improves the analysis by considering a beta PDF distribution of the uncertain parameters tuned against specific clinical data.

6.1 Results for the Guessed Uniform PDF Distributions

In this first stochastic analysis, the PDF distributions were assumed to be uniform in the interval SV $\in [61.5, 85.6]$ml, $T \in [0.71, 0.99]$s (see Sect. 3). The stochastic analysis was carried out from the results of 16 deterministic simulations. We start by analyzing the results of the single deterministic simulations. Figure 5 shows the surface fraction of the TAWSS value levels obtained for SV fixed and equal to 69.5 ml and varying the cardiac cycle period T (increasing from left to right). It can be seen that the TAWSS values tend to increase when decreasing T, which is particularly evident in the descending part of the aorta. This can be again explained in terms of losses inside the aorta: for fixed values of SV, decreasing T gives rise to a higher value of cycle-averaged flow rate, which in turn produces a higher cycle-averaged pressure difference between the inlet and the outlets. This results in an overall increase of TAWSS values. Figure 6 shows the same quantities as Fig. 5, this time for T equal to 0.8 s and varying the inlet stroke volume (SV increases going from left to right). It can be seen that the TAWSS values increase with increasing stroke volume. Once again, this effect is more important in the descending part

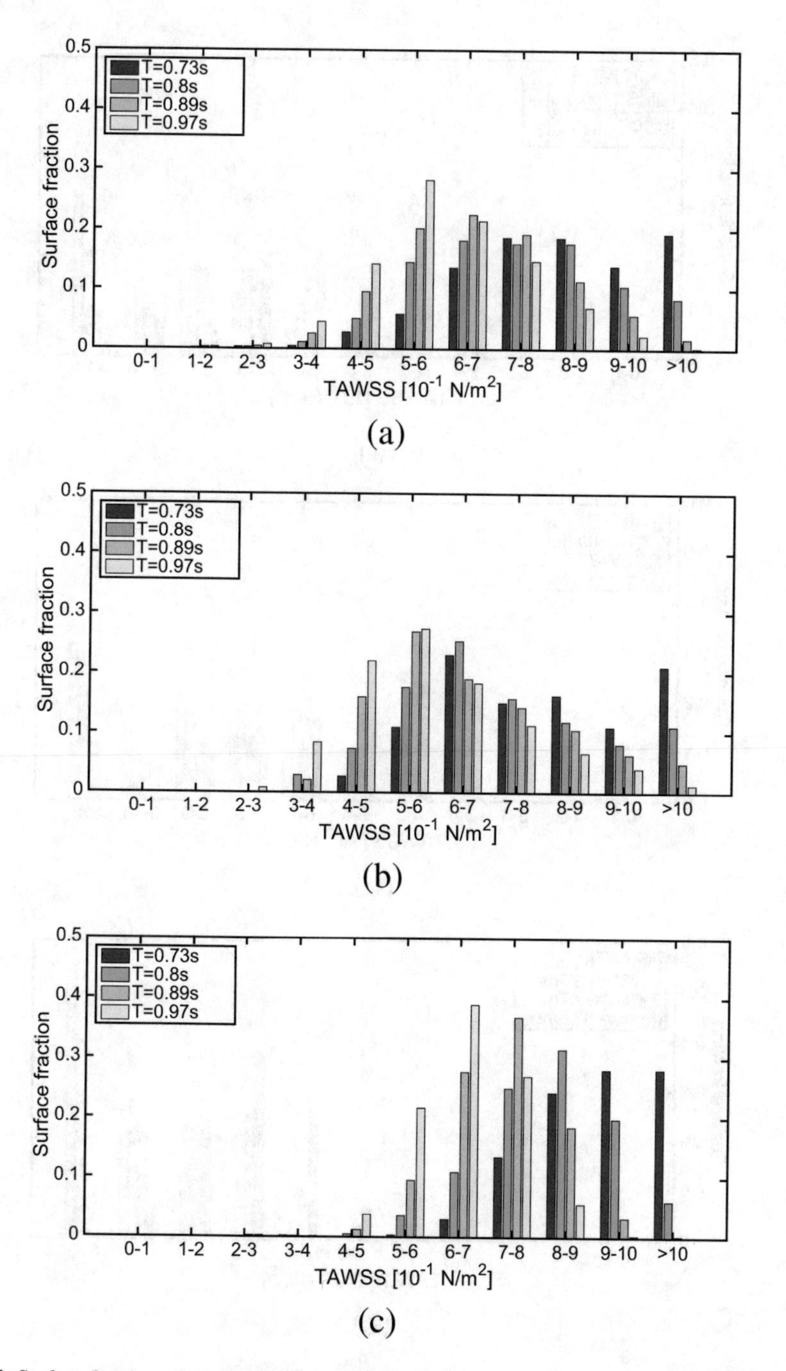

Fig. 5 Surface fraction of the TAWSS levels obtained for the deterministic simulations in which the uncertain parameter T varies (SV = 69.5 ml). (**a**) Ascending aorta. (**b**) Aortic arch. (**c**) Descending aorta

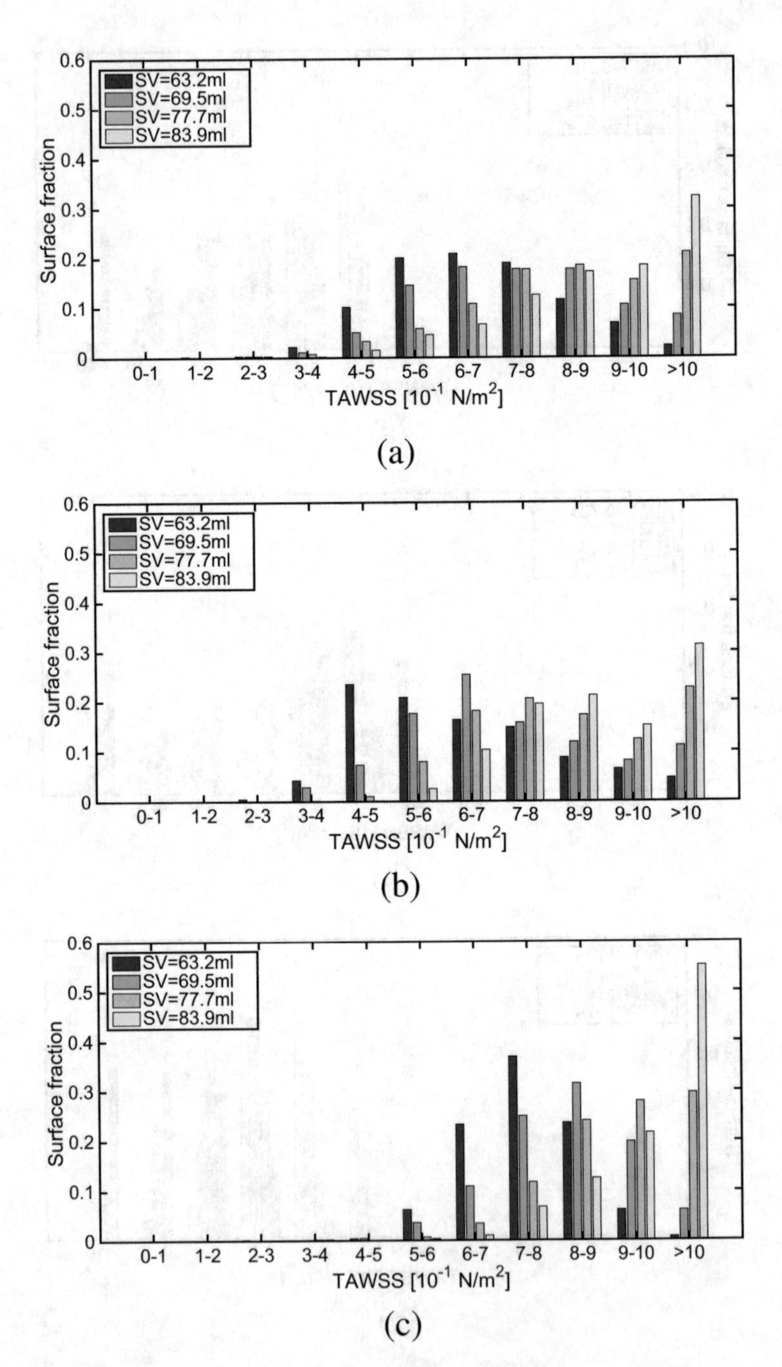

Fig. 6 Surface fraction of the TAWSS levels obtained for the deterministic simulations in which the uncertain parameter SV varies ($T = 0.8$ s). (**a**) Ascending aorta. (**b**) Aortic arch. (**c**) Descending aorta

of the aorta. This behavior can be intuitively justified by using an analogy with a Poiseuille flow, in which the wall shear stresses are given by:

$$|\boldsymbol{\tau}(t)| = \frac{4\mu Q_i(t)}{\pi r_i^3}$$

where $Q_i(t)$ and r_i are the flow rate and the radius at the considered vessel section, respectively. In particular, it can be verified that in the present case the decrease in vessel radius in the descending aorta outweighs the one in flow rate, giving rise to larger resulting wall shear stresses compared to those found in the other two regions. The previous relation indicates that also the variation of $|\boldsymbol{\tau}(t)|$ with $Q(t)$ is larger in the descending aorta: the same trend is thus expected also for the corresponding cycle-averaged quantities, i.e. TAWSS, which is consistent with Figs. 5 and 6.

A more quantitative appraisal of the variability of TAWSS with the considered inflow parameters can be obtained from Figs. 7 and 8, which show the isocontours and the fraction surfaces of the stochastic standard deviation of TAWSS in the parameter space, obtained with the gPC approach. It can be seen that the variability is almost uniform everywhere, the higher values being assumed in correspondence of the distal ends of the small branches and in the proximal part of the aortic arch. Also the region of flow impingement on the posterior part of the ascending aorta

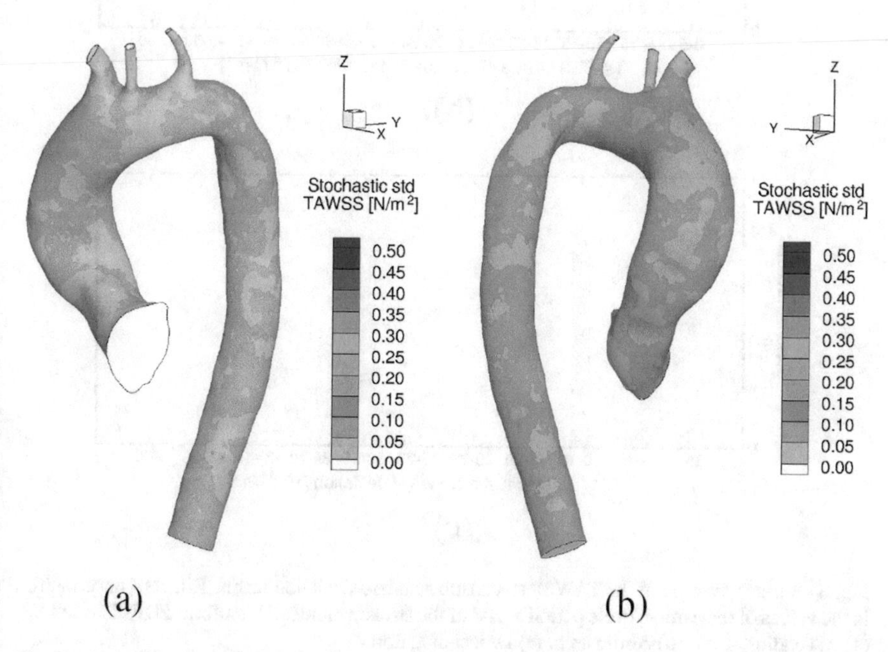

Fig. 7 Stochastic standard deviation of TAWSS. Effect of uncertainties in the values of the cardiac cycle period T and of the stroke volume SV, uniform PDF distributions. (**a**) Front view. (**b**) Back view

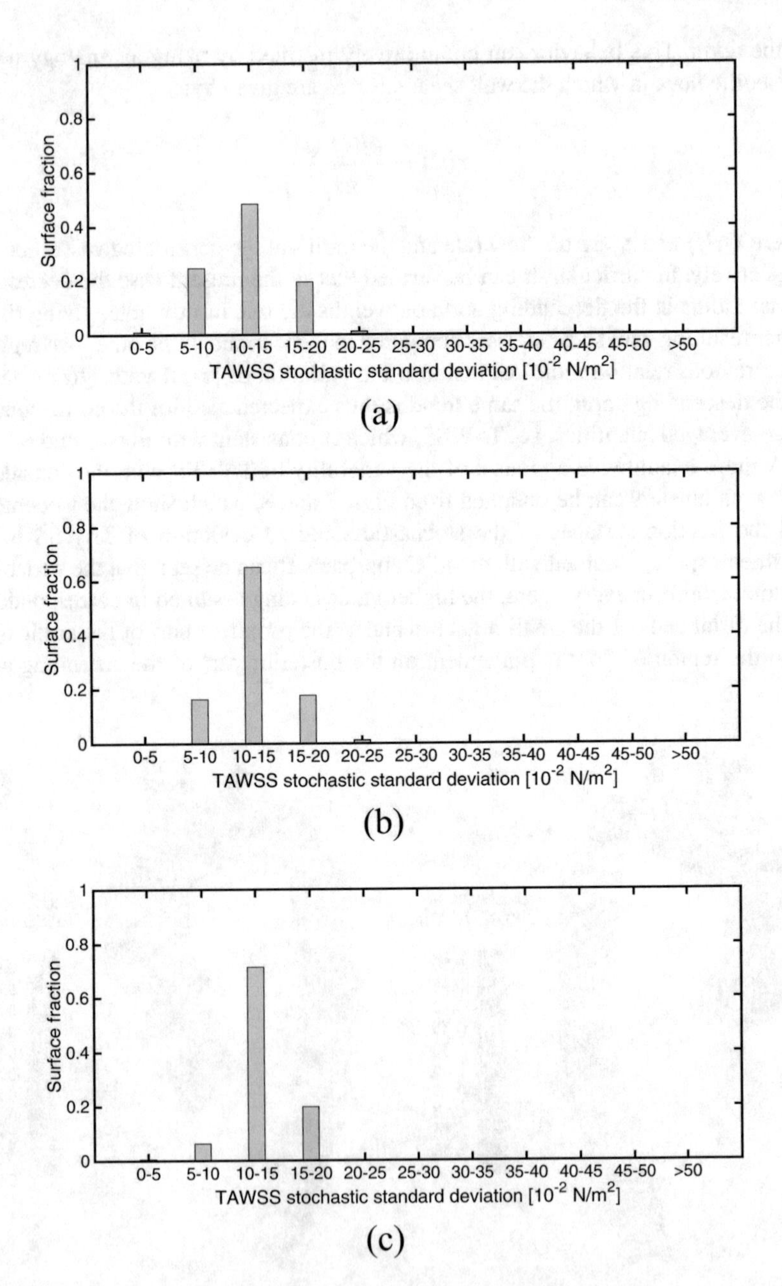

Fig. 8 Surface fraction of the TAWSS stochastic standard deviation levels. Effect of uncertainties in the values of the cardiac cycle period T and of the stroke volume SV, uniform PDF distributions. (**a**) Ascending aorta. (**b**) Aortic arch. (**c**) Descending aorta

takes values of stochastic standard deviation larger than average. However, Fig. 8c shows that the region with the largest overall sensitivity to the uncertainties in the inlet flow rate parameters is the descending aorta, in which about 95% of the surface has a stochastic standard deviation higher than 0.1 Pa.

Finally, Fig. 9 shows the isocontours of the partial variances, or Sobol indices, which quantify the sensitivity of TAWSS to the single input parameters, namely T (Fig. 9a, b) and SV (Fig. 9c, d), and to the interaction between these two parameters (Fig. 9e, f). It can be seen that in the ascending and descending parts of the aorta the most important parameter is the cardiac cycle period T, while in the upstream region of the aortic arch the stroke volume becomes predominant. On the other hand, the interaction between the uncertainties in T and SV seems to have a negligible impact on the variability of TAWSS.

6.2 *Results for the Clinical Beta PDF Distributions*

In this section, the PDF distributions of the two input parameters are no longer assumed uniform with a desired standard deviation, but rather based on the clinical data regarding 23 patients as explained in Sect. 3. The resulting PDF shape and variation interval are different from those guessed in the previous analysis; in particular, the "realistic" variation interval of SV is significantly larger. Also for these distributions a total amount of 16 deterministic simulations was needed.

As previously, we analyze first the results of the single deterministic simulations. Figure 10a shows the surface fraction of the TAWSS levels obtained for the deterministic simulations in which the uncertain parameter T varies between 0.67 s and 1.14 s for SV $=$ 69.1 ml (see Table 2). For the sake of brevity, in the present case we only show the TAWSS levels computed on the whole aorta wall surface instead of dividing it in three regions. Anyway, as in Sect. 6.1, TAWSS clearly decreases in average when T is increased. As previously explained, this is due to the fact that, for fixed SV, a reduction of T would lead during the systolic phase to higher instantaneous flow rates, which corresponds to larger instantaneous WSS. Figure 10b shows the same quantities as in Fig. 10a for fixed $T = 0.67$ s and SV varying between 41.4 ml and 133.8 ml. Once again, it is evident that, as expected, TAWSS increases with increasing SV.

Figure 11 shows the distribution of the stochastic standard deviation of TAWSS. Compared to the case of guessed uniform PDF distributions (Fig. 7), we can notice a significantly increased variability all over the aorta. This is confirmed by the histograms in Fig. 12, which also show that the descending aorta is again the region with the largest stochastic standard deviation.

A significant difference with the previous case is also found in the partial sensitivity of TAWSS to the input parameters, reported in Fig. 13. Indeed, the variability of the wall shear stresses is now mainly produced by the uncertainty in the stroke volume SV, in all the regions of the aorta (Fig. 13c, d). Predominance of the cardiac cycle period T can only be seen in a small anterior portion of the

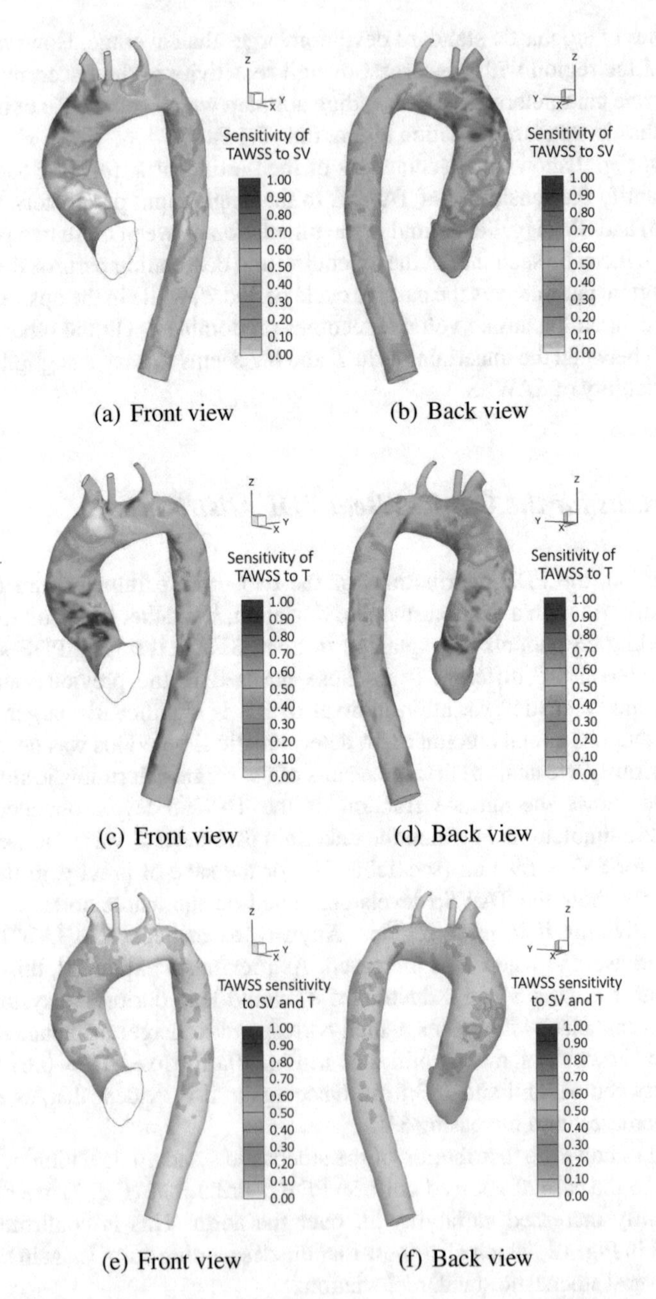

Fig. 9 Partial sensitivities of TAWSS to uncertainties in T (**a**, **b**), to uncertainties in SV (**c**, **d**), and to the interaction between T and SV (**e**, **f**). Stochastic analysis with uniform PDF distributions

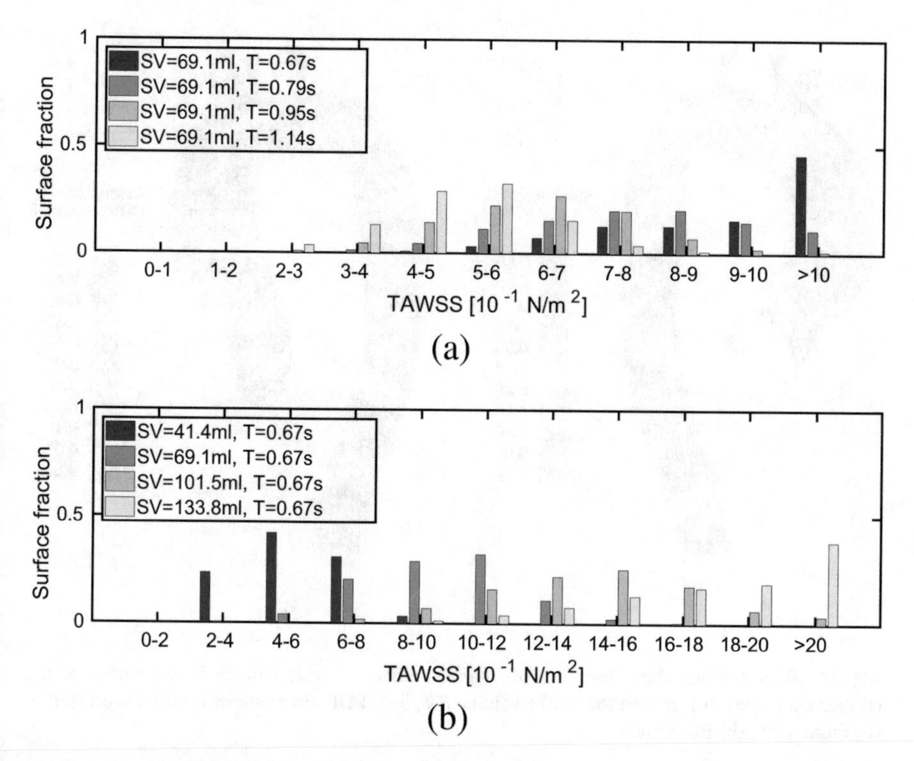

Fig. 10 Surface fraction of the TAWSS levels obtained for the deterministic simulations in which (**a**) T varies (SV $= 69.1$ ml), (**b**) SV varies ($T = 0.67$ s)

ascending aorta (Fig. 13a), which however is a location where the total variability of TAWSS was found to be very low, as shown in Fig. 11.

7 Conclusions

The aim of the present work was to evaluate the impact of inlet boundary conditions on the results of hemodynamic simulations of a patient-specific ATAA geometry. We carried out a stochastic analysis of the sensitivity of the computed wall shear stresses to the inlet flow waveform, parameterized by varying the stroke volume and the cardiac cycle period. We used gPC to build the response surface in the parameter space. As a first step, uniform PDF of the input parameters was initially assumed, as it is usually done in the most common situation in which specific clinical information of patient real variability is lacking. Also the range of variability of the parameters was guessed. It was found that uncertainties in these two parameters lead to a significant variability of computed TAWSS, especially in the descending aorta. The cardiac cycle period has the largest impact in the ascending part of the aorta,

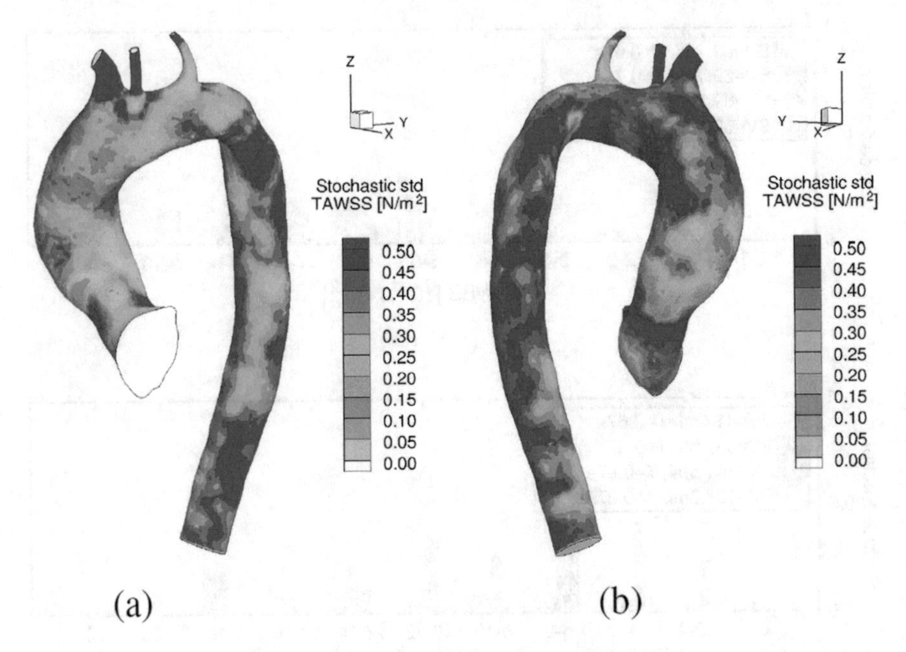

Fig. 11 Stochastic standard deviation of TAWSS. Effect of uncertainties in the values of the cardiac cycle period T and of the stroke volume SV, beta PDF distributions (case of rigid walls). (**a**) Front view. (**b**) Back view

while the variability in the upstream region of the aortic arch is dominated by the stroke volume and both parameters affect the TAWSS values in the descending aorta. As a second step, the PDFs of the inflow parameters were obtained from inflow waveform data available from 23 patients. The resulting PDF shape and variation interval are different from those guessed in the previous analysis; in particular, the "realistic" variation interval of SV is significantly larger. As a result, the variability of TAWSS was found to be significantly larger than in the previous analysis and this variability appeared to be mostly due to uncertainties in SV. The present study indicates thus that patient-specific inlet conditions should be used for quantitative accurate predictions of WSS and TAWSS. A way to obtain these patient-specific inlet conditions is clearly through *in vivo* measurements. These measurements have however a limited resolution and may be affected by noise. The stochastic methodology used in the present work could be useful also to quantify the impact of uncertainties in patient-specific measurements and this could be the object of future work.

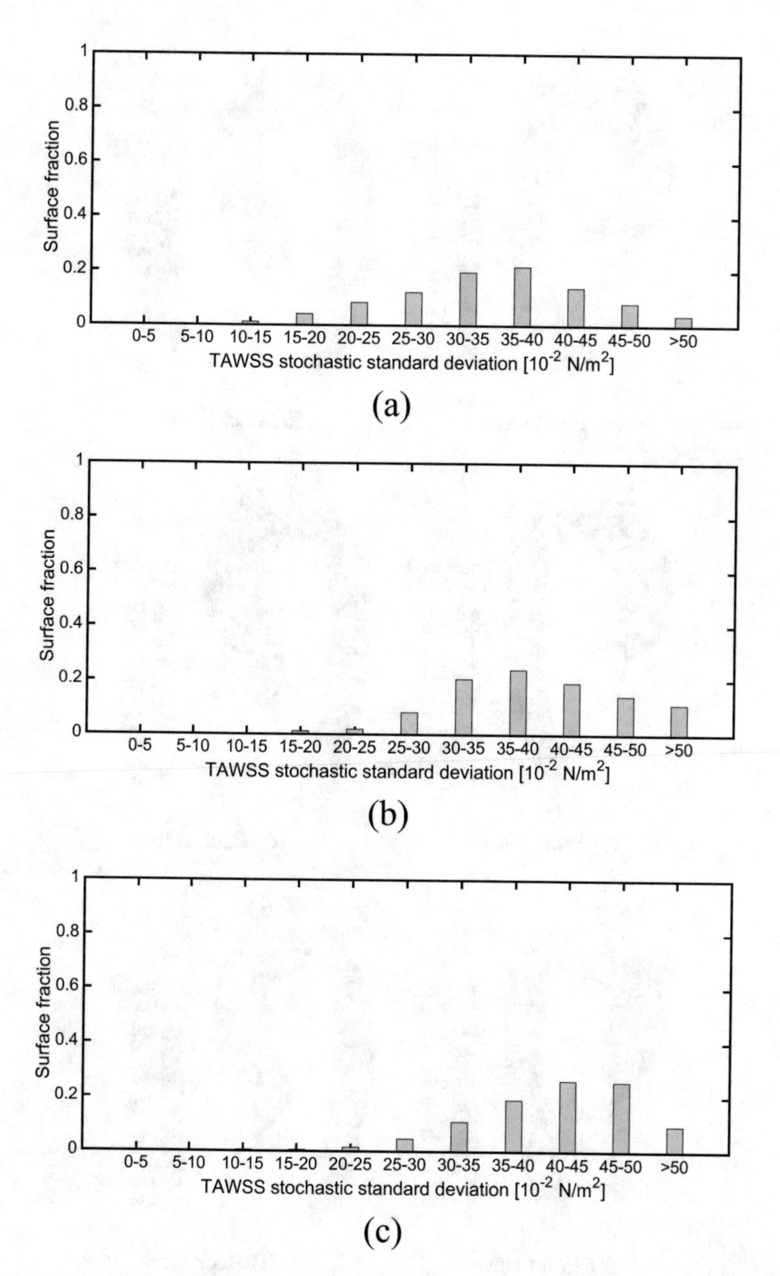

Fig. 12 Surface fraction of the TAWSS stochastic standard deviation levels. Effect of uncertainties in the values of the cardiac cycle period T and of the stroke volume SV, beta PDF distributions (case of rigid walls). (**a**) Ascending aorta. (**b**) Aortic arch. (**c**) Descending aorta

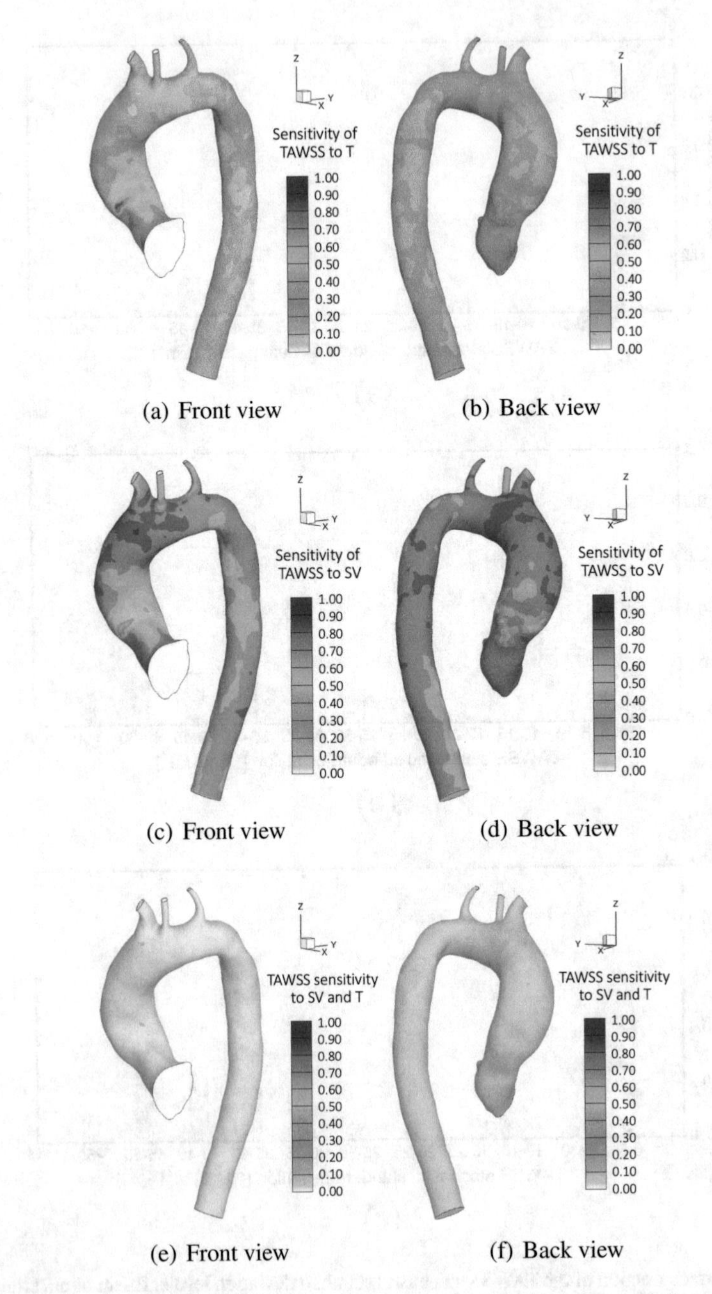

(a) Front view (b) Back view

(c) Front view (d) Back view

(e) Front view (f) Back view

Fig. 13 Partial sensitivities of TAWSS to uncertainties in T (**a**, **b**), to uncertainties in SV (**c**, **d**), and to the interaction between T and SV (**e**, **f**). Stochastic analysis with beta PDF distributions (case of rigid walls)

References

1. Boccadifuoco, A., Mariotti, A., Celi, S., Martini, N., Salvetti, M.V.: Uncertainty quantification in numerical simulations of the flow in thoracic aortic aneurysms. In: ECCOMAS Congress 2016 - Proceedings of the 7th European Congress on Computational Methods in Applied Sciences and Engineering, vol. 3, pp. 6226–6249 (2016)
2. Boccadifuoco, A., Mariotti, A., Celi, S., Martini, N., Salvetti, M.V.: Impact of uncertainties in outflow boundary conditions on the predictions of hemodynamic simulations of ascending thoracic aortic aneurysms. Comput. Fluids **165**, 96–115 (2018)
3. Boccadifuoco, A., Mariotti, A., Capellini, K., Celi, S., Salvetti, M.V.: Validation of numerical simulations of thoracic aorta hemodynamics: comparison with *in vivo* measurements and stochastic sensitivity analysis. Cardiovasc. Eng. Technol. **9**(4), 688–706 (2018)
4. Capellini, K., Vignali, E., Costa, E., Gasparotti, E., Biancolini, M.E., Landini, L., Positano, V., Celi, S.: Computational fluid dynamic study for aTAA hemodynamics: an integrated image-based and radial basis functions mesh morphing approach. J. Biomech. Eng. **140**(11), 111007 (2018)
5. Celi, S., Berti, S.: Biomechanics and FE modelling of aneurysm: review and advances in computational models. In: Murai, Y. (ed.) Aneurysm, pp. 3–26. IntechOpen, London (2012)
6. Celi, S., Berti, S.: Three-dimensional sensitivity assessment of thoracic aortic aneurysm wall stress: a probabilistic finite-element study. Eur. J. Cardiothorac. Surg. **45**(3), 467–475 (2014)
7. Coady, M.A., Rizzo, J.A., Hammond, G.L., Mandapati, D., Darr, U., Kopf, G.S., Elefteriades, J.A., Isom, O.W., Robicsek, F., Griepp, R.B.: What is the appropriate size criterion for resection of thoracic aortic aneurysms? J. Thorac. Cardiovasc. Surg. **113**(3), 476–491 (1997)
8. Gallo, D., De Santis, G., Negri, F., Tresoldi, D., Ponzini, R., Massai, D., Deriu, M.A., Segers, P., Verhegghe, B., Rizzo, G., Morbiducci, U.: On the use of *in vivo* measured flow rates as boundary conditions for image-based hemodynamic models of the human aorta: implications for indicators of abnormal flow. Ann. Biomed. Eng. **40**(3), 729–741 (2012)
9. Lantz, J., Renner, J., Karlsson, M.: Wall shear stress in a subject specific human aorta - influence of fluid-structure interaction. Int. J. Appl. Mech. **3**, 759–778 (2011)
10. Markl, M., Wallis, W., Harloff, A.: Reproducibility of flow and wall shear stress analysis using flow-sensitive four-dimensional MRI. J. Magn. Reson. Imaging *33*(4), 988–994 (2011)
11. Martufi, G., Gasser, T.C., Appoo, J.J., Di Martino, E.S.: Mechano-biology in the thoracic aortic aneurysm: a review and case study. Biomech. Model. Mechanobiol. **13**(5), 917–928 (2014)
12. Morbiducci, U., Ponzini, R., Rizzo, G., Cadioli, M., Esposito, A., Montevecchi, F.M., Redaelli, A.: Mechanistic insight into the physiological relevance of helical blood flow in the human aorta: an *in vivo* study. Biomech. Model. Mechanobiol. **10**(3), 339–355 (2011)
13. Morbiducci, U., Ponzini, R., Gallo, D., Bignardi, C., Rizzo, G.: Inflow boundary conditions for image-based computational hemodynamics: impact of idealized versus measured velocity profiles in the human aorta. J. Biomech. **46**, 102–109 (2013)
14. Nobile, F., Tempone, R., Webster, C.G.: A sparse grid stochastic collocation method for partial differential equations with random input data. SIAM J. Numer. Anal. **46**(5), 2309–2345 (2008)
15. Pape L.A., Tsai T.T., Isselbacher E.M., Oh J.K., O'gara P.T., Evangelista A., Fattori R., Meinhardt G., Trimarchi S., Bossone E., Suzuki T., Cooper J.V., Froehlich J.B., Nienaber C.A., Eagle K.A. Aortic diameter ≥5.5 cm is not a good predictor of type A aortic dissection: Observations from the International Registry of Acute Aortic Dissection (IRAD). Circulation **116**(10), 1120–1127 (2007)
16. Shang, E.K., Nathan, D.P., Fairman, R.M., Bavaria, J.E., Gorman, J.H., Jackson, B.M.: Use of computational fluid dynamics studies in predicting aneurysmal degeneration of acute type B aortic dissections. J. Vasc. Surg. **62**(2), 279–284 (2015)
17. Sobol, I.M.: Global sensitivity indices for nonlinear mathematical models and their Monte Carlo estimates. Math. Comput. Simul. **55**, 271–280 (2001)
18. Trimarchi, S., Nienaber, C.A., Rampoldi, V., Myrmel, T., Suzuki, T., Mehta, R.H., Bossone, E., Cooper, J.V., Smith, D.E., Menicanti, L., Frigiola, A., Oh, J.K., Deeb, M.G., Isselbacher,

E.M., Eagle, K.A.: Contemporary results of surgery in acute type A aortic dissection: The International Registry of Acute Aortic Dissection experience. J. Thorac. Cardiovasc. Surg. **129**(1), 112–122 (2005)
19. Xiu, D., Karniadakis, G.: The Wiener-Askey polynomial chaos for stochastic differential equations. SIAM J. Sci. Comput. **24**, 619–644 (2003)
20. Youssefi, P., Gomez, A., Arthurs, C., Sharma, R., Jahangiri, M., Figueroa, C.A.: Impact of patient-specific inflow velocity profile on hemodynamics of the thoracic aorta. J. Biomech. Eng. **140**, 011002 (2018)
21. Updegrove, A., Wilson, N.M., Merkow, J., Lan, H., Marsden, A.L., Shadden S.C.: Simvascular: an open source pipeline for cardiovascular simulation. Ann. Biomed. Eng. **45**, 525–541 (2017)

Cavitation Model Parameter Calibration for Simulations of Three-Phase Injector Flows

Alessandro Anderlini, Maria Vittoria Salvetti, Antonio Agresta, and Luca Matteucci

Abstract A stochastic sensitivity analysis and calibration of the cavitation model parameters in the URANS simulations of a configuration representative of high-pressure injectors for automotive applications is carried out. A popular homogeneous-flow cavitation model is considered, in which the mass transfer due to cavitation is given by the Schnerr–Sauer model together with the classical Rayleigh–Plesset equation. A stochastic approach based on the generalized Polynomial Chaos (gPC) expansion is adopted, which allows continuous response surfaces of the quantities of interest in the parameter space to be obtained starting from a few deterministic simulations. The considered uncertain parameters are the so-called scaling factors. The calibration of these parameters is carried out by using the gPC response surfaces for a axisymmetric simplified geometry against the experimental value of the critical cavitation point, i.e. the condition at which the injector is choked. The procedure is carried out for two different turbulence models, viz. the $k - \omega$ SST and RSM models. The so-obtained optimal parameter set-ups are then validated for the real three-dimensional geometry. The $k - \omega$ SST optimal set-up gives very accurate predictions also in the three-dimensional case. Finally, the results obtained with this optimal set-up are compared to those given by standard values, confirming that the predictions of the different flow regimes occurring in high-pressure injectors are highly sensitive to cavitation model parameters.

A. Anderlini (✉) · M. V. Salvetti
University of Pisa, Department of Civil and Industrial Engineering, Pisa, Italy
e-mail: alessandro.anderlini@ing.unipi.it; mv.salvetti@ing.unipi.it

A. Agresta · L. Matteucci
Vitesco Technologies Italy Srl, San Piero a Grado, Pisa, Italy
e-mail: antonio.agresta@continental-corporation.com;
Luca.2.Matteucci@continental-corporation.com

M. D'Elia et al. (eds.), *Quantification of Uncertainty: Improving Efficiency and Technology*, Lecture Notes in Computational Science and Engineering 137,
https://doi.org/10.1007/978-3-030-48721-8_9

1 Introduction

This work is focused on the numerical simulation of real size injector flows for automotive applications. It has been largely demonstrated that the spray behavior in the combustion chamber deeply influences the ignition process and, consequently, the particulate emission level (see for instance [9, 19]). A desired spray must be low-penetrating (in order to avoid phenomena like the 'piston impingement' [9]) with a high cone angle, such that the fuel is uniformly distributed all over the chamber. In turn, the spray configuration is strongly dependent on the flow characteristics inside the injector. Several authors have indeed demonstrated that turbulence and cavitation play a primary role in spray atomization (see, e.g., [3, 4, 7, 12, 16, 17]). When an injector is subjected to a pressure difference, the local pressure near the inlet corner of the channel may go under the saturation value, and cavitation starts to occur. Keeping constant the inlet pressure and decreasing the outlet value, the cavitating region becomes more extended along the channel length while, at the same time, the pressure becomes constant all over the channel inlet cross-section and equal to the saturation value. Once this last condition is reached, the speed of sound of the flow drops drastically and the disturbances can not propagate upstream. From this condition, denoted by an outlet pressure called Critical Cavitation Point (CCP), the injector (as well as the Mass-Flow-Rate MFR) is considered chocked since further decreases of the outlet pressure do not influence the upper part of the injector. This is true till the inception of the hydraulic flip, i.e. a back-flow of air from the outlet reservoir to inside the orifice that progressively replaces cavitation.

In a pioneering work, Soteriou et al. [16] carried out experiments and visualizations on a wide range of both large scale and real size simplified cylindrical nozzles. They found a correlation between the orifice internal flow and the characteristics of the spray; in particular cavitation occurring inside the channel improves the spray atomization and the cone angle, and this is particularly true when the cavitating region approaches the outlet section, also referred to 'super-cavitation' (desired spray condition). However, when hydraulic flip occurs, the spray pattern changes drastically emerging in a highly-penetrating narrow liquid column. It was also noticed in [16] that between super-cavitation and hydraulic flip, there is a pressure drop interval in which the transition from the two opposite spray patterns was 'not clear'. This condition has been analyzed by Chaves et al. [7] who noticed a regime of transition between cavitating/not cavitating flow.

From a computational point of view, the numerical modeling must take into account all the previously listed phenomena, and it is known that modeling choices influence the simulation results [1, 13]. For what concerns turbulence closure, it has been largely demonstrated that turbulent scale resolving approaches, as Large-Eddy Simulations (LES), seem to be able to accurately describe the flow dynamics. However, the use of LES implies a very large computational cost, unaffordable in an industrial context. Focusing on an industrial viewpoint, in fact, the required compromise between accuracy and computational time makes the 'cheaper' Unsteady-Reynalods-Averaged-Navier–Stokes (URANS) equations be very attractive. Another key issue is the modeling of cavitation. In the literature, several models have been proposed with different levels of complexity (see, for instance, [8]), but it is not still clear which is the best choice. Once again, in a industrial context, a compromise between accuracy and computational complexity must be accepted and achieved. In this work, focused on industry issues, we adopted the classical *one-fluid* or *homogeneous-flow* approach, in which the phase mixture is treated as a single fluid whose properties are a weighted sum of the properties of the pure phases. This approach, widely used also in commercial codes, requires additional transport equations for $N - 1$ phase fractions, being N the number of phases simulates. The mass-transfer due to cavitation is modeled in a source term by the Schnerr–Sauer model [15] together with the classical Rayleigh–Plesset equation [5]. This model contains a certain number of free-parameters that must be a-priori assigned, and once again their values can affect the simulation results. Thus, a suitable cavitation model parameter set-up has to be found; this set-up should be able to obtain accurate predictions of the conditions at which the major injector flow changes occur, such as CCP or hydraulic flip inception.

The aim of the present work is to investigate the sensitivity of URANS predictions of an injector flow to some of the parameters of the cavitation model and to carry out a model calibration in order to match the available reference data. This analysis has been performed for a simplified cylindrical nozzle geometry for which the CCP measurement was carried out in-house by CONTINENTAL AUTOMOTIVE. To this purpose, we used a stochastic analysis, in particular generalized Polynomial Chaos (gPC) [21]. Since the costs of each single simulation of the considered problem is large, the stochastic approach is interesting since it permits to reconstruct the response surfaces of the quantities of interest in the parameter space starting from a small number of deterministic simulations. In order to investigate the effect of the turbulence closure, we repeated the analysis for the $k - \omega$ SST [11] and RSM [10] models.

2 Physical Modeling and Numerical Discretization

In addition to the working liquid and to the vapor generated by cavitation of the liquid phase, a third gaseous phase, namely air, immiscible with the other phases is considered. The governing equations are herein the Unsteady Reynolds-Averaged Navier–Stokes (URANS) equations, meaning that all the flow quantities are time-averaged. A single-fluid approach has been adopted for the multi-phase flow model; thus, a single set of URANS equations is solved for a homogeneous mixture whose properties depend on its composition according to the following relations:

$$\rho = \alpha_l \rho_l + \alpha_v \rho_v + \alpha_g \rho_g \tag{1}$$

$$\mu = \alpha_l \mu_l + \alpha_v \mu_v + \alpha_g \mu_g \tag{2}$$

where ρ and μ denote the density and the molecular viscosity of the mixture, the subscripts l, v and g indicate the liquid, vapor and gas phases respectively and α_i is the volume fraction of phase i, defined as:

$$\alpha_i = \lim_{\delta V \to 0} \frac{\delta V_i}{\delta V} \tag{3}$$

being δV an infinitesimal volume and δV_i the volume part occupied by the phase i. Clearly, we have:

$$\delta V_l + \delta V_v + \delta V_g = \delta V \tag{4}$$

that leads to the obvious relation

$$\alpha_l + \alpha_v + \alpha_g = 1 \tag{5}$$

In this work, all the pure phase properties are considered constant.

Under the previous assumptions, the continuity equation for the mixture can be written as follows:

$$\frac{\partial u_j}{\partial x_j} = (\rho_v^{-1} - \rho_l^{-1})\dot{m} \tag{6}$$

where u_j is the time-averaged velocity in the j direction. Since cavitation implies a mass transfer between liquid and vapor phases, it is necessary to introduce the source term $\dot{m}$, which is the mass transfer rate per unit volume due to cavitation. This term must be closed by the cavitation model, as it will be shown in the following.

The momentum equation is the following:

$$\frac{\partial \rho u_i}{\partial t} + \frac{\partial \rho u_i u_j}{\partial x_j} = -\frac{\partial p}{\partial x_i} + \frac{\partial}{\partial x_j}(\mu(\frac{\partial u_i}{\partial x_j} + \frac{\partial u_j}{\partial x_i})) + \frac{\partial \tau_{ij}^t}{\partial x_j} \tag{7}$$

where ρ and μ are the mixture properties as defined in Eqs. (1) and (2), p is the time-averaged pressure and τ_{ij}^t is the so-called Reynolds-stress tensor that contains the effects of turbulence on the mean flow field. In order to close the last term, a turbulence model is needed. In this work the sensitivity to the turbulence closure has investigated by considering two different turbulence models for the Reynolds-stress tensor: the eddy-viscosity $k - \omega SST$ model [11] and the Reynolds-Stress-Model (RSM) (see e.g. [10]).

The previous equation system must be completed with the transport equations for the volume fraction of the gaseous and the vapor phases:

$$\frac{\partial \alpha_g}{\partial t} + \frac{\partial \alpha_g u_j}{\partial x_j} = 0 \tag{8}$$

$$\frac{\partial \alpha_v}{\partial t} + \frac{\partial \alpha_v u_j}{\partial x_j} = \frac{\dot{m}}{\rho_v} \tag{9}$$

As anticipated, the source term $\dot{m}$ introduced in Eqs. (6) and (9) accounts for the mass transfer between liquid and vapor phases due to cavitation. It is modeled herein by using the Schnerr–Sauer model [15]:

$$\dot{m} = 4\pi n_0 \alpha_l \rho_v R^2 \frac{dR}{dt} \tag{10}$$

in which n_0 is the "seed-density" defining the concentration of cavitation nuclei per unit liquid volume, while the cavitation bubbles assumed to be spherical of radius R. R is inferiorly bounded by R_0, also called 'seed-radius', which represents the radial dimension of the cavitation nuclei in the unperturbed flow.

The cavitation bubble dynamics, expressed by the term $\frac{dR}{dt}$, is described by a simplified form of the classical Rayleigh–Plesset equation [5], the so-called inertia controlled bubble growth, in which the viscous and the surface-tension terms are neglected:

$$\frac{dR}{dt} = SF^{+/-} sgn(p_{sat} - p)\sqrt{\frac{2}{3}\frac{|p_{sat} - p|}{\rho_l}} \tag{11}$$

p_{sat} appearing in Eq. (11) is the saturation pressure of the working liquid, and the expression $sgn(p_{sat} - p)$ accounts for the different sign in the bubble dynamics: if the pressure p is larger than the saturation pressure, the bubble is collapsing ($\frac{dR}{dt} < 0$), and the condensation rate is directly multiplied by the factor SF^-; on the contrary, if p is lower than p_{sat}, the bubble is growing (i.e. $\frac{dR}{dt} > 0$) and the vaporization rate can be tuned by the factor SF^+. The two multiplicative parameters SF^+ and SF^-, also known as 'scaling factors', were not present in the original formulation of the Schnerr–Sauer cavitation model, but they are usually introduced in current CFD implementations (see e.g. [1, 6]).

The four values n_0, R_0, SF^+ and SF^- are free parameters that must be a-priori specified in order to close the cavitation model. Difficulties arise from the fact that their 'correct' values are unknown in most practical applications and must be thus tuned to fit empirical/numerical reference data. Consequently, a stochastic sensitivity analysis seems particularly suitable to investigate the sensitivity of the numerical predictions by reducing the number of deterministic simulations. The aim is to find and validate a possible "optimum" setup.

The simulations have been carried out by using the commercial code SIEMENS STAR-CCM+ V 12.04 based on a finite-volume method. The convective terms are discretized through a second-order accurate upwind scheme, while a second-order central scheme is used for the diffusive terms; the Venkatakrishnan limiter [20] is implemented to limit the reconstructed gradients. The time-advancing is carried out by using a second-order accurate implicit scheme (see [18]), while the velocity-pressure coupling is dealt with the SIMPLE algorithm.

3 Uncertainty Quantification Methodology

In the previous section, we pointed out that the cavitation model requires that the values of a certain number of free-parameters to be a-priori assigned, whose values are unknown in most practical cases. On the other hand, the numerical predictions can be significantly influenced by the parameter set-up. The use of stochastic methodologies, often called also Uncertainty Quantification (UQ) methods, permits to quantify how much the uncertain parameters may affect the simulation at significantly reduced computational costs compared with deterministic approaches. Indeed, continuous response surfaces of a certain quantity of interest cane be obtained by using a limited number of deterministic simulations is very attractive in parameter calibration and data-fitting. In this work we use the Generalized Polynomial Chaos (gPC) expansion method.

3.1 *Generalized Polynomial Chaos Expansion*

The non-intrusive gPC approach is an interpolant method which allows a given random process to be projected over a known orthogonal basis [21]. This projection can be written as follows:

$$\begin{aligned} R(\omega) = a_0 B_0 + \sum_{i_1=1}^{M} a_{i_1} B_1(\xi_{i_1}) + \sum_{i_1=1}^{M}\sum_{i_2=1}^{i_1} a_{i_1 i_2} B_2(\xi_{i_1}, \xi_{i_2}) + \\ + \sum_{i_1=1}^{M}\sum_{i_2=1}^{i_1}\sum_{i_3=1}^{i_2} a_{i_1 i_2 i_3} B_3(\xi_{i_1}, \xi_{i_2}, \xi_{i_3}) + \dots \end{aligned} \tag{12}$$

where $R(\omega)$ denotes a random process that is function of $\xi(\omega) = [\xi_1, \xi_2, \dots, \xi_M]$, a random vector contained in the M-dimensional parameter space Ω. B_k is the polynomial of order k containing the interaction of a set of k parameters among M. The same expression can be simplified by using a term-based indexing as follows:

$$R(\omega) = \sum_{j=0}^{+\infty} \beta_j \Psi_j(\xi(\omega)) \tag{13}$$

where $\Psi_j(\xi)$ is the gPC polynomial base of generic index j and β_j is the related Galerkin projection coefficient. Clearly, a bijective relation exists between β_j and $a_{i_1 i_2 \dots}$.

Each coefficient β_j can be easily calculated taking advantage of the orthogonality properties of the basis as follows:

$$\beta_j = \frac{\langle R, \Psi_j \rangle}{\langle \Psi_j, \Psi_j \rangle} \tag{14}$$

where

$$\langle f, g \rangle = \int_{\Omega} w(\xi) f(\xi) g(\xi) d\xi \tag{15}$$

denotes the scalar product between functions f and g. The two functions are multiplied by a particular weight function w that is associated with the chosen polynomial family.

From a practical point of view, since the exact expansion of process $R(\omega)$ would imply the evaluation of an infinite number of terms, it is necessary to introduce a term-limit T such that the spectral decomposition (13) is truncated to a finite number of elements:

$$T = \sum_{j=1}^{M} (P + 1) - 1 \tag{16}$$

where P is the maximum polynomial degree for each single parameter (supposed here the same for sake of simplicity) and M is the already introduced dimension of the random parameter space. The truncation here defined is also called tensor-product and contains all the multi-dimensional polynomials till order P and some of the polynomials of higher order till order MP.

The truncated gPC expansion is thus the following:

$$R_{gPC}(\omega) = \sum_{j=0}^{T} \beta_j \Psi_j(\xi(\omega)) \tag{17}$$

and requires the evaluation of $T + 1$ terms.

Each integral involved in this methodology is evaluated through Gaussian Quadrature Formula (GQF) using $m_i = P + 1$ quadrature points per parameter whose tensor-product grid define a set of input parameters for which deterministic simulations must be carried out. The locations of such nodes depend on the chosen polynomial family, indicating that this choice is another key-point of the UQ methodology. The same number of terms appearing in Eq. (17), in fact, may lead to very different level of accuracy depending on which polynomial base is used. In particular, a suitable polynomial family needs a few number of terms to reduce the error of the quadrature formula under a certain threshold tolerance, thus influencing the speed of convergence of the gPC method. The optimal polynomial family is that having a weight function similar to the measure of the random variable, and thus to the Probability Density Function (PDF) of the uncertain parameter [21].

4 Circular Channel

The test case taken into consideration is, formally, a cylindrical channel 400 µm long with a diameter of 200 µm. In practice, however, the measurements of the real internal geometry highlighted that the actual dimensions are slightly different and are summarized in Table 1.

The experiments on the aforementioned geometry were carried out in-house by CONTINENTAL AUTOMOTIVE using the Oil ISO4113 as working fluid. It is a liquid-into-air injection and provides the measurements of the Critical Cavitation Point (CCP), i.e. the pressure in the outlet reservoir at which the flow inside the channel can be considered chocked. The pressure in the inlet reservoir has been kept constant at 100 bar while the outlet pressure has been adjusted to obtain a desired pressure difference. From a practical point of view, the methodology followed to evaluate the CCP (see [14]) is based on the relation between the Mass Flow Rate (MFR) and the cavitation number CN, here defined as:

$$CN = \frac{p_{in} - p_{out}}{p_{in} - p_{sat}} \simeq \frac{p_{in} - p_{out}}{p_{in}} \tag{18}$$

where the simplification $p_{in} - p_{sat} \simeq p_{in}$ is justified by the fact that the saturation pressure is much lower than the inlet pressure. Several studies (see for instance [12, 14]) have shown that the MFR-CN curves for cavitating orifices (before hydraulic flip appearance) can be divided in a not-cavitating region in which MFR is linearly dependent on $\sqrt{CN}$, and a cavitating region in which MFR is not dependent on CN (chocked condition). Under the previous hypothesis, the CCP is evaluated intersecting the two lines shown in Fig. 1: the black line (not-cavitating condition) passes through the points $CN = 0.4$ and $CN = 0.55$ corresponding to 60 and 45 bar of outlet pressure respectively, while the horizontal blue line (chocked flow) is computed by the flow conditions at CN=0.7, i.e. 30 bar of outlet pressure. The CCP expression can be easily computed as:

$$CCP = p_{in}(1 - CN_{CCP}) \tag{19}$$

where CN_{CCP} is the cavitation number corresponding to CCP.

Table 1 Nominal vs. real channel geometry dimensions

	Nominal [µm]	Real [µm]
Inlet diameter D_{in}	200	234
Outlet diameter D_{out}	200	221
Length L	400	384

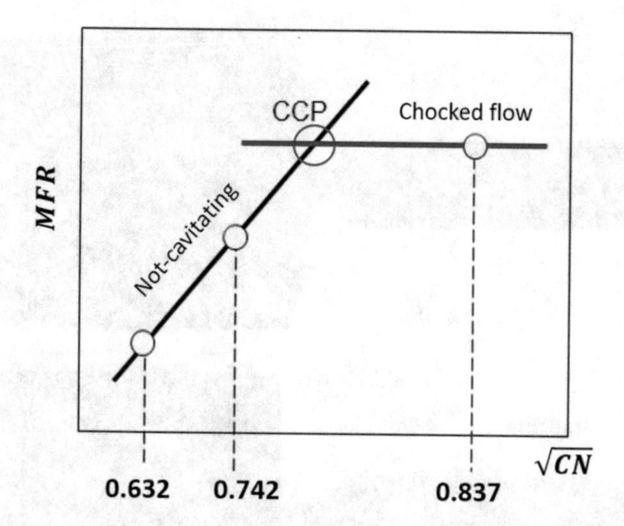

Fig. 1 Geometrical interpretation of the critical cavitation point

Table 2 Experimental measurements of the critical cavitation point

Inlet pressure [bar]	CCP [bar]	MFR @CCP [g/s]
100	37.15	3.475

The experimental measurements are reported in Table 2.

4.1 Computational Setup

The details of the computational geometry used in the simulations are shown in Fig. 2. A stagnation pressure condition of 100 bar, allowing the entrance of the pure liquid phase, has been imposed at the inlet surface, while a static pressure condition (whose value depend on the desired pressure difference) is used at the outlet; on the remaining surfaces of the computational domain a no-slip condition is imposed. In order to follow the same CCP evaluation methodology as used in the reference experiment, three pressure differences are considered herein: $\Delta p = 40$ bar, $\Delta p = 55$ bar and $\Delta p = 70$ bar.

As previously anticipated, the working fluid is the Oil ISO4113; the liquid and vapor properties, together with those of the air present in the outlet reservoir, are reported in Table 3.

The computational grid used to discretize the domain is composed of hexahedral cells and counts about 12 million nodes. The maximum grid spacing inside the channel is about 2.8 μm in all directions. Since no wall-functions have been used, the grid resolution should be such that $y^+ < 1$ at the wall. In order to obtain this condition by increasing the near-wall resolution, a structured prism layer is present at each wall: in particular, 11 layers, whose wall-normal size increases with

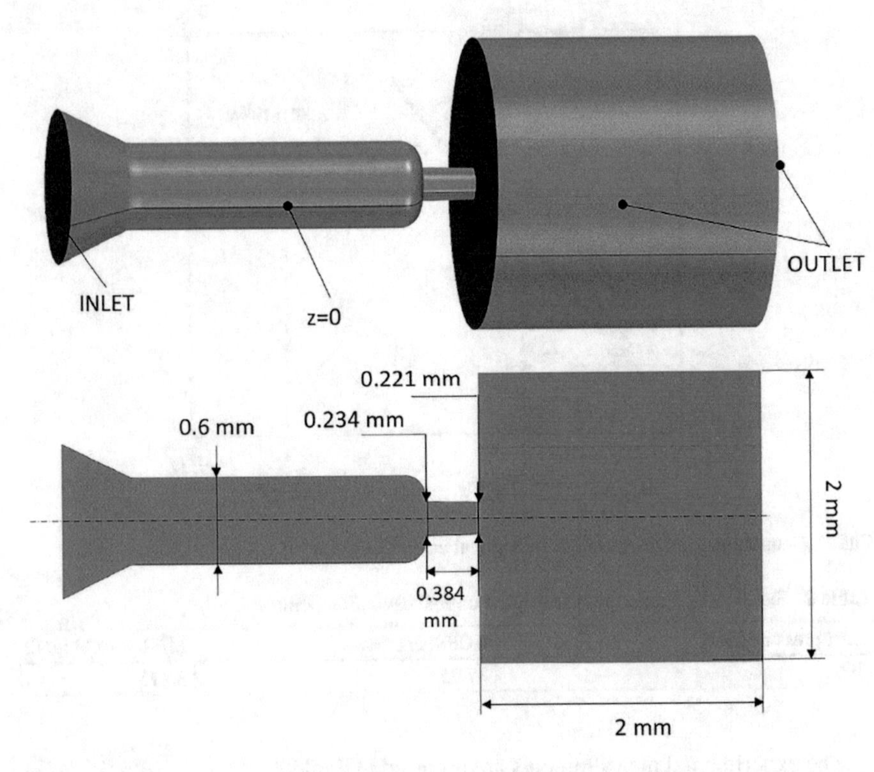

Fig. 2 Computational domain and boundary condition details

Table 3 Fluid properties

	Oil Liquid	Oil Vapor	Air
Density ρ [$\frac{kg}{m^3}$]	812.81	4.04	1.18
Viscosity μ [$\frac{kg}{ms}$]	$2.48 \cdot 10^{-3}$	$7.0 \cdot 10^{-6}$	$1.86 \cdot 10^{-5}$
Saturation pressure p_{sat}(40 °C) [Pa]	28.2		

a geometrical progression with a constant growth-rate equal to 1.2, are distributed over a total height of 2.5 µm. The condition $y^+ < 1$ has been verified a-posteriori. As it will described in Sect. 5.1, the cavitation model calibration has been performed on a axisymmetric grid, which was extracted from the three-dimensional geometry by taking the plane $z = 0$ (see Fig. 2). Finally, since the simulations cover very different flow configurations (pure liquid, weakly/massively cavitating flow and hydraulic flip), the time-advancing was performed with a pressure drop-dependent time step, whose expression is:

$$\Delta t = \frac{0.0000122}{\sqrt{\frac{2\Delta p}{\rho_L}}} \tag{20}$$

that corresponds to a maximum CFL number of about 30 near the inlet corner of the channel in all the simulations.

For what concerns the stochastic sensitivity analysis, we showed, in a previous work [2] containing a cavitation model parameter calibration in two-phase flow condition (liquid-into-liquid injection), that the parameters having the deepest impact on numerical predictions of cavitating flows are the two scaling factors, that directly influence the rate of vaporization (SF^+) and condensation (SF^-). The optimum setup obtained in [2] was tested in this new flow configuration, but the agreement between experiment and numerical predictions was not very good. The disagreement can be reasonably explained by the introduction of the third phase that adds different phenomena compared with the 2-phase flow, as, for instance, the hydraulic flip. For these reasons, it has been decided to perform a new cavitation model calibration considering as uncertain only the scaling factors, fixing the Seed Radius R_0 and the concentration of cavitation nuclei n_0 to the same values as in [2], i.e. $R_0 = 1.9\,\mu\text{m}$ and $n_0 = 4 \cdot 10^{12}\,\text{m}^{-3}$.

Regarding the ranges of variation of the uncertainties, we carried out first a few explorative deterministic simulations which indicated that the parameter ranges for which the numerical results include the experimental measurements depend on the turbulence model used. In light of these preliminary simulations, we investigate the following variation ranges: $SF^+ \in]700; 1200[$ and $SF^- \in]1; 100[$ for the RSM and $SF^+ \in]15; 250[$ and $SF^- \in]45; 60[$. As it can be seen, the intervals of parameter values in which the optimal set-up must be searched are significantly different for the two considered turbulence models. Although turbulence does not explicitly appears in the cavitation model, the interplay between turbulence and cavitation modeling is strong and complex. Indeed, the turbulence model affects the mean velocity and pressure fields which in turn affect cavitation. On the other hand, the cavitation model, and in particular the values of its parameters, obviously influences the mean flow field and turbulence. This explains why the values of the cavitation parameters that match the reference data can be very different for different turbulence models. Finally, since there is no information on the probability distribution of the uncertain parameter values inside these ranges, a constant PDF has been assigned to each parameter. Consequently we used Legendre polynomials on a Gauss–Legendre tensor-product grid.

5 Results

The computational cost of a three-dimensional simulation is quite large. On a workstation having 36 CPU-cores at 2.3 GHz (3.3 GHz with TurboBoost), the order of magnitude of a typical 3D simulation is about 5 days to reach statistical convergence on a computational grid of about 12 million nodes. Since a single evaluation of the CCP implies three deterministic simulations (one for 40, 55 and 70 bar of pressure drop respectively), a fully stochastic analysis with 2 uncertain parameters on the real 3D geometry appears excessively expensive. Conversely, a

2D simulation, for which the grid counts about 35,000 nodes, needs about 1 h to reach statistical convergence. For this reason, taking advantage of the geometrical shape of the orifice, it has been decided to perform the cavitation model parameter calibration on a axisymmetric grid. The 'optimal' setup obtained by the calibration has been, thus, tested for the real 3D geometry.

5.1 Cavitation Model Calibration

The gPC analysis has been performed on a tensor-product grid; it means that, since the number of uncertain parameters is 2, the total number of deterministic point to be simulated is $(p_{max} + 1)^2$ for each pressure drop and turbulence model, being p_{max} the maximum complete order of the multi-dimensional polynomials present in the expansion of Eq. (17). Since the parameter space to be analyzed is significantly different for the two turbulence models, it has been decided to use different truncation orders: $p_{max} = 3$ for the $k-\omega$ SST and a higher order $p_{max} = 6$ for the RSM. The truncation order of the gPC expansion for RSM was increased in order to have roughly the same variation of the parameter values between the different deterministic simulations as for the SST model. However, the convergence of the gPC coefficients (not shown here for sake of brevity) highlighted that the speed of convergence is similar, and, concerning the RSM expansion, the terms higher than order 3 have a negligible impact on the results of the gPC analysis. For each quadrature point in the parameter space, i.e. for a given combination of parameter value, three simulations were carried out for the three considered pressure drops and the CCP was determined following the procedure explained in Sect. 4; then the gPC expansion was applied to the computed CCP. As previously said, the procedure was repeated for the two considered turbulence models.

In Fig. 3a, b, the PDF of the difference between the numerical and experimental CCP normalized with its peak value (such as the maximum value is 1) is shown for the $k - \omega$ SST and RSM respectively. As anticipated in Sect. 4.1, the results obtained for the considered parameter variation ranges include the experimental CCP measurements, denoted by the 0 value on the x-axis, indicating that a proper parameter set-up is able to match the reference data,. However, it can also be seen that the most probable value, i.e. the prediction obtained by most of the parameter values, corresponds to a difference with the experiments of about 3 bars. Finally, both the PDFs are asymmetric with not-negligible tails on the left, and this is particularly true for the $k - \omega$ SST turbulence model.

The response surfaces of the difference between the numerical and experimental CCP in the parameter space is shown in Fig. 4a, b for the two different turbulence models. Despite the differences between the two responses, some common features can be identified. As it can be noticed, in fact, the largest underestimation of the numerical CCP is concentrated on the bottom-left corner of figures, while the overestimated predictions are located on the opposite side. This observation suggests that increasing both parameters means increasing the predicted CCP value.

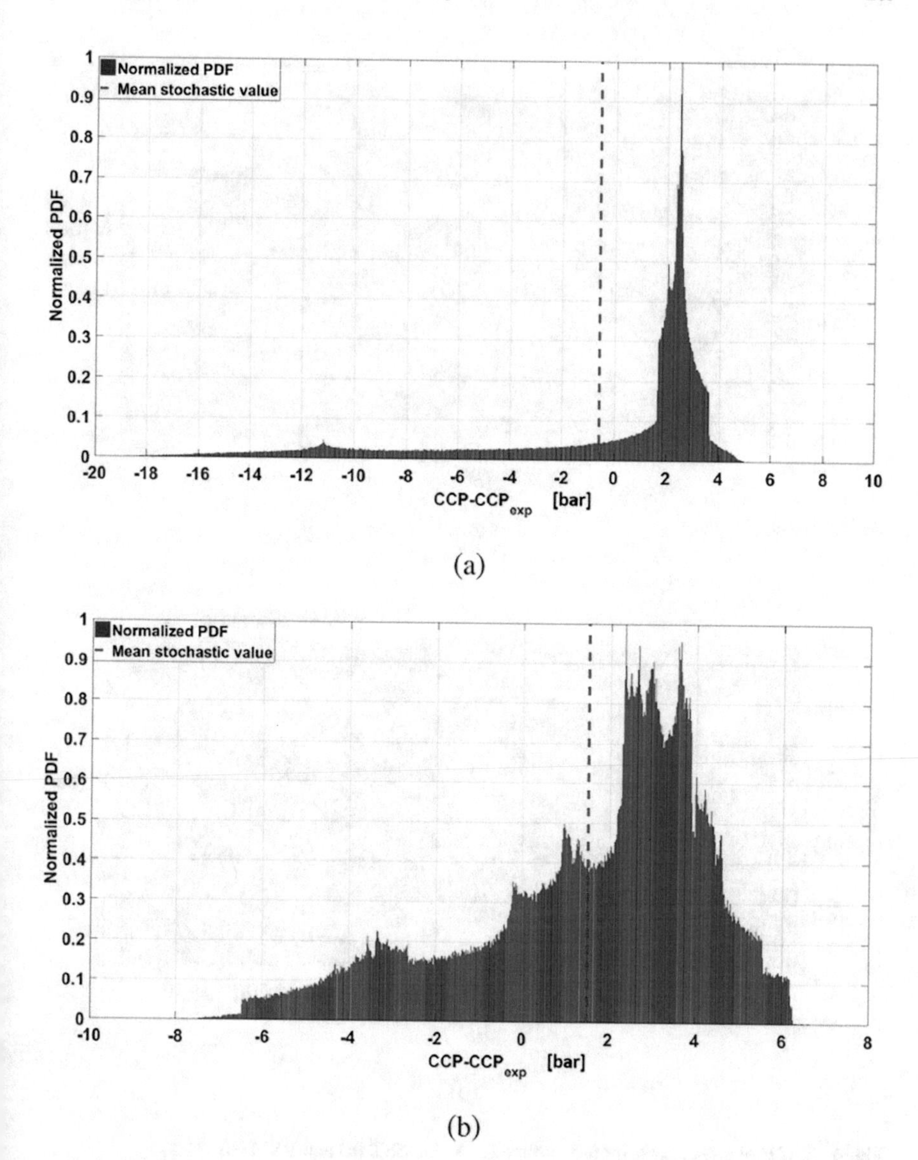

Fig. 3 Normalized PDF of $(CCP - CCP_{exp})$ for $k-\omega$ SST (**a**) and RSM (**b**). The mean stochastic value is represented with the red line

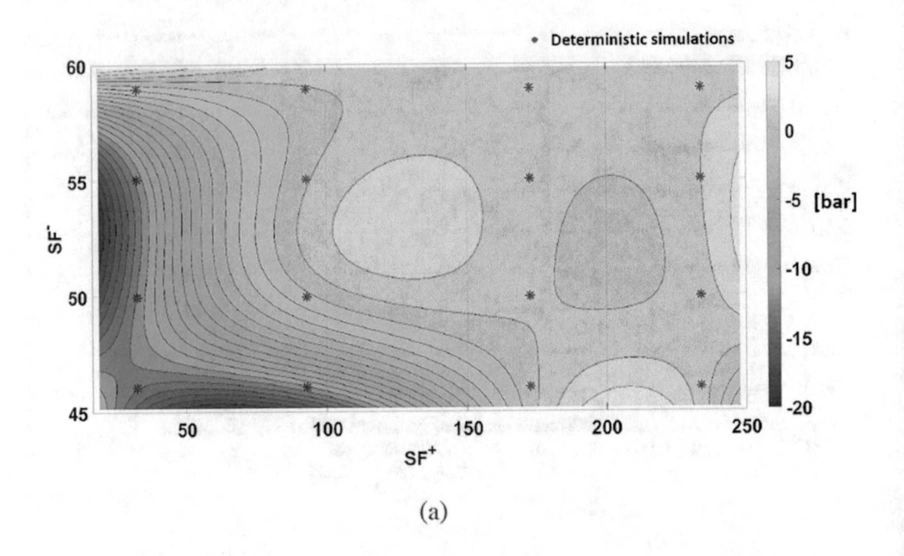

(a)

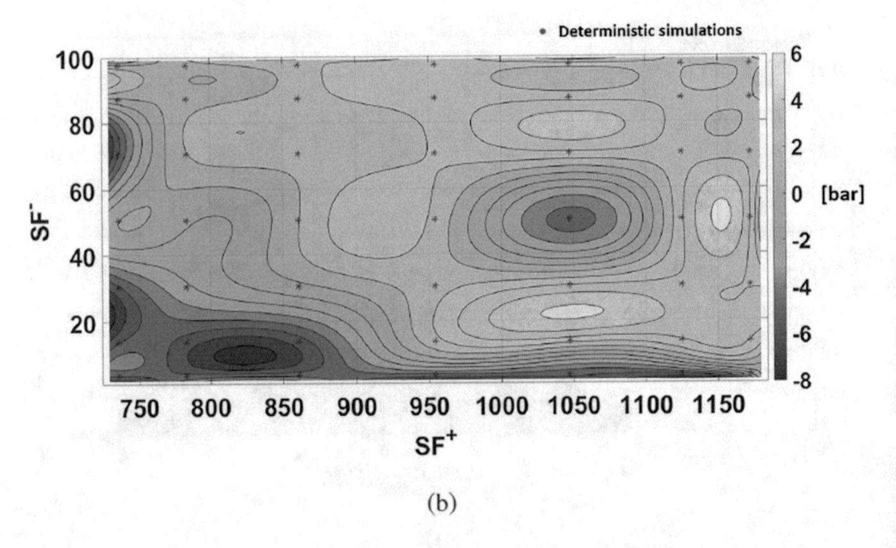

(b)

Fig. 4 $(CCP - CCP_{exp})$ response surface for $k - \omega$ SST (**a**) and RSM (**b**)

In order, thus, to find an 'optimum' cavitation model parameter set-up, we performed a model calibration starting from the response surfaces of CCP and $MFR_{@CCP}$. The procedure is divided in two steps:

1. We extracted a parameter sub-domain $\Omega_{reduced}$ in which the absolute difference $|CCP(\omega) - CCP_{exp}|$ is lower than a threshold value. In this work, the threshold value has been fixed to 0.5 bar.

2. Inside the parameter space $\Omega_{reduced}$, we found the minimum of the percent difference on $MFR_{@CCP}$, defined as:

$$\%MFR(\Omega_{reduced}) = 100 \cdot \frac{MFR_{@CCP}(\Omega_{reduced}) - MFR_{@CCP_{exp}}}{MFR_{@CCP_{exp}}} \tag{21}$$

The sub-domains that verify the condition on the CCP for both turbulence models (see Figs. 5a and 6a) are very similar, with a narrow region dividing the parameter space: this indicates that the "optimal" region is inside a high-gradient zone, and the most part of the parameter set-ups cannot give predictions inside the prescribed error tolerance. The percent differences on MFR, together with the "optimum" set-up, are shown in Figs. 5b and 6b; as a remark, the absolute minimum of $\%MFR(\Omega_{reduced})$ with the $k-\omega$ SST turbulence model (see Fig. 5b) is on a very flat region of the response surface. Since the difference between the actual minimum and the nearest deterministic simulated point inside the acceptable sub-domain is lower that 0.15%, it has been decided to consider that deterministic set-up as the "optimum". The two optimized set-ups are the following:

- $k-\omega$ SST
  - $SF^{+} = 93$
  - $SF^{-} = 58.96$
- RSM
  - $SF^{+} = 860.9$
  - $SF^{-} = 30.41$

The optimized set-ups have thus been used to characterize this injector flow for other intermediate pressure drops. Figure 7 shows the results for both the $k-\omega$ SST and RSM. Focusing on the curve obtained with the $k-\omega$ SST, the liquid/weakly cavitating flow can be recognized by the linear dependence between $\sqrt{(CN)}$ and MFR on the left of the figure (corresponding thus to the lowest pressure drops). At a certain point, the slope of the curve starts to strongly decrease and becomes horizontal: this is the chocked flow condition, corresponding to a highly cavitating flow. For higher pressure drops the curve has an abrupt decrease, meaning that the flow is in a steady hydraulic flip condition (see Fig. 8), i.e a back-flow of air from the outlet reservoir that replaces cavitation inside the channel. The MFR reduction can be explained by the fact that, in steady condition, the interface between air and liquid phases (that are immiscible) is a stream-line; it means that the air back-flow is similar to a modification of the internal geometry of the channel, and the liquid entering the channel "sees" a narrower orifice. When steady hydraulic flip occurs, the flow is liquid and not chocked anymore; for this reason, after the onset of the hydraulic flip, MFR recovers the linear dependence with $\sqrt{CN}$ and it can increase again. The results for the optimized set-up with RSM are very similar and have not been here described in detail for the sake of brevity.

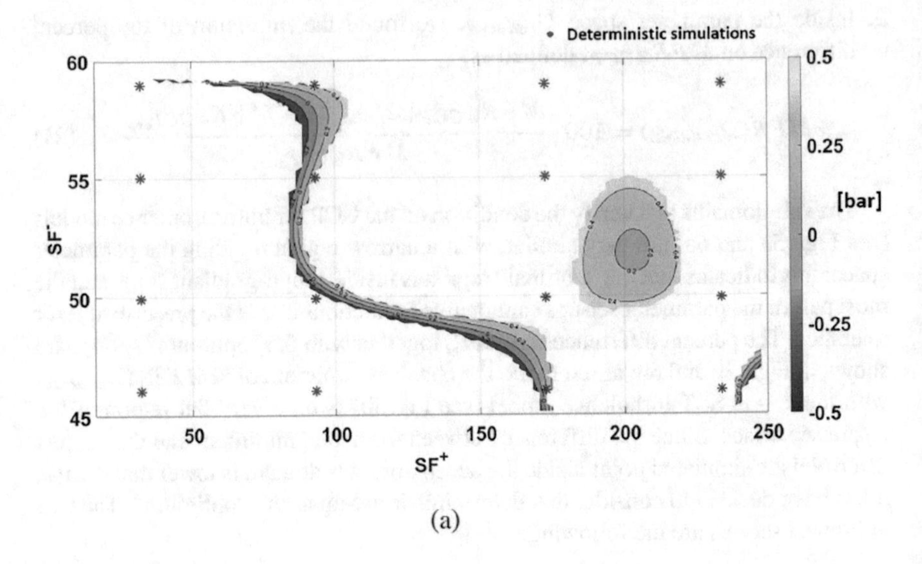

(a)

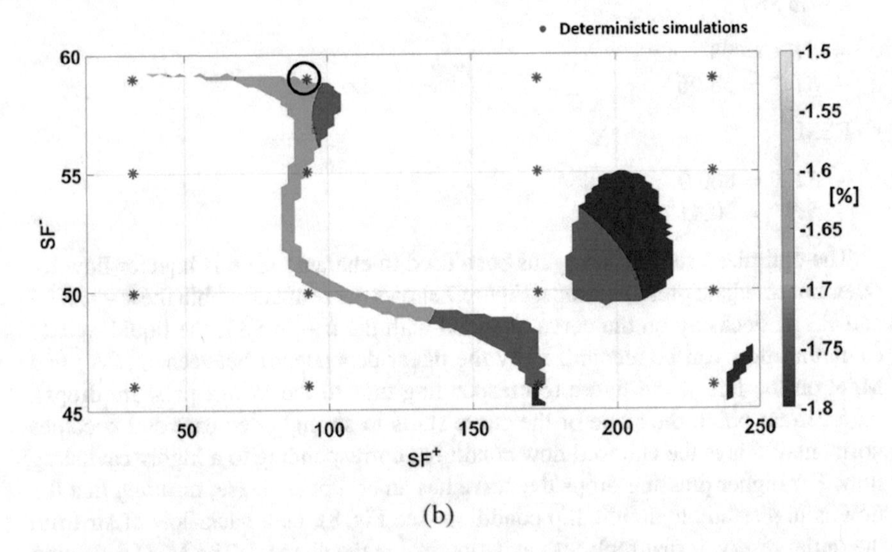

(b)

Fig. 5 Acceptable parameter space for the $(CCP - CCP_{exp})$ tolerance (**a**) and percent difference on MFR (**b**) for the $k - \omega$ SST. In the black circle, the 'optmimum' setup has been highlighted

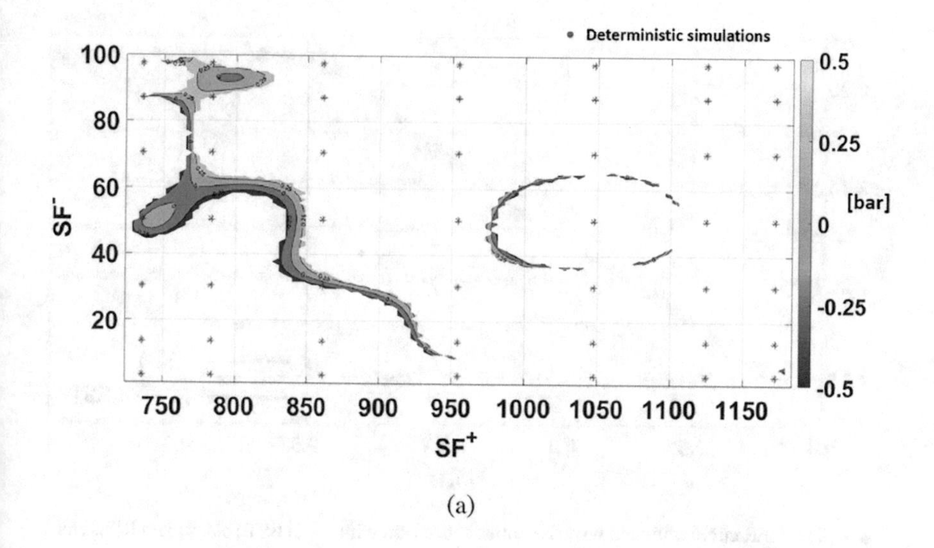

(a)

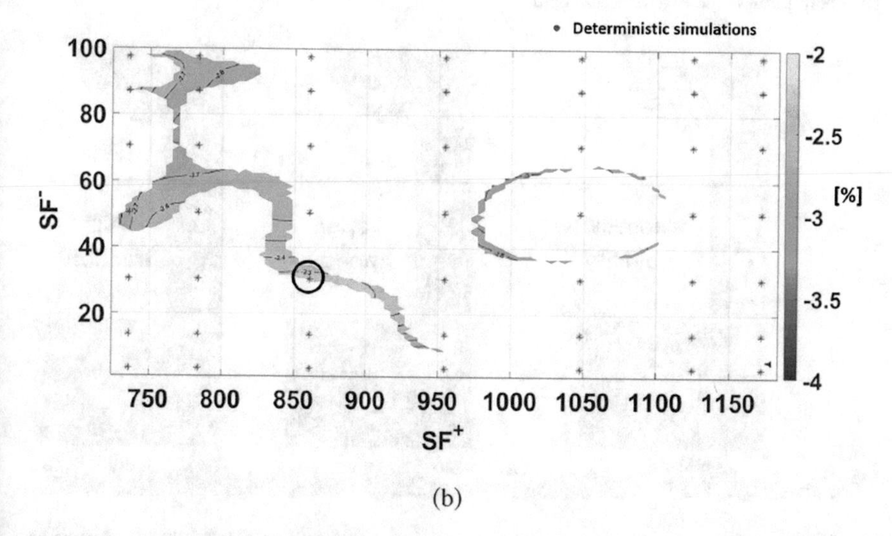

(b)

Fig. 6 Acceptable parameter space for the $(CCP - CCP_{exp})$ tolerance (**a**) and percent difference on MFR (**b**) for the RSM. In the black circle, the 'optmimum' setup has been highlighted

5.2 3D Validation

The two 'optimum' set-ups, found for the $k - \omega$ SST and the RSM models, are the result of a calibration on an axisymmetric geometry. However, to be considered real 'optima' set-ups, the results of the axisymmetric calibration must be validated over the more realistic 3D geometry.

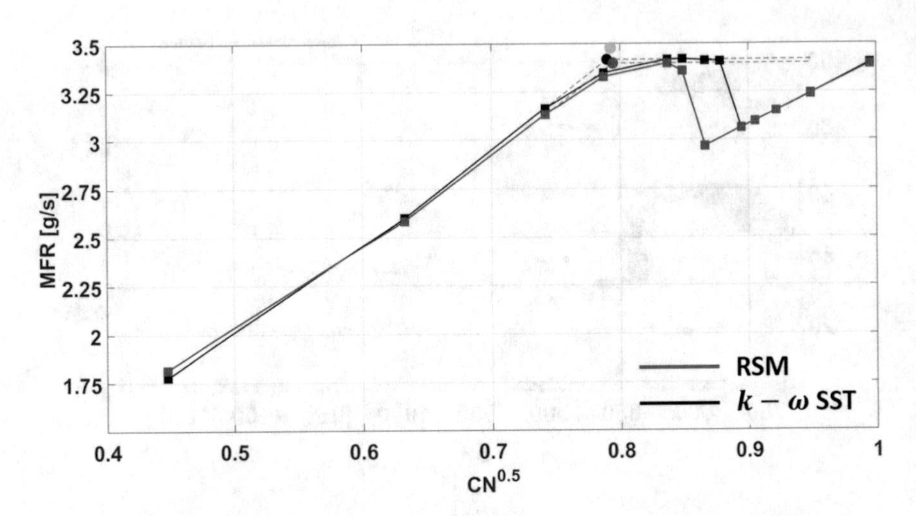

Fig. 7 CN-MFR curve obtained with the optimized set-up with $k-\omega$ SST (black) and RSM (red). The green point is the experimental data

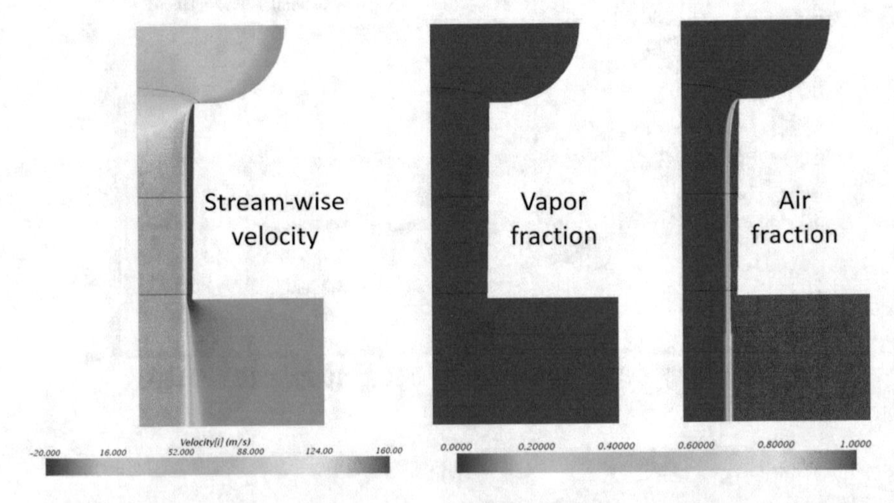

Fig. 8 From left to right, stream-wise velocity, vapor and air fraction fields for 82 bar of pressure drop with the optimized $k-\omega$ SST set-up, corresponding to stable hydraulic flip condition

Figure 9 and Table 4 shown the major results of the validation. The $k-\omega$ SST turbulence model preserves the good prediction of the CCP obtained in axysimmetric condition also in the real 3D model; in particular the difference $CCP_{k-\omega} - CCP_{exp}$ is about 0.76 bar, very close to the tolerance used in the calibration procedure, while the percent difference on MFR is -2.88%. Conversely, the prediction with the RSM is much worse than that obtained in axisymmetric condition since the difference between numerical and experimental CCP is greater than 4 bar and thus far from the threshold tolerance of 0.5 bar. This result suggests

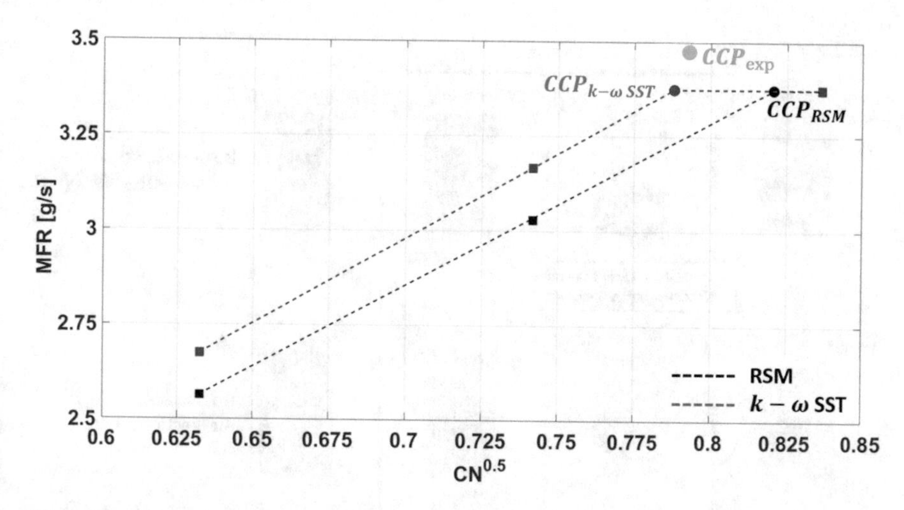

Fig. 9 Numerical prediction of the CCP for both turbulence models with the 3D geometry. The experimental measurements is the green circle

Table 4 Numerical prediction of the CCP and MFR(@CCP) for the 3D geometry

	CCP [bar]	MFR(@CCP) [g/s]
Experiment	37.15	3.475
$k-\omega$ SST	37.91	3.375
RSM	32.64	3.374

that RSM requires a new calibration to be carried out on the 3D geometry. This new potential calibration would require huge computational costs and it is far beyond the purpose of this paper; for these reasons it has been decided to neglect the set-up with the RSM focusing only on the more promising $k-\omega$ SST. As a remark, however, it must be noticed that the MFR for $\Delta p = 70$ bar ($CN = 0.7$), thus for highly-cavitating flow, is very low-dependent on the turbulence model chosen; this fact was also observed in [2].

The flow configuration for $\Delta p = 70$ bar requires a deeper investigation. In Fig. 10, the iso-surfaces of streamwise velocity, air and vapor fraction are shown at different time-instants. As it can be noticed looking at the first panel on the left of Fig. 10, in correspondence of the end of the cavitating flow, there is a recirculation area originated by the adverse pressure gradient (due to the cavity collapse [16]) between the vapor cavity, that is at the saturation pressure, and the liquid, that has a pressure similar to the outlet. If the pressure drop is such that the vapor approaches the channel outlet, it happens that the recirculation reaches the exit area sucking air inside the orifice (second panel from the left). Once entered, the air moves toward the inlet corner progressively replacing the vapor cavity (see Fig. 10). The pressure drop $\Delta p = 70$ is however not sufficient to allow the inception of a stable hydraulic flip condition. In this flow configuration, indeed, once the air reaches the channel inlet, it is pushed out of the orifice and the phenomenon starts again from the beginning. This 'unstable' hydraulic flip may explain the 'atypical' behavior

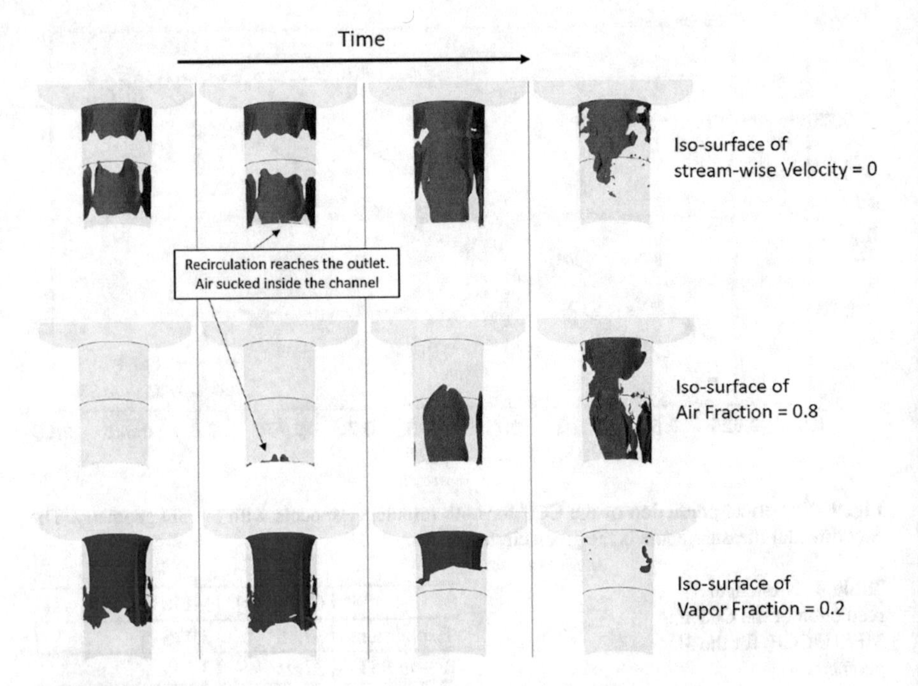

Fig. 10 Flow evolution in unstable hydraulic flip condition ($\Delta p = 70$)

Table 5 'Optimum' vs. default parameter set-up predictions

	CCP [bar]
Experiment	37.15
'Optimum' set-up	37.91
Default set-up	32.46

noticed in the experiments on real size cylindrical nozzles in [7, 16] in the range of cavitation numbers between super-cavitation and stable hydraulic flip conditions.

In Sect. 5.1, it has been demonstrated that, with axisymmetric conditions, the 'optimum' set-up improves the numerical prediction of the CCP compared to the default cavitation model parameter setting provided by STARCCM+ (i.e. $SF^+ = 1$ and $SF^- = 1$). The default set-up, indeed, has been excluded from the stochastic model calibration since it lays on a parameter space region where the CCP prediction is underestimated (see Fig. 4a). It has been then shown how the 'optimum' set-up preserves a very good accuracy even with the real 3D geometry. As last check, however, it is interesting to verify that the prediction improvement between default and optimized set-up observed with axisymmetric conditions is maintained also in 3D. In Table 5 it has been thus shown the comparison between the 'optimum' and the default prediction of CCP where it is obvious to observe that the absolute difference respect to the experimental measurement passes from 4.7 bar (default) to the already mentioned 0.76 bar ('optimum').

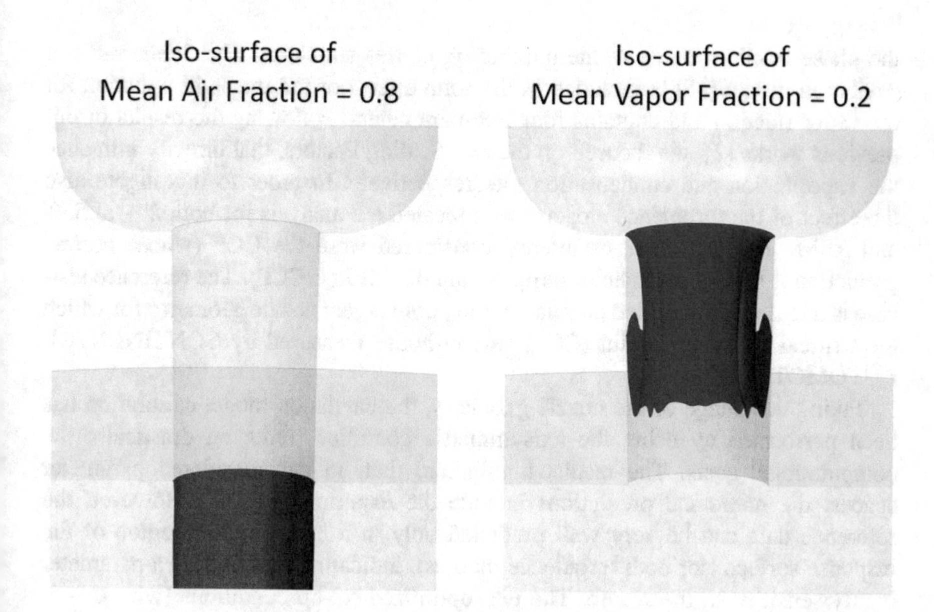

Fig. 11 ISO-surfaces of time-averaged air and vapor fraction for $\Delta p = 70$ bar using the default parameter set-up

Moreover, taking a deeper look at the flow configuration for $\Delta p = 70$ bar predicted by the default set-up (see Fig. 11), a time-averaged vapor region extending till about the 70% of the channel length that is not able to initiate the hydraulic flip condition; the time-averaged air region demonstrates, indeed, that the air phase never enters the orifice. Since the 'optimum' and the default parameter set-ups predict different flow behaviors for $\Delta p = 70$ bar and different accuracy in matching the experimental data, this confirms once again the need of calibration of the cavitation model parameters to correctly capture the flow phenomena occurring in high-pressure injectors.

6 Conclusions

In this work, focused on URANS simulations of high-pressure injectors for automotive applications, we performed a calibration of two parameters contained in a cavitation model implemented in the commercial code STARCCM+ through a stochastic approach. In particular, the generalized Polynomial Chaos method has been used to reconstruct the continuous response surface of some quantities of interest and to minimize the differences with experimental reference data. The phase mixture has been handled with the classical *one-fluid* model, i.e. a single fluid whose properties are an weighted sum of the properties of the pure phases. The multi-phase flow is then closed by $N - 1$ additional transport equation for the phase fraction, being N the number oh phases simulates. The Schnerr–Sauer cavitation model [15] is included in the source term of the transport equation for the vapor fraction and contains four free-parameters. Following the results of our previous works [2], we focused on the two Scaling Factors, that directly influence the vaporization and condensation rate respectively. In order to investigate also the effect of the turbulence closure, we repeated the analysis for both $k - \omega$ SST and RSM. The quantities of interest considered were the CCP (whose correct prediction is crucial for injector designs) and the MFR(@CCP). The reference test-case is a real-size simplified circular slightly convergent nozzle geometry for which the Critical Cavitation Point (CCP) was in-house measured by CONTINENTAL AUTOMOTIVE.

Taking advantage of the nozzle geometry, the cavitation model calibration has been performed by using the axisymmetric condition, reducing drastically the computational costs. The results highlighted that, in the considered parameter spaces, the numerical predictions include the experimental CCP. However, the reference data can be very well predicted only in a high-gradient region of the response surface (for both turbulence models), indicating that CCP is a parameter highly sensitive to the set up. The two optimized set-up conditions (with $k - \omega$ SST and RSM models) have then been used to compute the entire Cavitation number(CN)-MFR curve for the axisymmetric case. We observed that the flow behaviors predicted by both the set-ups are very similar, with some negligible differences on the pressure drop at which stable hydraulic flip occurs.

The optimized set-ups have been then validated on the realistic 3D geometry. The $k - \omega$ SST set-up demonstrated to preserve the good prediction accuracy even in the three-dimensional case. In particular, the difference $|CCP - CCP_{exp}|$ is about 0.76 bar, very close to the tolerance imposed in the calibration procedure. Conversely, the accuracy of the predictions of the optimal set-up for RSM degrades for the 3D case. This is probably due to the fact that the flow field obtained in 3D simulations with RSM is characterized by significant 3D features and, hence, it is more different from the corresponding axisymmetric one than for the $k - \omega$ SST simulations. As a consequence, the values of the caviation parameters calibrated in 2D are no more valid and a new calibration on the real 3D geometry would be required. This is not easy in practice because of the huge computational costs required. Therefore, we decided to focus only on the optimal $k - \omega$ SST set-up.

As last check we showed that the optimized parameter set-up improves the CCP prediction with respect to STARCCM+ default set-up also in 3D. Moreover, we noticed an important difference in the flow pattern for $\Delta p = 70$ bar. The default parameter setting, indeed, predicts a pure cavitating flow, with a cavitation region extending till about the 70% of the orifice length. On the other hand, the optimized set-up predictions suggest that the injector has already experienced the hydraulic flip inception, resulting in a roughly periodic transition between cavitation and air back-flow, i.e. an 'unstable hydraulic flip'. This confirms how the prediction of the different flow regimes occurring in high-pressure injectors are sensitive to cavitation model parameters.

References

1. Altimira, M., Fuchs, L.: Numerical investigation of throttle flow under cavitating conditions. Int. J. Multiphase Flow **75**, 124–136 (2015). https://doi.org/10.1016/j.ijmultiphaseflow.2015.05.006
2. Anderlini A., Salvetti, M.V., Agresta, A., Matteucci, L.: Stochastic sensitivity analysis of numerical simulations of high-pressure injectors to cavitation modeling parameters. In: Proceedings of ASME-FEDSM2017 (2017). https://doi.org/10.1115/FEDSM2017-69212
3. Bergwerk, W.: Flow pattern in diesel nozzle spray holes. Proc. Inst. Mech. Eng. **173**, 655–660 (1959)
4. Blessing, M., König, G., Krüger, C., Michels, U., Schwarz, V.: Analysis of flow and cavitation phenomena in diesel injection nozzles and its effect on spray and mixture formation. In: Proceedings of SAE 2003 World Congress and Exhibition (2003). https://doi.org/10.4271/2003-01-1358
5. Brennen, C.E.: Cavitation and Bubble Dynamics. Oxford University Press, Oxford (1995)
6. Cazzoli, G., Falfari, S., Bianchi, G.M., Forte, C., Catellani, C.: Assessment of the cavitation models implemented in OpenFOAM under DI-like conditions. Energy Proc. **101**, 638–645 (2016). https://doi.org/10.1016/j.egypro.2016.11.081
7. Chaves, H., Knapp, M., Kubitzek, A., Obermeier, F., Schneider, T.: Experimental study of cavitation in the nozzle hole of diesel injectors using transparent nozzles. SAE technical paper (1995). https://doi.org/10.4271/950290
8. Goncalves, E., Patella, R.F.: Numerical simulation of cavitating flows with homogeneous models. Comput. Fluids **38**, 1682–1696 (2009). https://doi.org/10.1016/j.compfluid.2009.03.001
9. Kiplimo, R., Tomita, E., Kawahara, N., Yokobe, S.: Effects of spray impingement, injection parameters, and EGR on the combustion and emission characteristics of a PCCI diesel engine. Appl. Thermal Eng. **37**, 165–175 (2012)
10. Launder, B.E., Reece, G.J., Rodi, W.: Progress in the development of a Reynolds-Stress turbulence closure. J. Fluid Mech. **68**, 537–566 (1975)
11. Menter, F.R.: Two-equation eddy-viscosity turbulence models for engineering applications. AIAA J. **32**, 1598–1605 (1994)
12. Nurick, W.H.: Orifice cavitation and its effect on spray mixing. J. Fluids Eng. Trans. ASME **98**, 681–687 (1976)
13. Orley, F., Trummler, T., Hickel, S., Mihatsch, M.S., Schmidt, S.J., Adams, N.A.: Large-Eddy simulation of cavitating nozzle flow and primary jet break-up. Phys. Fluids **27**, 086101 (2015). https://doi.org/10.1063/1.4928701
14. Payri, R., Guardiola, C., Salvador, F.J., Gimeno, J.: Critical cavitation number determination in diesel injection nozzles. Exp. Tech. **28**, 49–52 (2004)
15. Sauer, J., Schnerr, G.H.: Unsteady cavitating flow—a new cavitation model based on a modified front capturing method and bubble dynamics. In: Proceedings of the ASME-FEDSM2000, vol. 251, pp. 1073–1076 (2000)
16. Soteriou, C.C.E., Andrews, R., Smith, M.: Direct injection diesel sprays and the effect of cavitation and hydraulic flip on atomization. SAE technical paper 950080 (1995). https://doi.org/10.4271/950080
17. Sou, A., Tomiyama, A., Hosokawa, S., Nigorikawa, S., Maeda, T.: Cavitation in a two-dimensional nozzle and liquid jet atomization. JSME Int. J. B Fluids Thermal Eng. **49**, 1253–1259 (2006)
18. STAR-CCM+: Users Manual. http://www.cd-adapco.com/products/star-ccmdocumentation
19. Tree, D.R., Svensson, K.I.: Soot processes in compression ignition engines. Prog. Energy Combust. Sci. **33**, 272–309 (2007)
20. Venkatakrishnan, V.: On the convergence of limiters and convergence to steady state solutions. AIAA J. (1993) https://doi.org/10.2514/6.1993-880
21. Xiu, D., Karniadakis, G.E.: The Wiener–Askey polynomial chaos for stochastic differential equations. SIAM J. Sci. Comput. **24**, 619–644 (2002)

Non-intrusive Polynomial Chaos Method Applied to Full-Order and Reduced Problems in Computational Fluid Dynamics: A Comparison and Perspectives

Saddam Hijazi, Giovanni Stabile, Andrea Mola, and Gianluigi Rozza

Abstract In this work, Uncertainty Quantification (UQ) based on non-intrusive Polynomial Chaos Expansion (PCE) is applied to the CFD problem of the flow past an airfoil with parameterized angle of attack and inflow velocity. To limit the computational cost associated with each of the simulations required by the non-intrusive UQ algorithm used, we resort to a Reduced Order Model (ROM) based on Proper Orthogonal Decomposition (POD)-Galerkin approach. A first set of results is presented to characterize the accuracy of the POD-Galerkin ROM developed approach with respect to the Full Order Model (FOM) solver (OpenFOAM). A further analysis is then presented to assess how the UQ results are affected by substituting the FOM predictions with the surrogate ROM ones.

1 Introduction

Many methods have been developed to assess how uncertainties of input parameters propagate, through Computational Fluid Dynamics (CFD) numerical simulations, into the outputs of interest. The aim of this work is to carry out a study on the application of non-intrusive Polynomial Chaos Expansion (PCE) to CFD problems. The PCE method is a way of representing random variables or random processes in terms of orthogonal polynomials. One important feature of PCE is the possibility of decomposing the random variable into separable deterministic and stochastic components [20, 25]. By a computational stand point, the main problem in PCE consists in finding the deterministic coefficients of the expansion. In non-intrusive PCE, no changes are made in the simulations code, and the coefficients are computed in a post processing phase which follows the simulations. Thus, the

S. Hijazi · G. Stabile · A. Mola · G. Rozza (✉)
mathLab, Mathematics Area, SISSA, Trieste, Italy
e-mail: shijazi@sissa.it; gstabile@sissa.it; amola@sissa.it; grozza@sissa.it

M. D'Elia et al. (eds.), *Quantification of Uncertainty: Improving Efficiency and Technology*, Lecture Notes in Computational Science and Engineering 137,
https://doi.org/10.1007/978-3-030-48721-8_10

deterministic terms in the expansion are obtained via a sampling based approach such as the one used in [21, 32]. In this framework, samples of the input parameters are prescribed and then numerical simulations are carried out for each sample.

Once the output of the simulations corresponding to each sample is evaluated, it is used to obtain the PCE coefficients. In the projection approach, the orthogonality of the polynomials is exploited to compute the deterministic coefficients in the expansion through integrals in the sampling space. As the sampling points chosen are quadrature points for such integrals, the computational cost will grow exponentially as the parameter space dimension increases. This is of course quite undesirable, given the considerable computational cost of the CFD simulations associated to the output evaluation at each sampling point. To avoid such problem, in this work the PCE expansion coefficients are computed using a regression approach which is based on least squares minimization.

To explore even further reductions of the computational cost associated with sample points output evaluations, in the present work we apply the PCE algorithm both to the full order CFD model and to a reduced order model based on POD-Galerkin approach. In the last decade, there have been several efforts to develop reduced order models and apply them to industrial continuous mechanics problems governed by parameterized PDEs. We refer the interested readers to [11, 17, 30] for detailed theory on ROMs for parameterized PDE problems. More in particular, the solution of parameterized Navier–Stokes problem in a reduced order setting is discussed in [31]. In such work, the FOM discretization was based on Finite Element Method (FEM). Projection-based Reduced Order Methods (ROMs) have in fact been mainly developed for FEM, but in the last years many efforts have been dedicated to extend them to Finite Volume Method (FVM) and to CFD problems with high Reynolds numbers. Some examples of the application of ROMs based on the Reduced Basis (RB) method to a finite volume setting are found in [12, 15, 16]. In this work, we instead focus on POD-Galerkin methods applied to CFD computations based on FVM discretization. A large variety of works related to POD-Galerkin can be found in the literature, and here we refer only to some of them [2, 3, 6, 7, 24, 28], as examples. As for POD-Galerkin approach applied to Navier–Stokes flows discretized via FVM, we mention [26], in which the authors treat the velocity pressure coupling in the reduced model using the same set of coefficients for both velocity and pressure fields. In [35], the coefficients of velocity and pressure are instead different, and Poisson equation for pressure is added to close the system at reduced order level. In [36] a stabilization method for the finite volume ROM model is presented. In [8] a study on conservative reduced order model for finite volume method is discussed. For applications of ROMs to UQ problems we refer the readers to [9, 10, 14].

PCE is a tool that is independent of the output evaluator and in this work we will apply it to output parameters both obtained from the full order solution and to its POD-Galerkin reduced order counterpart. In this regard, the objective of the present work is to assess whether PCE results are significantly influenced by the use of a POD-Galerkin based model reduction approach. To this end, we will apply POD model reduction to CFD simulations based on incompressible steady Navier–

Stokes equations, and compare the PCE coefficients and sensitivities obtained for the reduced order solution to the ones resulting from the full order simulations.

This article is organized as follows: in Sect. 2 the physical problem under study is described at the full order level. In Sect. 3 the reduced order model is introduced. In particular, the most relevant notions on projection based methods are reported in Sect. 3.1, while boundary conditions treatment is discussed in Sect. 3.2. The theory of the non intrusive PCE is summarized in Sect. 4 with direct reference to the quantities of interest in the present work. Numerical results are presented in Sect. 5, starting with the ones of the reduced order model in Sect. 5.1 and then the PCE results in Sect. 5.2. Finally, conclusions and possible directions of future work are discussed in Sect. 6.

2 The Physical Problem

In this section, we describe the physical problem of interest which consists into the flow around an airfoil subjected to variations of the angle of attack and inflow velocity. In aerospace engineering, the angle of attack is the angle that lies between the flow velocity vector at infinite distance from the airfoil (U_∞) and the chord of the airfoil, see Fig. 1. We are interested in finding the angle of attack that produces the maximum lift coefficient before stall happens. Figure 2 depicts for the lift coefficient curve of the airfoil NACA $0012-64$ [1, 34] at a fixed Reynolds number of 10^6. The

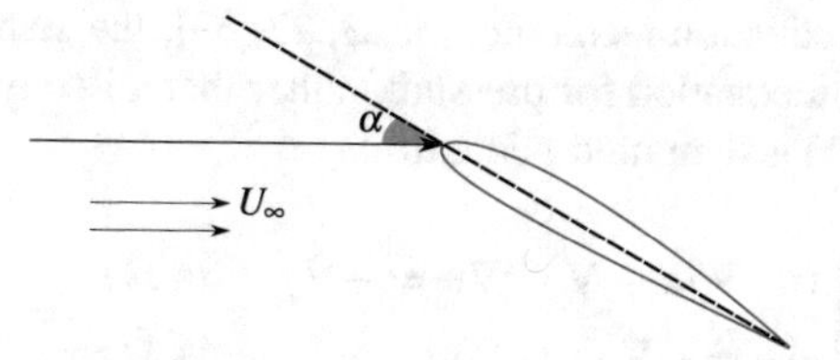

Fig. 1 The angle of attack on an airfoil

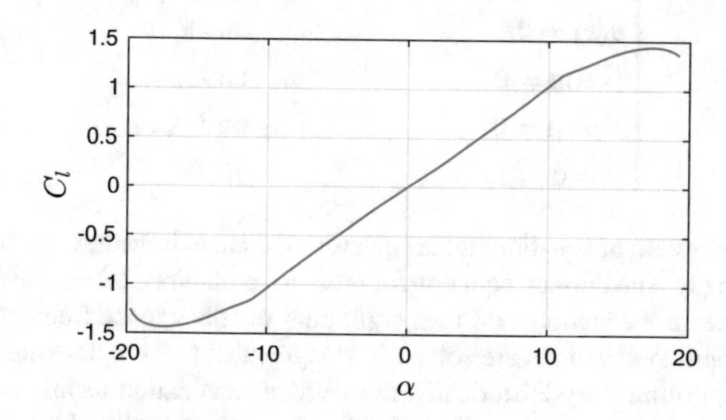

Fig. 2 The lift coefficient curve for the airfoil NACA0012

plot suggests that as the angle of attack increases, the lift coefficient grows until flow separation occurs leading to a loss of the lift force. At laminar flow regimes such as the ones that will be analyzed later in Sect. 5, such stall phenomenon happens in a mild fashion, as opposed to more abrupt stalls observed at higher velocities like the one in Fig. 2. It can be noticed from the plot that the lift coefficient reaches its maximum value when the angle is about 17° before stall happens. For lower Reynolds numbers, the maximum in the C_L-α curve is observed at higher angles of attack. The fluid dynamic problem is mathematically governed by the steady Navier-Stokes equations which read as follows:

$$\begin{cases} (\mathbf{u} \cdot \boldsymbol{\nabla})\mathbf{u} - \boldsymbol{\nabla} \cdot \nu \boldsymbol{\nabla} \mathbf{u} = -\boldsymbol{\nabla} p & \text{in } \Omega_f, \\ \boldsymbol{\nabla} \cdot \mathbf{u} = \mathbf{0} & \text{in } \Omega_f, \\ \mathbf{u}(x) = \mathbf{f}(\mathbf{x}, \boldsymbol{\mu}) & \text{on } \Gamma_{In}, \\ \mathbf{u}(x) = \mathbf{0} & \text{on } \Gamma_0, \\ (\nu \boldsymbol{\nabla} \mathbf{u} - p\mathbf{I})\mathbf{n} = \mathbf{0} & \text{on } \Gamma_{Out}, \end{cases} \tag{1}$$

where $\Gamma = \Gamma_{In} \cup \Gamma_0 \cup \Gamma_{Out}$ is the boundary of the fluid domain Ω_f and is composed by three different parts Γ_{In}, Γ_{Out} and Γ_0, indicating respectively inlet boundary, outlet boundary and wing. In the flow equations $\mathbf{u}$ is the flow velocity vector, ν is the fluid kinematic viscosity, and p is the normalized pressure, which is divided by the fluid density ρ_f. As for the boundary conditions in which $\mathbf{f}$ is a generic function that prescribes the value of the velocity on the inlet Γ_{In} and it is parameterized through the vector quantity $\boldsymbol{\mu}$. In the present work the problem is solved using a finite volume discretization technique [5, 23, 27, 39], the standard approach is to work with a Poisson equation for pressure, rather than directly with the continuity equation. System (1) is then modified into:

$$\begin{cases} (\mathbf{u} \cdot \boldsymbol{\nabla})\mathbf{u} - \boldsymbol{\nabla} \cdot \nu \boldsymbol{\nabla} \mathbf{u} = -\boldsymbol{\nabla} p & \text{in } \Omega_f, \\ \Delta p = -\nabla \cdot (\mathbf{u} \cdot \nabla)\mathbf{u} & \text{in } \Omega_f, \\ \mathbf{u}(x) = \mathbf{f}(x, \mu) & \text{on } \Gamma_{In}, \\ \mathbf{u}(x) = \mathbf{0} & \text{on } \Gamma_0, \\ (\nabla \mathbf{u})\mathbf{n} = \mathbf{0} & \text{on } \Gamma_{Out}, \\ \nabla p \cdot \mathbf{n} = 0 & \text{on } \Gamma \setminus \Gamma_{Out}, \\ p = 0 & \text{on } \Gamma_{Out}. \end{cases} \tag{2}$$

In the above system of equations all the quantities assume the same meaning of those presented in (1). The Poisson equation for pressure is obtained taking the divergence of the momentum equation, and then exploiting the divergence free constraint on velocity. The two equations are solved in a segregated fashion, making use of the SIMPLE algorithm [29]. Historically, the FVM discretization technique has been widely used in industrial applications and for flows characterized by higher values of the *Reynolds* number. One important feature of the FVM is that it ensures that conservative laws are satisfied at local level. In this work the Full Order Model (FOM) simulations are carried out making use of the finite volume open source C++ library OpenFOAM® (OF) [40].

3 The Reduced Order Model

The FOM simulations carried out by OpenFOAM present a high computational cost. In the framework of a many query problem such as the one associated with non-intrusive PCE employed in this work, the search for ways to reduce the computational cost becomes paramount. For this reason, we resort to reduce order modelling and we couple it with PCE in the next sections. In This section, we recall the notion of ROM and the POD approach to build the reduced order spaces. Here, only few details are addressed, while for further information on how to adapt ROM for finite volume discretization method the reader may refer to [26, 35, 36].

The key assumption of ROMs is that one can find a low dimensional space in which it is possible to express the solution of the full order problem with good approximation properties. That space is spanned by the reduced order modes [17]. The latter assumption translates to the following decomposition of the velocity and pressure fields:

$$\mathbf{u}(\mathbf{x}, \boldsymbol{\mu}) \approx \mathbf{u}_r(\mathbf{x}, \boldsymbol{\mu}) = \sum_{i=1}^{N_u} a_i(\boldsymbol{\mu}) \boldsymbol{\phi}_i(\mathbf{x}), \tag{3}$$

$$p(\mathbf{x}, \boldsymbol{\mu}) \approx p_r(\mathbf{x}, \boldsymbol{\mu}) = \sum_{i=1}^{N_p} b_i(\boldsymbol{\mu}) \chi_i(\mathbf{x}), \tag{4}$$

where $\mathbf{u}_r(\mathbf{x}, \boldsymbol{\mu})$ and $p_r(\mathbf{x}, \boldsymbol{\mu})$ are the reduced order approximations of velocity and pressure, respectively, a_i and b_i are scalar coefficients that depend on the parameter value $\boldsymbol{\mu}$, $\boldsymbol{\phi}_i$ and χ_i are the basis functions of the reduced basis spaces for velocity and pressure, respectively. N_u and N_p represent the dimension of the reduced basis spaces for velocity and pressure, respectively, obviously N_u and N_p are not supposed to have the same value.

The next step in constructing the reduced order model is to generate the reduced order space. For such step we resort to a POD approach. The POD space is constructed by solving the following minimization problem:

$$\mathbb{V}_{POD} = \arg\min \frac{1}{N_s} \sum_{n=1}^{N_s} ||\mathbf{u}_n - \sum_{n=1}^{N_s} (\mathbf{u}_n, \boldsymbol{\phi}_i)_{L^2(\Omega)} \boldsymbol{\phi}_i ||^2_{L^2(\Omega)}, \tag{5}$$

where $\mathbf{u}_n$ is a solution snapshot obtained for a certain parameter value $\boldsymbol{\mu}_\mathbf{n}$ and N_s is the total number of solution snapshots. One can see that the reduced order space (or $\mathbb{V}_{POD}$) is optimal in the sense that it is spanned by the modes that minimize the projection error between the fields and their projection into the modes. For further details on how the problem (5) is solved one can refer to [35].

3.1 Projection Based ROM

The next step in building the reduced order model (this procedure is referred as POD-Galerkin projection) is to project the momentum equation of (1) onto the POD space spanned by the POD velocity modes, namely:

$$\left(\boldsymbol{\phi}_i, (\mathbf{u} \cdot \nabla)\mathbf{u} - \nu \Delta \mathbf{u} + \nabla p\right)_{L^2(\Omega)} = 0. \tag{6}$$

Inserting the approximations (3) and (4) into (6) yields the following reduced system:

$$\nu \mathsf{B}\mathbf{a} - \mathbf{a}^T \mathsf{C}\mathbf{a} - \mathsf{H}\mathbf{b} = 0, \tag{7}$$

where **a** and **b** are the vectors of coefficients for reduced velocity and reduced pressure respectively, while B, C, H are the reduced discretized differential operators which are computed as follows:

$$B_{ij} = \left(\boldsymbol{\phi}_i, \Delta \boldsymbol{\phi}_j\right)_{L^2(\Omega)}, \tag{8}$$

$$C_{ijk} = \left(\boldsymbol{\phi}_i, (\boldsymbol{\phi}_j \cdot \nabla)\boldsymbol{\phi}_k)\right)_{L^2(\Omega)}, \tag{9}$$

$$H_{ij} = \left(\boldsymbol{\phi}_i, \nabla \chi_j\right)_{L^2(\Omega)}. \tag{10}$$

To solve the system (7), one needs N_p additional equations. The continuity equation cannot be directly used because the snapshots are divergence free and so are the velocity POD modes. The available approaches to tackle this problem are either the use of the Poisson equation [35, 36] or the use of the supremizer stabilization method [4, 33], which consists into the enrichment of the velocity space by the usage of supremizer modes. These modes are computed such that a reduced version of the inf-sup condition is fulfilled. The latter approach usually employed in a finite element context has been also extended to a FV formulations [36]. In this work we rely on the supremizer stabilization method. After a proper enrichment of the POD velocity space it is possible to project the continuity equation onto the space spanned by the pressure modes giving rise to the following system:

$$\begin{cases} \nu \mathsf{B}\mathbf{a} - \mathbf{a}^T \mathsf{C}\mathbf{a} - \mathsf{H}\mathbf{b} = 0, \\ \mathsf{P}\mathbf{a} = 0, \end{cases} \tag{11}$$

where the new matrix P, is computed as follows:

$$P_{ij} = \left(\chi_i, \nabla \cdot \boldsymbol{\phi}_j\right)_{L^2(\Omega)}. \tag{12}$$

System (11) can be solved respect to **a** and **b** in order to obtain the reduced order solution for velocity and pressure respectively.

3.2 *Treatment of Boundary Conditions*

The current problem involves non-homogeneous Dirichlet boundary conditions at the inlet Γ_{In}. The term $\mathbf{f}(\mathbf{x}, \boldsymbol{\mu})$ in (1) becomes $\mathbf{f}(\mathbf{x}, \boldsymbol{\mu}) = (\mu_x, \mu_y)$ where μ_x and μ_y are the components of the velocity at Γ_{In} along the x and y directions respectively. Non-homogeneous Dirichlet boundary conditions have been treated making use of the so called lifting control function. In this method the POD procedure is applied on a modified set of snapshots which have been homogenized in the following way:

$$\mathbf{u_i'} = \mathbf{u_i} - \mu_x \boldsymbol{\phi}_{\mathbf{c_x}} - \mu_y \boldsymbol{\phi}_{\mathbf{c_y}}, \quad \text{for} \quad i = 1, \ldots, N_s, \tag{13}$$

where $\boldsymbol{\phi}_{\mathbf{c_x}}$ and $\boldsymbol{\phi}_{\mathbf{c_y}}$ are the two lifting functions which have at the inlet Γ_{In} the following values $(1, 0)$ and $(0, 1)$ respectively. The approach used in this work to obtain the lifting functions involves solving two linear potential flow problems with the initial boundary conditions at Γ_{In} being $(1, 0)$ and $(0, 1)$ respectively for $\boldsymbol{\phi}_{\mathbf{c_x}}$ and $\boldsymbol{\phi}_{\mathbf{c_y}}$.

The POD is then applied to the snapshots matrix $\mathscr{U}' = [\mathbf{u_1'}, \mathbf{u_2'}, \ldots, \mathbf{u_{N_s}'}]$ that contains only snapshots with homogeneous boundary conditions. It has to be noted that the way the lifting functions have been computed assures that they are divergence free and thus the new set of snapshots has the same property.

At the reduced order level, it is then possible to deal with any boundary velocity at Γ_{In} (obviously the results will be more accurate if the prescribed velocity values are sufficiently close to the ones used during the training stage). If the new sample $\mu^\star$ (which has the new boundary velocity) is introduced in the online stage, one can compute the reduced velocity field as follows:

$$\mathbf{u}(\mathbf{x}, \boldsymbol{\mu}^\star) \approx \mu_x^\star \boldsymbol{\phi}_{\mathbf{c_x}} + \mu_y^\star \boldsymbol{\phi}_{\mathbf{c_y}} + \sum_{i=1}^{N_u} a_i(\boldsymbol{\mu}^\star) \boldsymbol{\phi}_\mathbf{i}(\mathbf{x}). \tag{14}$$

4 Non-intrusive PCE

According to Polynomial Chaos (PC) theory which was formulated by Wiener [41], real-valued multivariate Random Variables (RVs), such as the one considered in this work (the lift coefficient C_l) can be decomposed into an infinite sum of separable deterministic coefficients and orthogonal polynomials [22]. These polynomials are stochastic terms which depend on some mutually orthogonal Gaussian random variables. Once applied to our output of interest—the lift coefficient C_l—such decomposition assumption reads

$$C_l^{\star}(\boldsymbol{\zeta}) = \sum_{i=0}^{\infty} C_{li}\psi_i(\boldsymbol{\zeta}). \tag{15}$$

Here the random variable $\psi = (\alpha, U)$ is used to express the uncertainty in the angle of attack and inflow velocity. $\psi_i(\boldsymbol{\zeta})$ is the ith polynomial and C_{li} is the so-called ith stochastic mode. In practical application this series is truncated and only its first $P+1$ values are computed, namely

$$C_l^{\star}(\boldsymbol{\zeta}) = \sum_{i=0}^{P} C_{li}\psi_i(\boldsymbol{\zeta}). \tag{16}$$

In this work the orthogonal polynomial are called Hermite polynomials. These polynomials form an orthogonal set of basis functions in terms of Gaussian distribution [13]. In (16) $P+1$ is the number of Hermite polynomials used in the expansion and has to depend on the order of the polynomials chosen and the on dimension n of the random variable vector $\boldsymbol{\zeta} = \{\zeta_1, \ldots, \zeta_n\}$. More specifically, in an n-dimensional space, the number P of Hermite polynomials of degree p is given by $P+1 = \frac{(p+n)!}{p!n!}$ [13].

4.1 Coefficients Computation

The estimation of the coefficients $C_{li}(x)$ in (15) can be carried out in different ways. Among others, we mention the sampling based method and the quadrature method. The one here used is based on the sampling approach, following the methodology proposed by Hosder et al. [20]. The coefficient calculation algorithm starts from a discretized version of Eq. (16), namely

$$\begin{bmatrix} C_{l0}^* \\ C_{l1}^* \\ \vdots \\ C_{lN}^* \end{bmatrix} = \begin{bmatrix} \psi_1(\boldsymbol{\zeta}_0) & \psi_2(\boldsymbol{\zeta}_0) & \ldots & \psi_P(\boldsymbol{\zeta}_0) \\ \psi_1(\boldsymbol{\zeta}_1) & \psi_2(\boldsymbol{\zeta}_1) & \ldots & \psi_P(\boldsymbol{\zeta}_1) \\ \vdots & & \ddots & \\ \psi_1(\boldsymbol{\zeta}_N) & \psi_2(\boldsymbol{\zeta}_N) & \ldots & \psi_P(\boldsymbol{\zeta}_N) \end{bmatrix} \begin{bmatrix} C_{l0} \\ C_{l1} \\ \vdots \\ C_{lP} \end{bmatrix},$$

where N is the number of the samples taken. If N coincides with the number of Hermite polynomials $P+1$ needed for the PCE expansion, the system above presents a square matrix and can be solved to determine the coefficients C_{li} from the known output coefficients $C_{l_i}^{\star}$. In the most common practice, a redundant number of samples are considered and the system is solved in a least squares sense, namely

$$\mathbf{C_l} = (\boldsymbol{L}^T \boldsymbol{L})^{-1} \boldsymbol{L}^T \mathbf{C_l^*}, \tag{17}$$

where $\boldsymbol{L}$, $\mathbf{C_l}$ and $\mathbf{C_l^*}$ denote the rectangular matrix in (4.1), the PCE coefficients vector and output vector, respectively.

5 Numerical Results

This section presents the results for the simulations carried out with the POD-Galerkin ROM and PCE for UQ described in the previous sections. The first part of the analysis will be focused on the results obtained with the POD-Galerkin ROM. In the second part we will assess the performance of the UQ technique on the airfoil problem, both when FOM and ROM simulation results are used to feed the PCE algorithm. The overall objective of the present section is in fact twofold. The first aim is to understand the influence of the samples distribution used to train the ROM in the results of the POD-Galerkin ROM. The second aim is to compare between the PCE UQ results obtained using full order model to those obtained with POD Galerkin-ROM.

5.1 ROM Results

The FOM model used to generate the POD snapshots has been set up as reported in Sect. 2. Making use of the computational grid shown in Fig. 3, a set of simulations was carried out, selecting a Gauss linear numerical scheme for the approximation

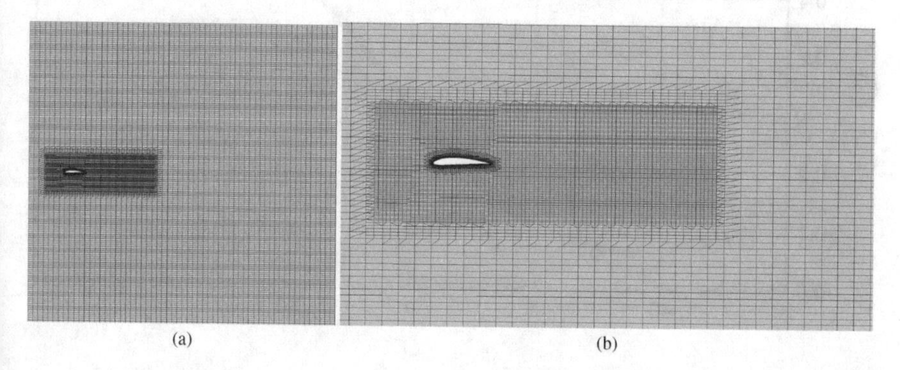

(a) (b)

Fig. 3 (**a**) The OpenFOAM mesh used in the simulations. (**b**) A picture of the mesh zoomed near the airfoil

of gradients and Laplacian terms, and a bounded Gauss upwind scheme for the convective term approximation.

As mentioned, the parameters considered in the ROM investigation are the airfoil angle of attack and the magnitude of the inflow velocity at the inlet. The training of the POD-Galerkin ROM requires a suitable amount of snapshots (FOM solutions) to be available. Thus, 520 samples have been produced and a single FOM simulation is launched for each sampling point. As for the distribution in the parameters space, the samples are obtained making use of the Latin Hyper Cube (LHC) [37] sampling algorithm. Figure 4 depicts the lift coefficient against the angle of attack curve obtained from a first FOM simulation campaign in which the 520 samples were generated imposing mean values of 100 m/s and 0° and variances of 20 m/s and 300, for velocity and angle of attack, respectively. As can be appreciated in the picture, the lift coefficient values do not significantly depend on the inflow velocity. In fact sampling points with equal α and different U_∞ values, result in practically identical output. For this reason, the input-output relationship appears like a curve in the C_l-α plane. We also point out that this is a consequence of considering a nondimensionalized force a C_l as our output, rather then the corresponding dimensional lift values.

The POD modes are generated after applying POD onto the snapshots matrices of the flow fields obtained in the simulation campaign. After such offline phase, the computation of the reduced order fields is performed in the online stage, as presented in Sect. 3. In this first reproduction test, we performed a single reduced simulation in correspondence with the velocity and the angle of attack used to generate each offline snapshot. This means that we used the same sample values both in the online stage and in the offline one. The ROM results of the reproduction test for the lift

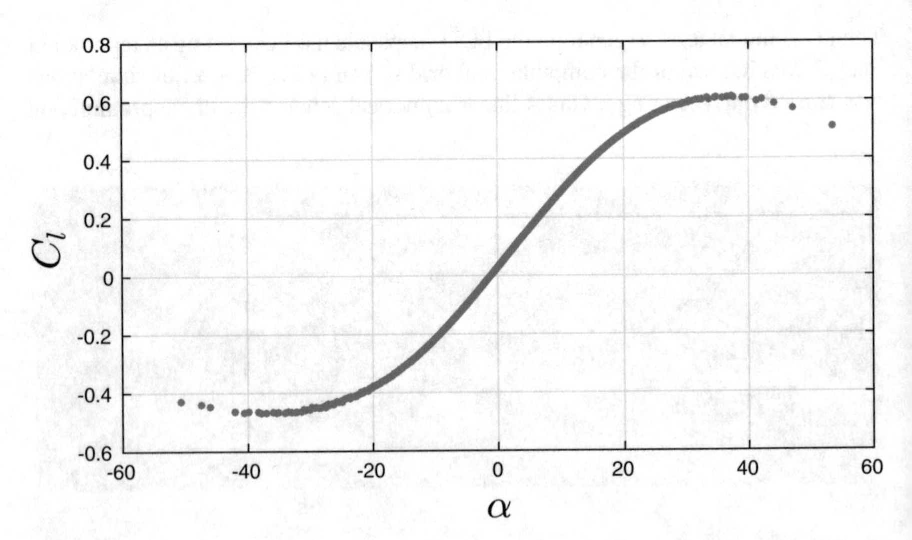

Fig. 4 The FOM lift coefficient for the first case, as it can be seen in the plot the vast majority of the samples is clustered around $\alpha = 0$

coefficient are reported in Fig. 5. The figure refers to the ROM results obtained considering 10 modes for the discretization of velocity, pressure and supremizer fields. The plot shows that the reconstruction of the lift coefficient is only accurate in the central region. In the lateral regions the lift coefficient computed with the ROM solution does not match the corresponding FOM solution. The poor quality of the ROM prediction on the lift coefficient, as well as of the forces acting on the airfoil, is a direct consequence of the fact that the fields were not reconstructed in an accurate way. For the particular physical phenomenon this inaccuracy may be even more undesirable, since the stall occurs in these regions. One might originally guess that the problem can be mitigated by increasing the amount of POD modes. Yet, as Fig. 6 clearly shows, even increasing the modes for velocity to 30 is not solving the problem. In this particular case adding more modes will not solve the problem since the energy added by considering more modes is negligible. An explanation of the poor performances of the ROM model in the stall region may be instead associated to the distribution of the offline samples used to generate the POD snapshots. In fact, the samples generated with LHC are distributed around a mean value of the angle of attack of 0° and their density is rather coarse in the stall regions.

To confirm such deduction, we tried to generate the snapshots by means of a different set of samples generated so as to be more dense in the stall regions. More specifically, we have generated thirteen different groups of samples in which the velocity mean and variance were kept fixed at values of 100 and 20 m/s respectively, while the mean and the variance for the angle of attack were varied in each set as summarized in Table 1. As illustrated in Figs. 7 and 8 for angle of attack and velocity respectively, the overall training data for the ROM offline phase has been generated

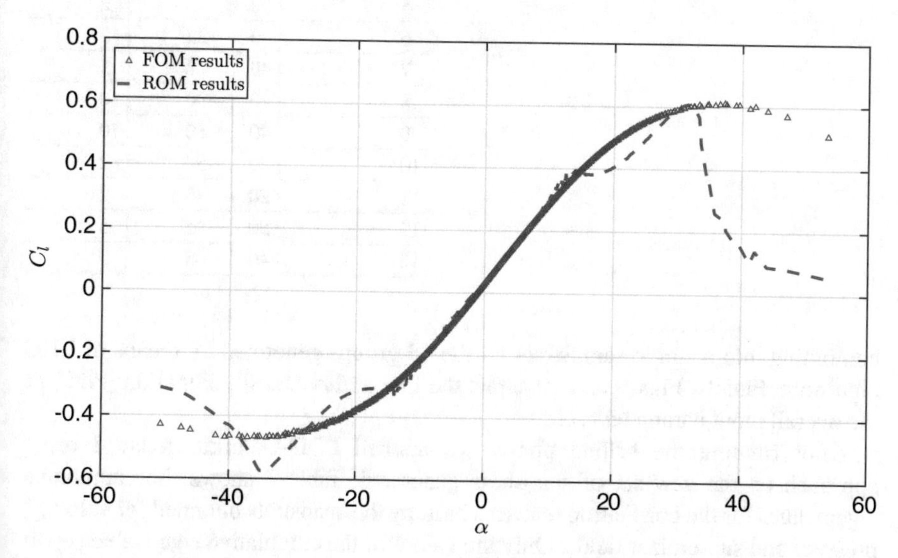

Fig. 5 The first sampling case: the full order lift coefficients curve versus the ROM reconstructed one with 10 modes used for each of velocity, pressure and supremizer fields

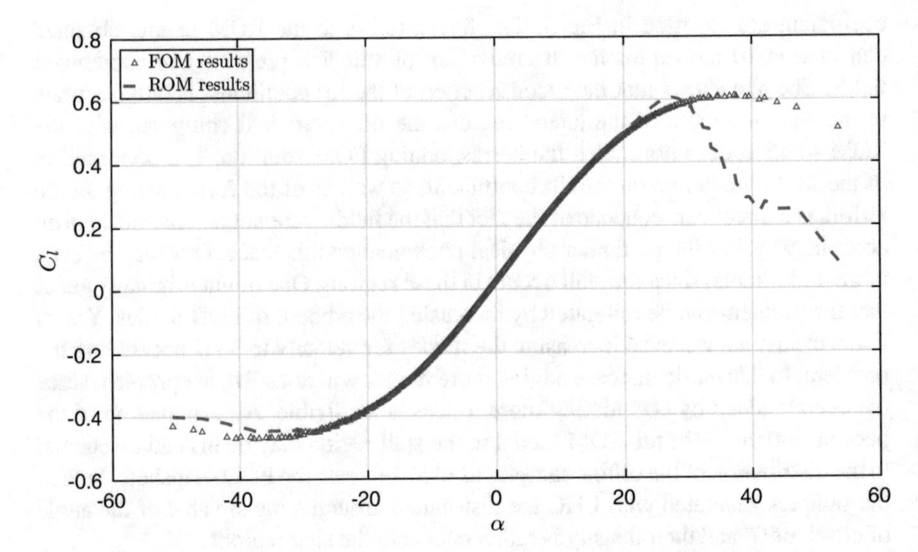

Fig. 6 The first sampling case: the full order lift coefficients curve versus the ROM reconstructed one with 30, 10 and 10 modes are used for velocity, pressure and supremizer fields respectively

Table 1 The mean and variance for the group of samples that form the training set in the second case

Group number	N	$E(\alpha)$ in °	$\sigma(\alpha)$ in °
1	90	0	20
2	20	−10	2
3	20	10	2
4	50	−15	2
5	50	15	2
6	40	−22	5
7	40	22	5
8	40	−30	10
9	40	30	10
10	20	−38	2
11	20	38	2
12	50	−45	5
13	40	45	5

combining into a single sample set all the 13 groups generated by means of LHC algorithm. Finally, Figs. 9 and 10 depict the Probability Density Function (PDF) of the overall input parameters set.

After running the offline phase, we applied POD-Galerkin reduced order approach on the new set of snapshots generated. Table 2 shows the cumulative eigenvalues for the correlation matrices built by the snapshots obtained for velocity, pressure and supremizer fields. Only the values of the cumulative eigenvalues up to

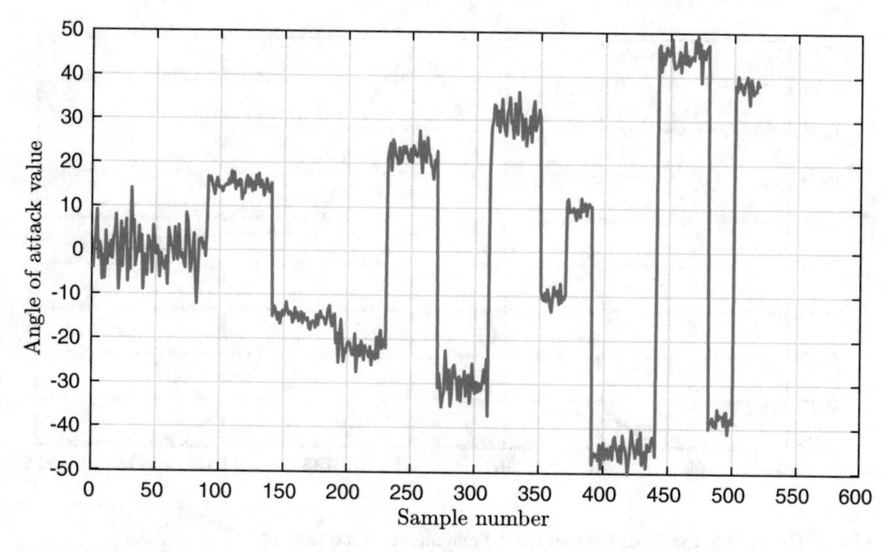

Fig. 7 The angle of attack samples for the second case

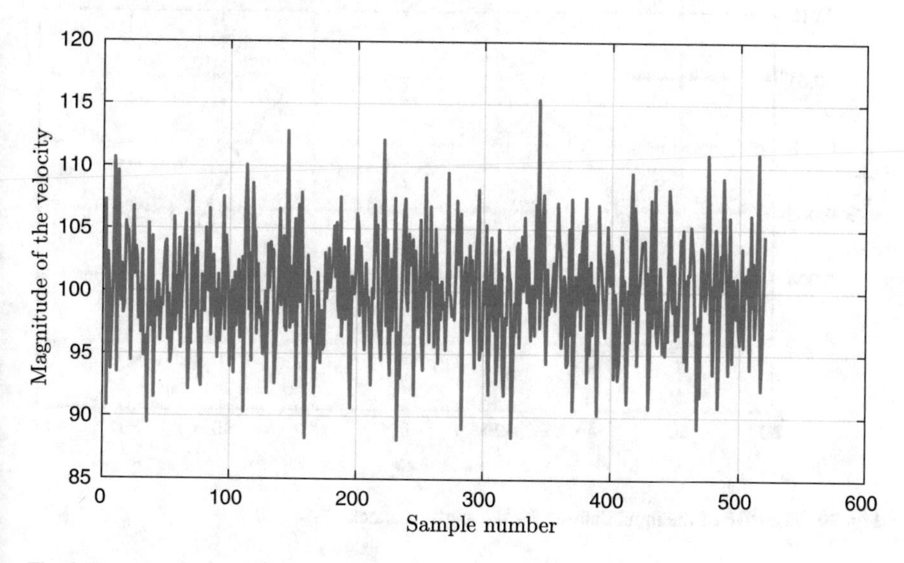

Fig. 8 The magnitude of velocity samples for the second case

the fifteenth mode are listed. Yet, such data indicate that by using 15 modes in the online phase, we can recover 99.9% of the energy embedded in the system.

Figure 11 displays the FOM lift coefficients of the airfoil corresponding to the set of angles of attack previously introduced in Table 1. Figure 12 shows the results obtained with the reduced order model trained with the samples summarized in Table 1. Here, the online phase was carried out using the same samples used in the offline stage. To provide a quantitative evaluation of the results, we used L^2

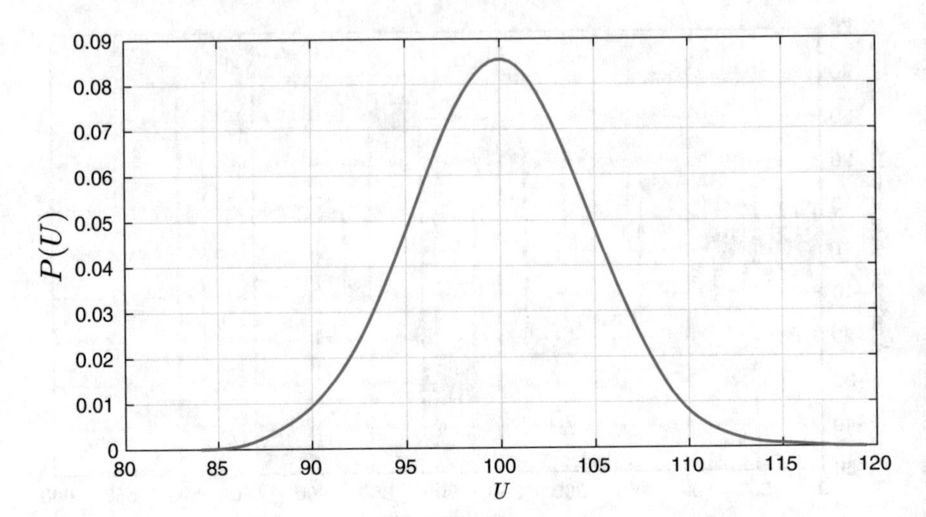

Fig. 9 The PDF of the input data set for the magnitude of the velocity

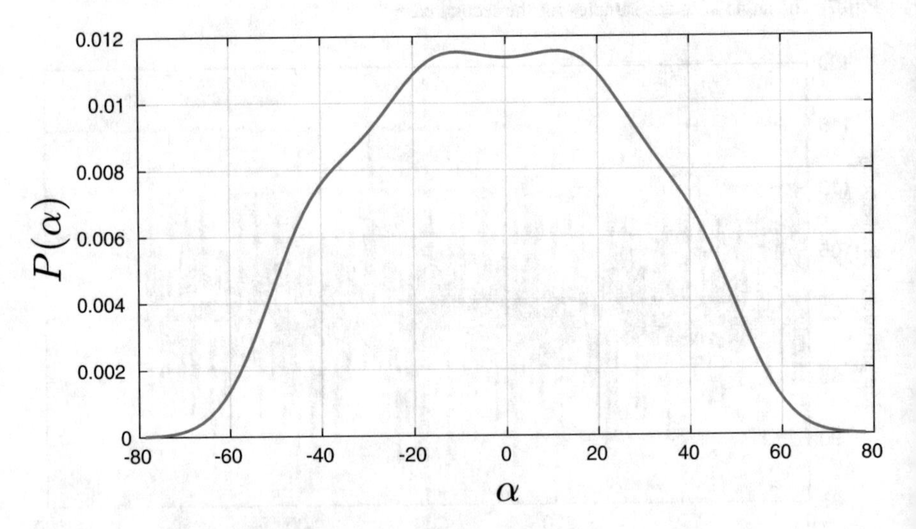

Fig. 10 The PDF of the input data set for the angle of attack

relative error, computed as follows:

$$\varepsilon = 100\frac{\sqrt{\sum_{t=1}^{n}(C_{l_t}^{FOM} - C_{l_t}^{ROM})^2}}{\sqrt{\sum_{t=1}^{n}(C_{l_t}^{FOM})^2}}\%, \tag{18}$$

where n is the number of sampling points, $C_{l_t}^{FOM}$ and $C_{l_t}^{ROM}$ are the t-th sample point lift coefficients for FOM and ROM, respectively. In this case we have a relative

Table 2 Cumulative Eigenvalues of the correlation matrices for velocity, pressure and supremizer fields

N modes	$\mathbf{u}$	p	$\mathbf{u}_{sup}$
1	0.679635	0.738828	0.557189
2	0.930038	0.960781	0.987862
3	0.955239	0.990746	0.995984
4	0.971768	0.998833	0.999228
5	0.981370	0.999730	0.999796
6	0.987603	0.999880	0.999927
7	0.992311	0.999945	0.999975
8	0.994793	0.999963	0.999983
9	0.996651	0.999976	0.999990
10	0.997914	0.999982	0.999993
11	0.998679	0.999987	0.999995
12	0.999165	0.999991	0.999997
13	0.999492	0.999993	0.999998
14	0.999700	0.999995	0.999999
15	0.999806	0.999996	0.999999

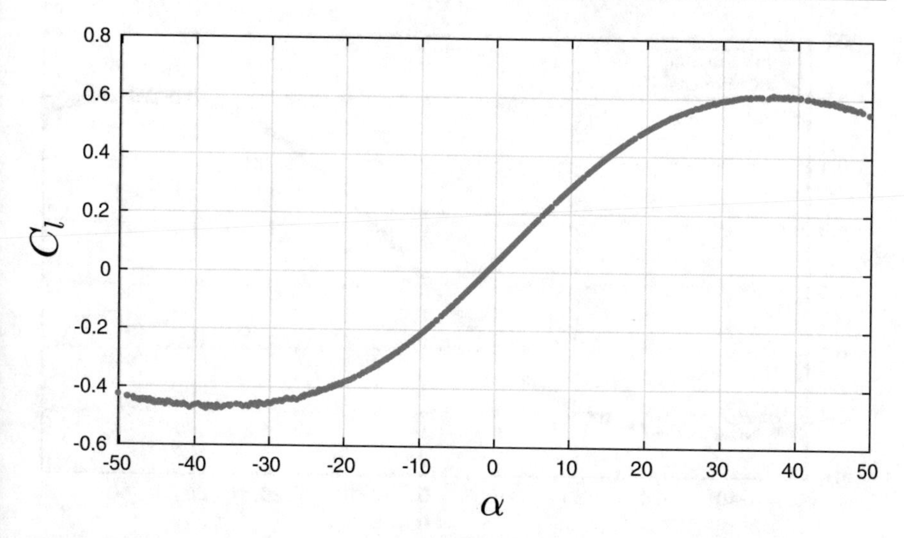

Fig. 11 The FOM lift coefficient as a function of the angle of attack α for the second case. Compared to Fig. 4, the samples span a wider range with sufficient accuracy sought by the ROM reconstruction

error of 6.69% in L^2 norm between the FOM and the ROM lift coefficients, when 10 modes have been used in the online phase for each of velocity, pressure and supremizer fields. Using 10 additional modes for velocity, results instead in a 3.75% error. The corresponding plots in Fig. 12 clearly suggest that the qualitative behavior of the ROM lift output was substantially improved with respect to the first case. This improvement in the prediction of the ROM lift coefficients is due to a more accurate reproduction of the ROM fields. This is highlighted by Fig. 13, which

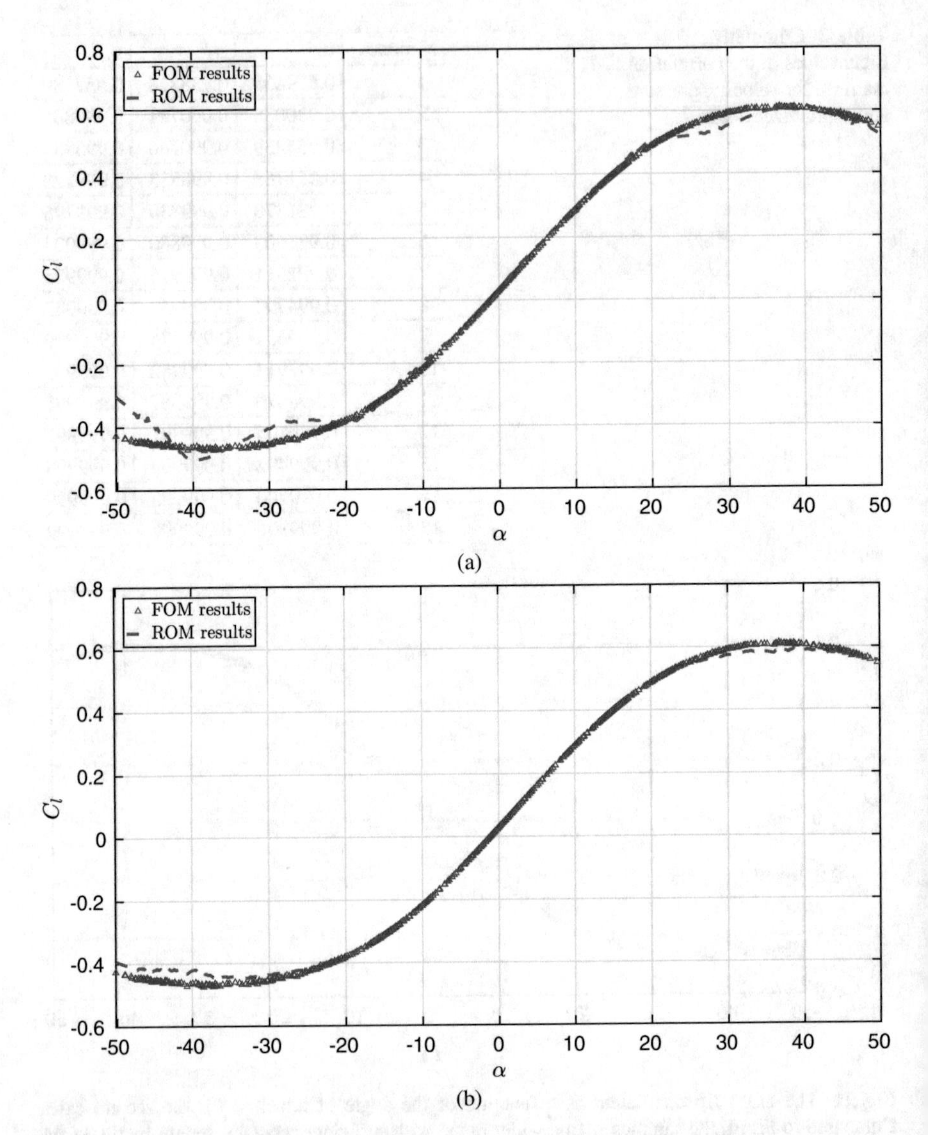

Fig. 12 A comparison between FOM and ROM reconstructed lift coefficients for the second case (**a**) 10 modes are used for each of velocity, pressure and supremizer fields. (**b**) 20, 10 and 10 modes are used for velocity, pressure and supremizer fields, respectively. In both figures we have online parameters set that coincide with the offline ones

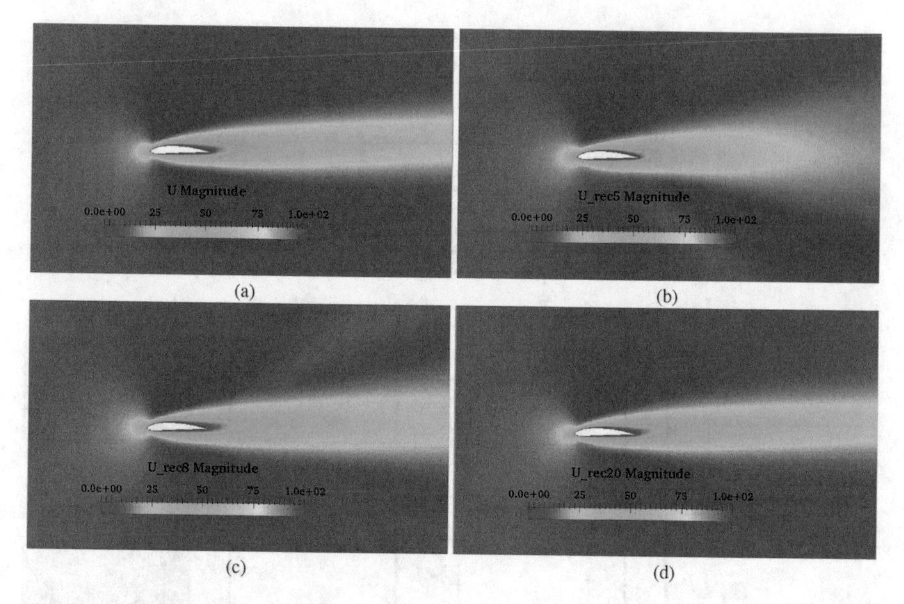

Fig. 13 The full order velocity field for the parameter $\bar{\mu} = (90.669, 5.4439\,\text{m/s})$ and a comparison with the reconstructed field by means of different number of modes for velocity, pressure and supremizer fields. (**a**) FOM field (**b**) ROM velocity field with 3, 10 and 10 modes used. (**c**) ROM velocity field with 8, 10 and 10 modes used. (**d**) ROM velocity field with 20, 10 and 10 modes used

shows the FOM velocity field along with different reconstructed surrogate fields obtained employing different number of modes at the projection stage.

5.2 PCE Results

The aim of the present section is to evaluate the performance of the PCE algorithm implemented for the fluid dynamic problem at hand. To better describe the amount of simulations carried out to both train and validate the UQ PCE model implemented, we present in Fig. 14 a conceptual scheme of the simulation campaign carried out in this work.

As mentioned, one of the main features of non intrusive PCE is that it can use any deterministic simulation software as a black box input source. We will then present different tests in which PCE has been fed with the output of fluid dynamic simulations based on models characterized by different fidelity levels. In a first test we in fact generated a PCE based on the FOM, and evaluate its performance in a prediction test. The second test consisted in generating a PCE based on the ROM described in the previous section. The latter test allows for an evaluation of how the PCE results are affected when the expansion is based on a surrogate ROM model rather than the FOM one. Given the relatively high number of samples required for the PCE setup, in fact it is interesting to understand if ROM can be used to reduce

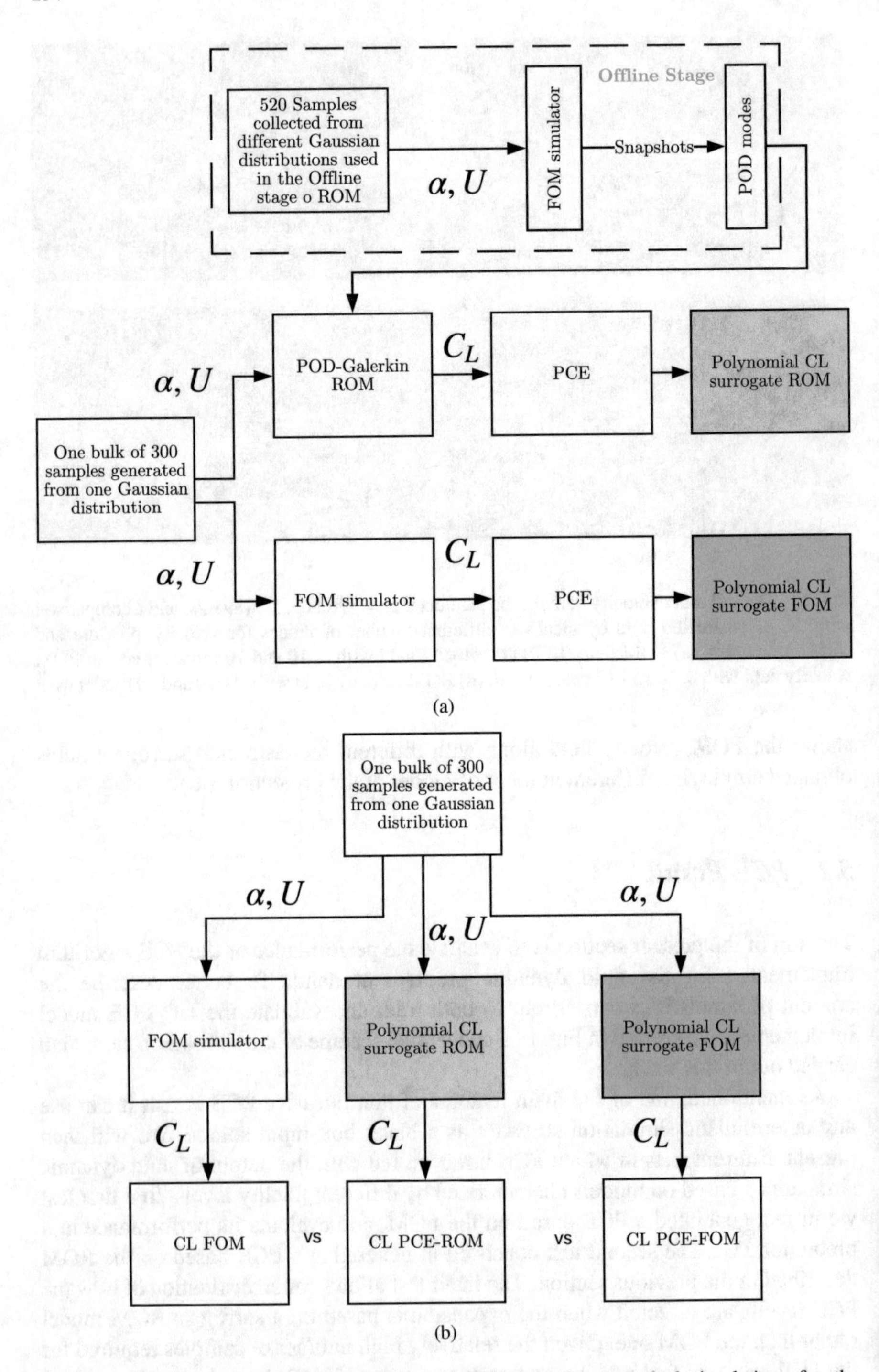

Fig. 14 The flowcharts describing procedure followed in the numerical simulations for the UQ model generation and validation campaign respectively. The top scheme in (**a**) focuses on the

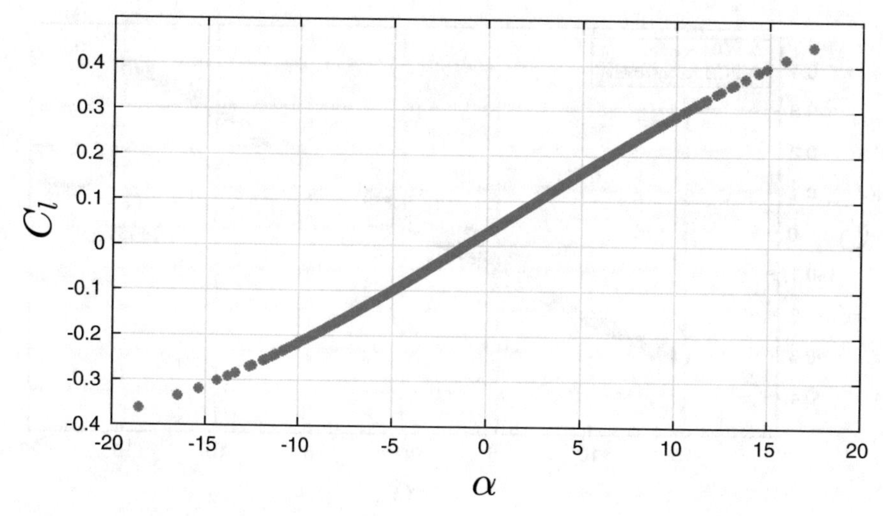

Fig. 15 The FOM lift coefficient for the UQ case

the computational cost for their generation of the system output at each sample, without a significant loss in terms of accuracy.

One of the main assumptions of the non intrusive PCE algorithm implemented is that of operating on Gaussian distributed input parameters. For such reason, the tests in the present section were generated with a set of 300 sampling points consisting of a single Gaussian distributed bulk. Making use of LHC algorithm, the samples have been randomly generated around angle of attack and velocity magnitude means of 0° and 100 m/s respectively. As for the variances, we prescribed 40° and 20 m/s, for angle of attack and velocity magnitude respectively. The FOM lift coefficient curve obtained with the input sampling points described can be seen in Fig. 15.

The samples, which have been just mentioned above, will be used as cross validation test for the ROM model developed in the previous subsection. As mentioned earlier, the first test will be to feed PCE with FOM output data and then to conduct a prediction test. PCE is used to predict the value of the lift coefficient

Fig. 14 (continued) procedure adopted for the generation of the UQ model, and in particular on the identification of the PCE coefficients. The polynomial surrogate based on the full order model (indicated in green) has been generated using 300 Gaussian distributed samples in the α, U space. The same samples have been used to obtain the polynomial surrogate input-output relationship for the POD-Galerkin ROM (denoted by the yellow box). Note that the ROM used in this simulation campaign has been trained by means of 520 samples in the α, U space, organized in 13 Gaussian distributed bulks, as reported in Table 1. Finally, the bottom flowchart in (**b**) illustrates the PCE validation campaign. Here, 300 sample points in the input space have been used to obtain the corresponding output with the full order model, with the polynomial UQ surrogate trained with the FOM simulations (green box), and with the polynomial UQ surrogate trained with the ROM simulations (yellow box)

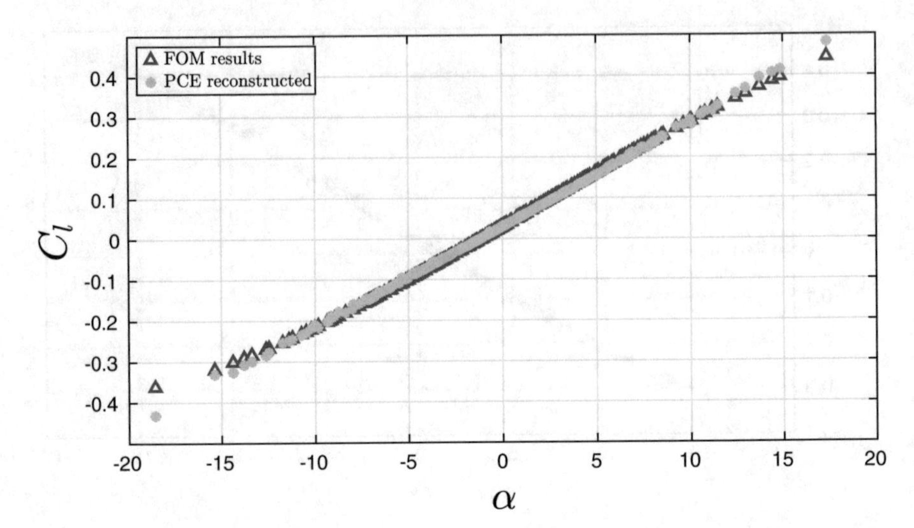

Fig. 16 PCE reconstructed lift coefficient versus FOM one, here polynomials of second degree have been used

for 200 samples which differ of the 100 samples used for the PCE coefficients evaluation. Thus the first 100 samples with their corresponding FOM lift coefficient values were used to build the matrix system (4.1), which has been solved in the least squares sense. Figure 16 displays the C_l values computed with both FOM and PCE in correspondence with all the samples used for check. The overall error in L^2 norm between FOM and PCE predications is 5.04%.

In the second test we have used ROM data as input for PCE. After using 100 samples to compute the PCE coefficients, we used the PCE to predict the lift coefficients at 200 additional samples used for check. We then compared the value of the predicted PCE coefficients in this case to both ROM values and FOM values.

The result of the aforementioned test are reported in Fig. 17. The figure includes comparison of the PCE predicted C_l curve with both its ROM and FOM counterparts. The plots show a similar behaviour of the PCE predictions obtained using ROM and FOM output data. By a quantitative standpoint, the PCE predictions present a 4.4% error with respect to the ROM predictions, while the L^2 norm of the error with respect to the FOM predictions is 5.14%. A summary of the comparisons made is reported in Table 3.

6 Conclusions and Future Developments

In this work, we studied two popular techniques which are used often in the fields of ROM and UQ which are the POD and PCE, respectively. The study aimed at comparing the accuracy of the two techniques in reconstructing the outputs of

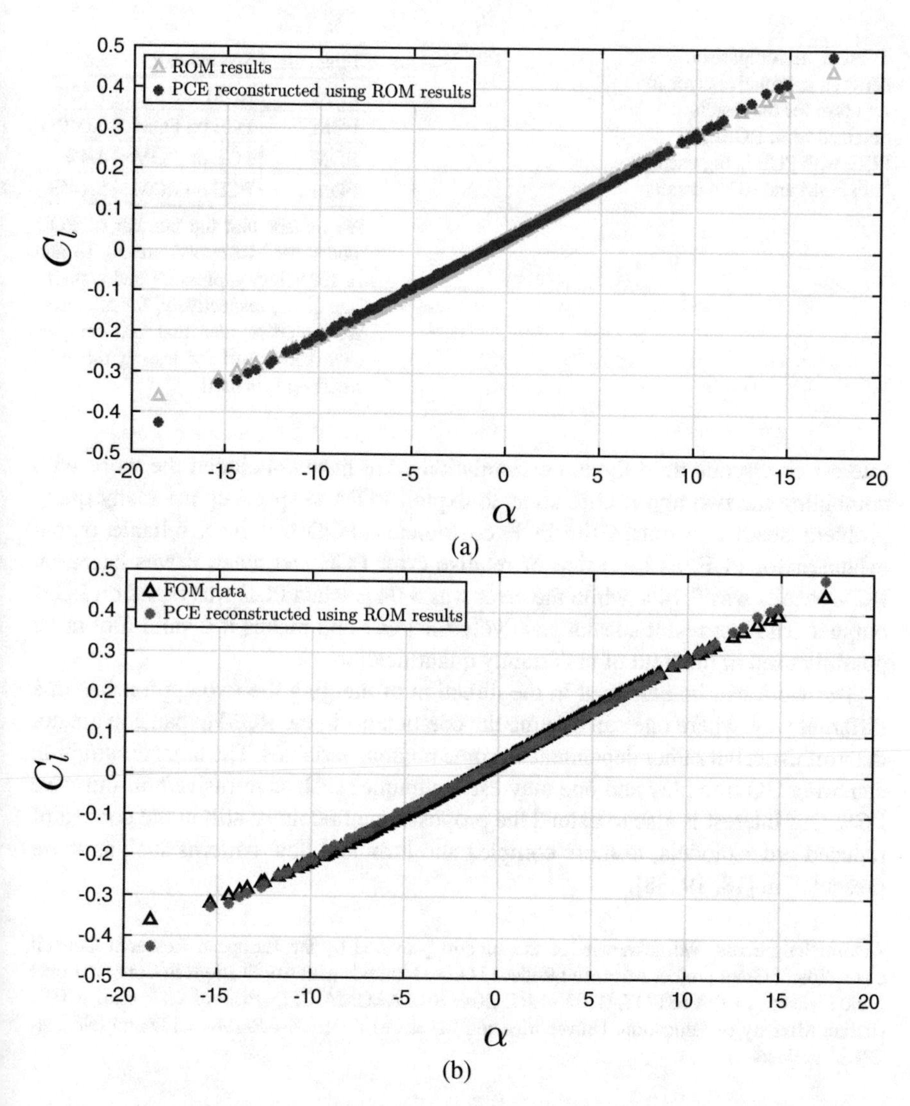

Fig. 17 (**a**) The ROM lift coefficient versus PCE lift coefficient curve when PCE has been applied on ROM output with 30, 10 and 10 modes used for velocity, pressure and supremizer fields respectively. (**b**) The FOM lift coefficient versus PCE lift coefficient curve when PCE has been applied on ROM output with same number of modes as in (**a**)

Table 3 A comparison between the relative error in L^2 norm for the results obtained from ROM and PCE, with PCE being used on both FOM and ROM results

First data	Second data	Error
FOM	ROM	1.53%
FOM	PCE on FOM	5.04%
ROM	PCE on ROM	4.4%
FOM	PCE on ROM	5.14%

We remark that the number of POD modes used (if apply) are 30, 10 and 10 for velocity, pressure and supremizer fields, respectively, for all cases. We underline also that 200 samples have been used for testing the PCE wherever it is used

interest of viscous fluid dynamic simulations. We have concluded the work with combining the two approaches so as to exploit ROM to speed up the many query problem needed to obtain the PCE coefficients. POD can be a reliable output evaluator for PCE, as the value of relative error PCE had when it was based on ROM results was 5.14% while the error was 5.04% when PCE was based on FOM outputs. The last result speaks positively for POD and makes it a valid tool to be possibly used in the field of uncertainty quantification.

The work can be extended in the direction of merging the two approaches in a different way, where one can assume the coefficients in the ROM expansion are not deterministic, but rather dependent on some random variables. The latter assumption can bring UQ into play and one may use techniques such as intrusive/non-intrusive PCE. Our interest is also to extend the proposed methodology, still in the context of reduced order models, to more complex and turbulent flow patterns such as those presented in [18, 19, 38].

Acknowledgments We acknowledge the support provided by the European Research Council Consolidator Grant project Advanced Reduced Order Methods with Applications in Computational Fluid Dynamics—GA 681447, H2020-ERC COG 2015 AROMA-CFD (PI: Prof. G. Rozza), MIUR (Italian Ministry of Education, Universities and Research) FARE-X-AROMA-CFD and INdAM-GNCS projects.

References

1. Abbott, I.: Theory of Wing Sections: Including a Summary of Airfoil Data. Dover Publications, Mineola (1999)
2. Akhtar, I., Nayfeh, A.H., Ribbens, C.J.: On the stability and extension of reduced-order Galerkin models in incompressible flows. Theor. Comput. Fluid Dyn. **23**(3), 213–237 (2009). https://doi.org/10.1007/s00162-009-0112-y
3. Baiges, J., Codina, R., Idelsohn, S.: Reduced-order modelling strategies for the finite element approximation of the incompressible Navier-Stokes equations. Comput. Methods Appl. Sci. **33**, 189–216 (2014). https://doi.org/10.1007/978-3-319-06136-8_9

4. Ballarin, F., Manzoni, A., Quarteroni, A., Rozza, G.: Supremizer stabilization of POD-Galerkin approximation of parametrized steady incompressible Navier-Stokes equations. Int. J. Numer. Methods Eng. **102**(5), 1136–1161 (2014). https://doi.org/10.1002/nme.4772
5. Barth, T., Ohlberger, M.: Finite Volume Methods: Foundation and Analysis. Wiley, Hoboken (2004). https://doi.org/10.1002/0470091355.ecm010
6. Bergmann, M., Bruneau, C.H., Iollo, A.: Enablers for robust POD models. J. Comput. Phys. **228**(2), 516–538 (2009). https://doi.org/10.1016/j.jcp.2008.09.024
7. Burkardt, J., Gunzburger, M., Lee, H.C.: POD and CVT-based reduced-order modeling of Navier-Stokes flows. Comput. Methods Appl. Mech. Eng. **196**(1–3), 337–355 (2006). https://doi.org/10.1016/j.cma.2006.04.004
8. Carlberg, K., Choi, Y., Sargsyan, S.: Conservative model reduction for finite-volume models. J. Comput. Phys. **371**, 280–314 (2018). https://doi.org/10.1016/j.jcp.2018.05.019
9. Chen, P., Quarteroni, A., Rozza, G.: Comparison between reduced basis and stochastic collocation methods for elliptic problems. J. Sci. Comput. **59**(1), 187–216 (2014)
10. Chen, P., Quarteroni, A., Rozza, G.: Reduced basis methods for uncertainty quantification. SIAM/ASA J. Uncertain. Quantif. **5**(1), 813–869 (2017)
11. Chinesta, F., Huerta, A., Rozza, G., Willcox, K.: Model reduction methods. Encyclopedia of Computational Mechanics Second Edition, pp. 1–36 (2017)
12. Drohmann, M., Haasdonk, B., Ohlberger, M.: Reduced basis approximation for nonlinear parametrized evolution equations based on empirical operator interpolation. SIAM J. Sci. Comput. **34**(2), A937–A969 (2012). https://doi.org/10.1137/10081157X
13. Ghanem, R.G., Spanos, P.D.: Stochastic Finite Elements: A Spectral Approach. Courier Corporation, North Chelmsford (2003)
14. Gunzburger, M.D., Peterson, J.S., Shadid, J.N.: Reduced-order modeling of time-dependent PDEs with multiple parameters in the boundary data. Comput. Methods Appl. Mech. Eng. **196**(4), 1030–1047 (2007)
15. Haasdonk, B., Ohlberger, M.: Reduced basis method for finite volume approximations of parametrized linear evolution equations. Math. Model. Numer. Anal. **42**(2), 277–302 (2008). https://doi.org/10.1051/m2an:2008001
16. Haasdonk, B., Ohlberger, M., Rozza, G.: A reduced basis method for evolution schemes with parameter-dependent explicit operators. ETNA Electron. Trans. Numer. Anal. **32**, 145–161 (2008)
17. Hesthaven, J.S., Rozza, G., Stamm, B.: Certified Reduced Basis Methods for Parametrized Partial Differential Equations. Springer, Berlin (2016). https://doi.org/10.1007/978-3-319-22470-1
18. Hijazi, S., Ali, S., Stabile, G., Ballarin, F., Rozza, G.: The effort of increasing Reynolds number in projection-based reduced order methods: from laminar to turbulent flows. In: Lecture Notes in Computational Science and Engineering, Springer International Publishing, pp. 245–264 (2020). https://doi.org/10.1007/978-3-030-30705-9_22
19. Hijazi, S., Stabile, G., Mola, A., Rozza, G.: Data-driven POD–Galerkin reduced order model for turbulent flows. J. Comput. Phys. **416**, 109513 (2020)
20. Hosder, S., Walters, R., Perez, R.: A non-intrusive polynomial chaos method for uncertainty propagation in CFD simulations. In: 44th AIAA Aerospace Sciences Meeting and Exhibit, p. 891 (2006)
21. Isukapalli, S.S.: Uncertainty analysis of transport-transformation models. Unpublished Ph.D. Disseration, New Brunswick, NJ: Rutgers, The State University of New Jersey, Department of Chemical and Biochemical Engineering (1999)
22. Janya-Anurak, C.: Framework for Analysis and Identification of Nonlinear Distributed Parameter Systems Using Bayesian Uncertainty Quantification Based on Generalized Polynomial Chaos, vol. 31. KIT Scientific Publishing, Karlsruhe (2017)
23. Jasak, H.: Error analysis and estimation for the finite volume method with applications to fluid flows. Ph.D. Thesis, Imperial College, University of London (1996). http://powerlab.fsb.hr/ped/kturbo/OpenFOAM/docs/HrvojeJasakPhD.pdf

24. Kunisch, K., Volkwein, S.: Galerkin proper orthogonal decomposition methods for a general equation in fluid dynamics. SIAM J. Numer. Anal. **40**(2), 492–515 (2002). https://doi.org/10.1137/S0036142900382612
25. Loeve, M.: Probability Theory vol. II (Graduate Texts in Mathematics). Springer, Berlin (1994)
26. Lorenzi, S., Cammi, A., Luzzi, L., Rozza, G.: POD-Galerkin method for finite volume approximation of Navier-Stokes and RANS equations. Comput. Methods Appl. Mech. Eng. **311**, 151–179 (2016). http://dx.doi.org/10.1016/j.cma.2016.08.006
27. Moukalled, F., Mangani, L., Darwish, M.: The Finite Volume Method in Computational Fluid Dynamics: An Advanced Introduction with OpenFOAM and Matlab, 1st edn. Springer, Berlin (2015)
28. Noack, B.R., Eckelmann, H.: A low-dimensional Galerkin method for the three-dimensional flow around a circular cylinder. Phys. Fluids **6**(1), 124–143 (1994). https://doi.org/10.1063/1.868433
29. Patankar, S., Spalding, D.: A calculation procedure for heat, mass and momentum transfer in three-dimensional parabolic flows. Int. J. Heat Mass Transf. **15**(10), 1787–1806 (1972). http://dx.doi.org/10.1016/0017-9310(72)90054-3
30. Quarteroni, A., Manzoni, A., Negri, F.: Reduced Basis Methods for Partial Differential Equations. Springer, Berlin (2016). https://doi.org/10.1007/978-3-319-15431-2
31. Quarteroni, A., Rozza, G.: Numerical solution of parametrized Navier–Stokes equations by reduced basis methods. Numer. Methods Partial Differ. Equ. **23**(4), 923–948 (2007). https://doi.org/10.1002/num.20249
32. Reagana, M.T., Najm, H.N., Ghanem, R.G., Knio, O.M.: Uncertainty quantification in reacting-flow simulations through non-intrusive spectral projection. Combust. Flame **132**(3), 545–555 (2003)
33. Rozza, G., Veroy, K.: On the stability of the reduced basis method for Stokes equations in parametrized domains. Comput. Methods Appl. Mech. Eng. **196**(7), 1244–1260 (2007). https://doi.org/10.1016/j.cma.2006.09.005
34. Sarkar, S., Witteveen, J., Loeven, A., Bijl, H.: Effect of uncertainty on the bifurcation behavior of pitching airfoil stall flutter. J. Fluids Struct. **25**(2), 304–320 (2009). https://doi.org/10.1016/j.jfluidstructs.2008.06.006
35. Stabile, G., Hijazi, S., Mola, A., Lorenzi, S., Rozza, G.: POD-Galerkin reduced order methods for CFD using finite volume discretisation: vortex shedding around a circular cylinder. Commun. Appl. Ind. Math. **8**(1), (2017). https://doi.org/10.1515/caim-2017-0011
36. Stabile, G., Rozza, G.: Finite volume POD-Galerkin stabilised reduced order methods for the parametrised incompressible Navier–Stokes equations. Comput. Fluids **173**, 273–284 (2018). https://doi.org/10.1016/j.compfluid.2018.01.035
37. Stein, M.: Large sample properties of simulations using latin hypercube sampling. Technometrics **29**(2), 143–151 (1987). https://doi.org/10.1080/00401706.1987.10488205
38. Tezzele, M., Demo, N., Mola, A., Rozza, G.: An integrated data-driven computational pipeline with model order reduction for industrial and applied mathematics. Submitted, Special Issue ECMI (2018)
39. Versteeg, H.K., Malalasekera, W.: An Introduction to Computational Fluid Dynamics. The Finite Volume Method. Longman, London (1995)
40. Weller, H.G., Tabor, G., Jasak, H., Fureby, C.: A tensorial approach to computational continuum mechanics using object-oriented techniques. Comput. Phys. **12**(6), 620–631 (1998)
41. Wiener, N.: The homogeneous chaos. Am. J. Math. **60**(4), 897–936 (1938)

A Practical Example for the Non-linear Bayesian Filtering of Model Parameters

Matthieu Bulté, Jonas Latz, and Elisabeth Ullmann

Abstract In this tutorial we consider the non-linear Bayesian filtering of static parameters in a time-dependent model. We outline the theoretical background and discuss appropriate solvers. We focus on particle-based filters and present Sequential Importance Sampling (SIS) and Sequential Monte Carlo (SMC). Throughout the paper we illustrate the concepts and techniques with a practical example using real-world data. The task is to estimate the gravitational acceleration of the Earth g by using observations collected from a simple pendulum. Importantly, the particle filters enable the adaptive updating of the estimate for g as new observations become available. For tutorial purposes we provide the data set and a Python implementation of the particle filters.

1 Introduction

An important building block of uncertainty quantification is the statistical estimation and sustained learning of parameters in mathematical and computational models. In science and engineering models are used to emulate, predict, and optimise the behaviour of a system of interest. Examples include the transport of contaminants by groundwater flow in hydrology, the price of a European option in finance, or the motion of planets by mutual gravitational forces in astrophysics. The associated mathematical models for these examples are an elliptic partial differential equation (PDE), the parabolic Black-Scholes PDE, and a system of ordinary differential equations (ODEs) describing the N-body dynamics.

Assuming that we have observational data of the system of interest, it is now necessary to calibrate the model with respect to these observations. This means that we identify model parameters such that the model output is close to the observations

M. Bulté · J. Latz · E. Ullmann (✉)
Zentrum Mathematik, Technische Universität München, München, Germany
e-mail: matthieu.bulte@tum.de; jonas.latz@ma.tum.de; elisabeth.ullmann@ma.tum.de

M. D'Elia et al. (eds.), *Quantification of Uncertainty: Improving Efficiency and Technology*, Lecture Notes in Computational Science and Engineering 137,
https://doi.org/10.1007/978-3-030-48721-8_11

in a suitable metric. In the examples above we need to calibrate the hydraulic conductivity of the groundwater reservoir, the volatility of the stock associated with the option, and the masses of the planets.

In this tutorial we focus on the next step following the model calibration, namely the updating of the estimated parameters as additional observations become available. This is an important task since many systems are only partially observed. Thus it is often unlikely to obtain high quality estimates of underlying model parameters by using only a single data set. Moreover, it is often very expensive or impossible to restart the parameter estimation with all data sets after a new data set becomes available. The problem of combining a parameter estimate with a new set of observations to update the estimate based on all observations is called *filtering* in statistics. Filtering can be considered as a *learning process*: a certain state of knowledge based on previous observations is combined with new observations to reach an improved state of knowledge.

Throughout this tutorial we consider a practical example for filtering. We study the periodic motion of a pendulum. The underlying mathematical model is an ODE. The model parameters are the length of the pendulum string ℓ, and the gravitational acceleration of the Earth g. We assume that ℓ is known, however, we are uncertain about g. Our goal is to estimate and update the estimate for g based on real-world observational data. Importantly, the pendulum experiment can be carried out without expensive equipment or time-consuming preparations. Moreover, the mathematical model is simple and does not require sophisticated or expensive numerical solvers. However, the filtering problem is non-linear and non-Gaussian. It does not have an analytic solution, and an efficient approximate solution must be constructed. We use particle filters for this task.

The simple pendulum setting allows us to focus on the statistical aspects of the estimation problem and the construction of particle filters. The filters we discuss are well known in the statistics and control theory communities, and textbooks and tutorials are available, see e.g. [14, 24, 33]. However, these works focus on filters for *state space estimation*. In contrast, we employ filters for *parameter estimation* in mathematical models, and within the Bayesian framework.

Bayesian inverse problems attracted a lot of attention in the applied mathematics community during the past decade since the work by Stuart [35] which laid out the mathematical foundations of Bayesian inverse problems. The design of efficient solvers for these problems is an active area of research, and particle filters offer attractive features which deserve further research. This tutorial enables interested readers to learn the building blocks of particle filters illustrated by a simple example. Moreover, we provide the source code so that the reader can combine the filters with more sophisticated mathematical models.

The remaining part of this tutorial is organised as follows. In Sect. 2 we give a precise formulation of the filtering problem and define a filter. We introduce the pendulum problem in Sect. 2.1, and review previous work on model calibration, filtering and the numerical approximation of these procedures in Sect. 2.2. In Sect. 3 we introduce the Bayesian solution to the filtering problem. Furthermore, we explain the statistical modeling of the pendulum filtering problem. In Sect. 4

we discuss particle-based filters, namely Sequential Importance Sampling and Sequential Monte Carlo. In Sect. 5 we apply both these methods to the pendulum filtering problem, and comment on the estimation results. Finally, we provide a discussion in Sect. 6.

2 The Filtering Problem

We motivated the filtering of model parameters in the preceding section. Now we give a rigorous introduction to filtering. Note that we first define the filtering problem in a general setting.

Let X and H be separable Banach spaces. X denotes the *parameter space*, and H denotes the *model output space*. We define a *mathematical model* $G : X \to H$ as a mapping from the parameter space to the model output space. Next, we observe the system of interest that is represented by the model. We collect measurements at $T \in \mathbb{N}$ points in time $t = 1, \ldots, T$. These observations are denoted by $y_1, y_2, \ldots, y_T$. Each observation y_t is an element of a *finite-dimensional* Banach space Y_t. The family of spaces $Y_1, \ldots, Y_T$ are the so called *data spaces*. We model the observations by *observation operators* $\mathcal{O}_t : H \to Y_t$, $t \geq 1$, that map the model output to the associated observation. Furthermore, we define a family of *forward response operators* $\mathcal{G}_t := \mathcal{O}_t \circ G$, $t \geq 1$, that map from the parameter space directly to the associated data space. We assume that the observations are noisy and model this fact by randomness. The randomness is represented on an underlying probability space $(\Omega, \mathcal{A}, \mathbb{P})$. Each observation y_t is the realisation of a random variable $\widetilde{y}_t : \Omega \to Y_t$. Moreover,

$$\widetilde{y}_t \sim L_t(\cdot|\theta^\dagger) \tag{1}$$

where $L_t : Y_t \times X \to [0, \infty)$, $t \geq 1$, is a parameterised probability density function (w.r.t. the Lebesgue measure). $\theta^\dagger$ denotes the true parameter associated with the observations.

Example 1 (Additive Gaussian Noise) A typical assumption is that the measurement noise is Gaussian and additive. In that case y_t is a realisation of the random variable $\widetilde{y}_t = \mathcal{G}_t(\theta^\dagger) + \eta_t$, where $\eta_t \sim \mathrm{N}(0, \Gamma_t)$ and $\Gamma_t : Y_t \to Y_t$ is a linear, symmetric, positive definite covariance operator, $t \geq 1$. It holds

$$L_t(y_t|\theta) = \exp\left(-\frac{1}{2}\|\Gamma^{-1/2}(\mathcal{G}_t(\theta) - y_t)\|^2\right).$$

The *inverse* or *smoothing problem* at a specific timepoint $t \geq 1$ is the task to identify the unknown true parameter $\theta^\dagger$ given the data set $(y_1, \ldots, y_t) =: y_{1:t}$. We denote the estimate for the parameter by $\widehat{\theta}(y_{1:t})$. Hence, a formal expression for smoothing is the map

$$y_{1:t} \mapsto \widehat{\theta}(y_{1:t}).$$

The *filtering problem*, on the other hand, is the task to update the estimate $\widehat{\theta}(y_{1:t})$ after the observation y_{t+1} is available. Hence, a formal expression for a *filter* is the map

$$\{\widehat{\theta}(y_{1:t}), y_{t+1}\} \mapsto \widehat{\theta}(y_{1:t+1}).$$

Filtering can be considered as a *learning process* in the following sense. Our point of departure is a current state of knowledge represented by the parameter estimate $\widehat{\theta}(y_{1:t})$. This involves all observations up to the point in time t. The data set y_{t+1} is then used to arrive at a new state of knowledge represented by the updated parameter estimate $\widehat{\theta}(y_{1:t+1})$. We depict this learning process in Fig. 1.

Next we describe two practical filtering problems. The pendulum filtering problem is used for illustration purposes, and the tumor filtering problem highlights a more involved application of filtering.

Example 2 (Tumor) The tumor inverse and filtering problem has been discussed extensively in the literature, see e.g. [8, 18, 25] and the references therein. In this problem we model a tumor with a system of (partial) differential equations, for example, the Cahn-Hilliard or reaction-diffusion equations, or, alternatively, an atomistic model. The goal is to predict the future growth of the tumor. Moreover, we wish to test, compare and select suitable therapeutical treatments. To do this we need to estimate model parameters, e.g. the tumor proliferation and consumption rate, and chemotaxis parameters. These model parameters are patient-specific and can be calibrated and updated using patient data. The data is given by tumor images obtained e.g. with *magnetic resonance imaging* (MRI) or with *positron emission tomography* (PET). The images are captured at different timepoints and monitor the progression of the tumor growth. Note that the data spaces are in general infinite-dimensional in this setting.

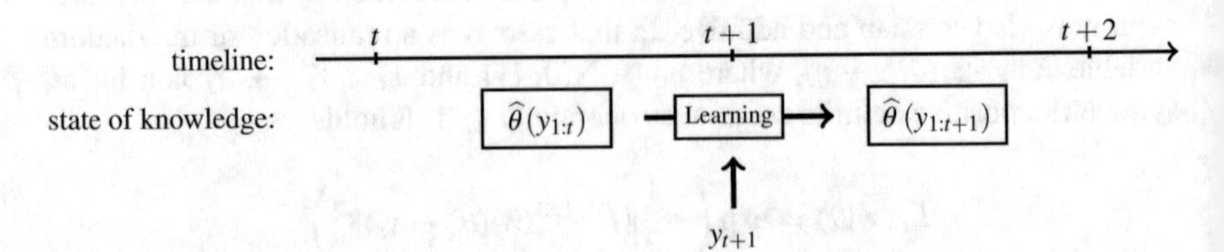

Fig. 1 The filtering problem. The starting point is the current state of knowledge $\widehat{\theta}(y_{1:t})$ at the point in time t. At the timepoint $t+1$ we observe y_{t+1}. We want to use these observations to improve our knowledge concerning $\theta^{\dagger}$. The new state of knowledge is given by the updated estimate $\widehat{\theta}(y_{1:t+1})$

2.1 Pendulum Example

In this section we describe a simple yet practical filtering problem that is associated with a real-world experiment and data. Throughout this tutorial we will come back to this problem to illustrate the filtering of model parameters.

The goal of the pendulum inverse problem is the estimation of the Earth's gravitational acceleration g using measurements collected from the periodic motion of a pendulum. Note that the gravitational acceleration at a particular location depends on the altitude and the latitude of this location. We use measurements that were collected in Garching near Munich, Germany, where the height above mean sea level is $h = 482\,\mathrm{m}$ and the latitude is $\phi = 48°15' \mathrm{N} = 48.25° \mathrm{N}$. The formula (4.2) in [2] gives the gravitational acceleration in Garching:

$$\begin{aligned} g^\dagger =&\Big(9.780327\left(1 + 5.3024 \cdot 10^{-3}\sin^2(\phi) - 5.8 \cdot 10^{-6}\sin^2(2\phi)\right) \\ &- 1.965 \cdot 10^{-6} hm^{-1}\Big)\frac{m}{s^2} \approx 9.808\frac{m}{s^2}. \end{aligned}$$

We use a simplified model to describe the dynamics of the pendulum. Specifically, we ignore friction, and assume that the pendulum movements take place in a two-dimensional, vertically oriented plane. In this case the state of the pendulum can be described by a single scalar that is equal to the angle enclosed by the pendulum string in its excited position and the stable equilibrium position. By using Newton's second law of motion and by considering the forces acting on the pendulum it is easy to see that this angle $x(\cdot; g)$ satisfies the parametrised non-linear initial value problem (IVP)

$$\begin{aligned} \ddot{x}(\tau; g) &= -\frac{g}{\ell}\sin(x(\tau; g)), \\ \dot{x}(0; g) &= v_0, \\ x(0; g) &= x_0, \end{aligned}$$

where ℓ denotes the length of the string that connects the two ends of the pendulum, and $\tau \in [0, \infty)$ denotes time. An illustration of the model is given in Fig. 2.

Following the framework presented in Sect. 2, we define the model $G(g) = x(\cdot; g)$ which maps the model parameter, here the gravitational acceleration, to the model output, here the time-dependent angle. For $t = 1, \ldots, 10$ we define the observation operator $\mathcal{O}_t$ by $\mathcal{O}_t(x(\cdot; g)) = x(\tau_t; g)$. This models the angle measurement of the pendulum at a fixed point in time τ_t. Note that in practise the measurement is reversed since we measure the time at a prescribed angle that is easy to identify. Mathematically, this can be interpreted as angle measurement at a specific timepoint. Finally, we define the forward response operators $\mathcal{G}_t = \mathcal{O}_t \circ G$ for $t = 1, \ldots, 10$. The data set $y_1, \ldots, y_t$ of angle measurements corresponds to realisations of the random variables

$$\widetilde{y}_t := \mathcal{G}_t(g^\dagger) + \eta_t$$

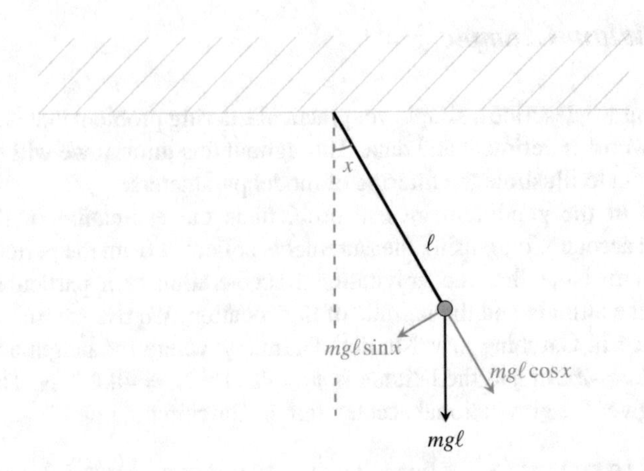

Fig. 2 Pendulum model and forces applied to the bob with mass m. The black vertical vector represents the gravitational force. It is decomposed into the gray vectors representing the components parallel and perpendicular to the motion of the pendulum. The dashed line represents the angle $x = 0$ where the pendulum is at rest. In this position the time measurements are taken. This figure is adapted from [7]

where $\eta_1, \ldots, \eta_t$ are independent and identically distributed Gaussian random variables according to $N(0, \sigma^2)$. The pendulum filtering problem consists of using time measurements (and the associated angle measurements) to sequentially improve the estimate of the true value of the Earth's gravitational acceleration $g^\dagger$.

Remark 1 It is possible to simplify the mathematical model of the pendulum motion. Assume that $v_0 = 0$. If $|x|$ is small, then $x \approx \sin(x)$. Hence, the nonlinear ODE $\ddot{x} = -(g/\ell)\sin(x)$ above can be replaced by the linear ODE $\ddot{x} = -(g/\ell)x$ with the analytical solution

$$x(\tau; g) := x_0 \cos(\tau\sqrt{g/\ell}).$$

However, the relation between the angle x and the model parameter g is still nonlinear and thus the filtering problem remains nonlinear with no analytical solution. For this reason we do not consider the linear pendulum dynamics.

2.2 *State of the Art*

Model calibration problems have often been approached with optimisation techniques, for example, the Gauss-Newton or Levenberg-Marquardt algorithm. These algorithms minimise a (possibly regularised) quadratic loss function which measures the distance between the data and the model output, see e.g. [29, §10]. Today's availability of high-performance computing resources has enabled statistical techniques for the calibration of computationally expensive models. A popular example is Bayesian inference. Here we consider the model parameters to be uncertain and model the associated uncertainty with a probability measure. By using Bayes' formula it is possible to include information from the observational data in this probability measure. In particular, the probability measure is conditioned with respect to the data (see [30]).

Recently, Bayesian inference for model parameters (so-called Bayesian inverse problems) has attracted a lot of attention in the literature. It was first proposed in [19] and thoroughly analysed in [9, 35]. A tutorial on Bayesian inverse problems is given by Allmaras et al. [2]; in fact this work inspired the authors of this article. However, most works on Bayesian inverse problems, including the works cited above, are concerned with the so-called *static* Bayesian learning where one uses a single set of observations and no filtering is carried out. Taking the step from including a single set of observations to iteratively including more observations is both practically important and non-trivial. We mention that the filtering of model parameters is closely related to the filtering of states in state space models. In the pendulum problem in Sect. 2.1 this task corresponds to the tracking and prediction of the pendulum motion. Filtering of states is a central problem in *data assimilation* (see e.g. [14, 24, 27]).

Linear-Gaussian filtering problems can be solved analytically with the Kalman filter, see e.g. [17]. For non-linear filtering problems several non-linear approximations to the Kalman filter have been proposed. Examples are the extended Kalman filter (EKF, see e.g. [17]) and the Ensemble Kalman Filter (EnKF, see [15]) which are essentially based on linearisations of the forward problem. The EKF uses a Taylor expansion for linearisation. The EnKF uses a probabilistic linearisation technique called *Bayes linear* (see e.g. [22, 34]). The Taylor expansion in the EKF can be evaluated analytically, however, for the Bayes linear approximation this is not possible. For this reason the EnKF uses a particle approximation—an *ensemble* of unweighted particles.

Another family of approximations, so-called *particle filters*, do not use linearisation strategies, but rely on importance sampling (see [31]). We will discuss two particle filters—Sequential Importance Sampling and Sequential Monte Carlo—in this tutorial. Note that these methods are not only used for filtering. Indeed, it is possible to sample from general sequences of probability measures by using particle filters. We refer to [10–12] for the theoretical background and further applications of particle filters. We mention the use of particle filters in static Bayesian inverse problems. SMC and SIS are here used together with a tempering of the likelihood (see e.g. [4, 5, 18, 20, 28]), with multiple resolutions of the computational model (see e.g. [6]) or with a combination of both (see e.g. [23]). An excellent overview of SMC and particle filters can be found on the webpage of Doucet [13]. Finally, note that SMC requires a Markov kernel that is typically given by a Markov Chain Monte Carlo (MCMC) method [26, 31].

3 Bayesian Filtering

In this section we explain the Bayesian approach to the filtering problem discussed in Sect. 2. We begin by introducing conditional probabilities and their construction using Bayes' formula. We conclude this section by revisiting the pendulum filtering problem presented in Sect. 2.1 and discuss implications from the Bayesian approach.

Consider some uncertain parameter θ. We model the uncertainty as a random variable; by definition this is a measurable map from the probability space $(\Omega, \mathscr{A}, \mathbb{P})$ to the parameter space X. The expected value $\mathbb{E}(\cdot)$ is the operator integrating a function with respect to $\mathbb{P}$. The probability measure of θ is called $\mu_0 = \mathbb{P}(\theta \in \cdot)$. The measure μ_0 reflects the knowledge concerning θ before we include any information given by observations. For this reason μ_0 is called *prior (measure)*.

In Sect. 2 we modeled the data generation at time t as an event that occurs and that we observe. This event is $\{\widetilde{y}_t = y_t\} \in \mathscr{A}$. The process of *Bayesian learning* consists in including the information "$\widetilde{y}_t = y_t$" into the probability distribution of θ. This is done with conditional probability measures. A good intuition about conditional probabilities can be obtained by consideration of discrete probabilities.

Example 3 (Conditional Probability) Let θ denote a uniformly distributed random variable on the parameter space $X := \{1, \ldots, 10\}$. The uniform distribution models the fact that we have no information whatsoever about θ. Next, an oracle tells us that "θ is about 4". We model this information by assuming that θ is equal to 3, 4 or 5 with equal probability. Our state of knowledge is then modeled by the following conditional probabilities:

$$\mathbb{P}(\theta = k|\theta \text{ is about } 4) := \begin{cases} 1/3, & \text{if } k = 3, 4, 5, \\ 0, & \text{otherwise.} \end{cases}$$

We revisit this example in the next subsection and illustrate the computation of the conditional probability measure. For now, we move back to the filtering problem. Having observed the first data set y_1 we replace the prior probability measure μ_0 by the conditional probability measure

$$\mu_1 = \mathbb{P}(\theta \in \cdot|\widetilde{y}_1 = y_1).$$

Analogously to Example 3 the measure μ_1 now reflects the knowledge about θ given the information that the event $\{\widetilde{y}_1 = y_1\}$ occurred. In the next step we observe $\widetilde{y}_2 = y_2$ and update $\mu_1 \mapsto \mu_2$, where

$$\mu_2 = \mathbb{P}(\theta \in \cdot|\widetilde{y}_1 = y_1, \widetilde{y}_2 = y_2) =: \mathbb{P}(\theta \in \cdot|\widetilde{y}_{1:2} = y_{1:2}).$$

This update models the Bayesian filtering from time point $t = 1$ to $t = 2$. More generally, we can define a Bayesian filter as a map

$$\{\mu_t, y_{t+1}\} := \{\mathbb{P}(\theta \in \cdot|\widetilde{y}_{1:t} = y_{1:t}), y_{t+1}\} \mapsto \mathbb{P}(\theta \in \cdot|\widetilde{y}_{1:t+1} = y_{1:t+1}) =: \mu_{t+1}.$$

Since μ_t reflects the knowledge about θ after seeing the data, this measure is called *posterior (measure) at time t* for every $t \geq 1$. The notions and explanations in Sect. 2 can be transferred to Bayesian filtering by replacing $\widehat{\theta}(y_{1:t})$ with μ_t.

3.1 Bayes' Formula

The Bayesian learning procedure—formalised by the map $\{\mu_t, y_{t+1}\} \mapsto \mu_{t+1}$—is fundamentally based on Bayes' formula. In order to define Bayes' formula, we make some simplifying, yet not necessarily restrictive assumptions.

In Sect. 2 the parameter space X can be infinite-dimensional. In the remainder of this tutorial we consider $X := \mathbb{R}^N$, a finite-dimensional parameter space. For a treatment of the infinite-dimensional case we refer to [35]. Moreover, we assume that μ_0 has a *probability density function (pdf)* $\pi_0 : X \to \mathbb{R}$ with respect to the Lebesgue measure. This allows us to represent μ_0 by

$$\mu_0(A) := \int_A \pi_0(\theta)\mathrm{d}\theta,$$

for any measurable set $A \subseteq X$. Furthermore, we assume that the *model evidence*

$$Z_{t+1}(y_{t+1}) := \mu_t\left(L_{t+1}(y_{t+1}|\cdot)\right) := \int_X L_{t+1}(y_{t+1}|\theta)\mathrm{d}\mu_t(\theta)$$

$$:= \int_X L_{t+1}(y_{t+1}|\theta)\pi_t(\theta)\mathrm{d}\theta, \quad t \geq 0,$$

is strictly positive and finite. Then it follows that the conditional measures $\mu_1, \mu_2, \ldots$ have pdfs $\pi_1, \pi_2, \ldots$ on X as well. The associated densities are given recursively by *Bayes' formula*

$$\pi_{t+1}(\theta) = \frac{1}{Z_{t+1}(y_{t+1})} L_{t+1}(y_{t+1}|\theta)\pi_t(\theta), \quad t \geq 0, \quad \text{a.e. } \theta \in X. \tag{2}$$

In some situations it is possible to use this formula to compute the densities $(\pi_t)_{t=1}^{\infty}$ analytically. In particular, this is possible if π_t is the pdf of a *conjugate prior* for the likelihood L_{t+1} for $t \geq 0$. However, in the filtering of parameters of nonlinear models it is typically impossible to find conjugate priors. In this case we construct approximations to the densities $(\pi_t)_{t=1}^{\infty}$ and the measures $(\mu_t)_{t=1}^{\infty}$, respectively. We discuss particle based approximations in Sect. 4.

Before moving on to the Bayesian formulation of the pendulum filtering problem, we briefly revisit Example 3 and explain Bayes' formula in this setting. Note that Bayes' formula holds more generally for probability density functions that are given w.r.t. to σ-finite measures, for example, *counting densities* on $\mathbb{Z}$; these are sometimes called *probability mass functions (pmf)*.

Example 3 (continued) Recall that we consider a uniformly distributed random variable θ taking values in $\{1, \ldots, 10\}$. Hence the counting density is given by $\pi_0 \equiv 1/10$. Furthermore, an oracle tells us that "θ is about 4" and we model this information by θ taking on values in $\{3, 4, 5\}$ with equal probability. Hence

$$L_1(y_1|k) = \mathbb{P}(\theta \text{ is about } 4|\theta = k) = \begin{cases} 1, & \text{if } k = 3, 4, 5, \\ 0, & \text{otherwise.} \end{cases}$$

Now we compute the posterior counting density using Bayes' formula. We arrive at

$$\begin{aligned} \pi_1(k) &= \frac{1}{Z_1(y_1)} \cdot \pi_0(k) \cdot L_1(y_1|k) \\ &= \begin{cases} \frac{1}{1 \cdot \frac{1}{10} + 1 \cdot \frac{1}{10} + 1 \cdot \frac{1}{10}} \cdot \frac{1}{10} \cdot 1, & \text{if } k = 3, 4, 5, \\ 0, & \text{otherwise.} \end{cases} \\ &= \begin{cases} 1/3, & \text{if } k = 3, 4, 5, \\ 0, & \text{otherwise.} \end{cases} \end{aligned}$$

Since the values of counting densities are identical to the probability of the respective singleton, we obtain $\pi_1(k) := \mathbb{P}(\theta = k|\theta \text{ is about } 4)$. This fits with the intuition discussed in Example 3.

3.2 Bayesian Filtering Formulation of the Pendulum Problem

Next, we revisit the pendulum filtering problem in Sect. 2.1, and reformulate it as a Bayesian filtering problem. This requires us to define a prior measure for the parameter g. The prior should include all information about the parameter before any observations are made. The first piece of information about g stems from the physical model which describes the motion of the pendulum. Indeed, since we know that gravity attracts objects towards the center of the Earth, we conclude that the value of g must be non-negative in the coordinate system we use. Furthermore, we assume that previous experiments and theoretical considerations tell us that $g \leq 20\,\text{m/s}^2$, and that g is probably close to the center of the interval $[0, 20]$. We model this information by a normal distribution with unit variance, centered at 10, and truncated support on the interval $[0, 20]$. Thus, the prior density is proportional to

$$\pi_0(g) \propto \begin{cases} \exp\left(-\frac{1}{2}(g - 10)^2\right), & 0 \leq g \leq 20, \\ 0, & \text{otherwise.} \end{cases}$$

Furthermore, the data generating distribution in Sect. 2.1 implies that the likelihoods are given by

$$L_t(y_t|g) \propto \exp\left(-\frac{1}{2\sigma^2}\,(y_t - \mathcal{G}_t(g))^2\right), \qquad t = 1, \ldots, 10, \tag{3}$$

where σ^2 denotes the noise variance. In the numerical experiments we use the estimate $\sigma^2 = 0.0025$. This is obtained by combining typical values for human reaction time and a forward Monte Carlo simulation as follows. Studies (see e.g. [36]) suggest that a typical visual reaction time for humans is 450 ms $\pm$ 100 ms. We now would like to use this information about the error in the time measurements for modelling the error in the angles. To this end we use the forward model parameterised by the mean of the prior distribution to compute a reference solution for a set of generated time measurements. We then compute a Monte Carlo estimate of the angle error resulting from adding a $N(0.45, 0.01)$ noise to the set of time measurements. The result of this numerical experiment indicates that $N(0, \sigma^2)$ with $\sigma^2 = 0.0025$ is a suitable model for the angle measurements error. Ideally, one would estimate the noise variance directly in the pendulum experiment.

By inserting the prior density and the definition of the likelihoods into Bayes' formula (2) we obtain a recursive expression for the densities $\pi_1, \ldots, \pi_{10}$ of the posterior measures $\mu_1, \ldots, \mu_{10}$ as follows:

$$\pi_{t+1}(g) := \frac{1}{Z_{t+1}(y_{t+1})}\exp\left(-\frac{1}{2\sigma^2}\,(y_{t+1} - \mathcal{G}_{t+1}(g))^2\right)\pi_t(g), \quad \text{for a.e. } g \in [0, 20],$$

$$Z_{t+1}(y_{t+1}) := \mu_t\left(\exp\left(-\frac{1}{2\sigma^2}\,(y_{t+1} - \mathcal{G}_{t+1}(g))^2\right)\right), \quad t \geq 0.$$

The family of posterior densities corresponding to the measurements in Table 1 is depicted in Fig. 3. As the time increases we see that the posterior densities become

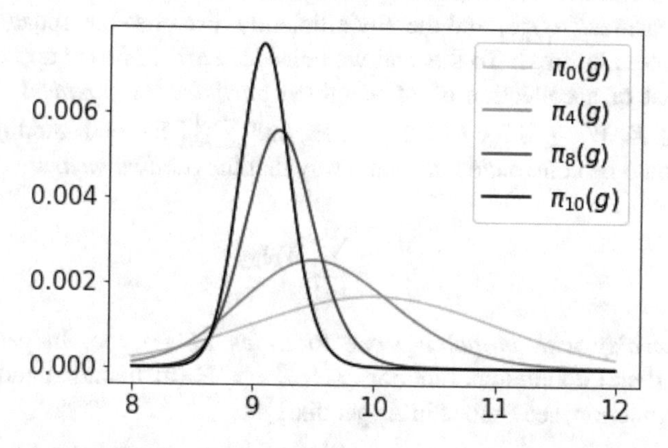

Fig. 3 Sequence of posterior densities for a given value of g for $t = 0$ (prior), 4, 8, 10 data points. The densities are computed with kernel density estimation based on a Sequential Monte Carlo particle approximation

more concentrated. Note that a more concentrated density indicates less uncertainty compared to a flat density. Thus, the picture is consistent with our intuition: The uncertainty in the parameter value for g is reduced as more data becomes available. We provide further comments on the estimation results in Sect. 5.2.

4 Particle Filters

It is typically impossible to compute the posterior measures $(\mu_t)_{t=1}^{\infty}$ analytically. In this section we discuss two particle filters to approximate the solution of the Bayesian filtering problem. In Sect. 4.1 we present the *Sequential Importance Sampling* (SIS) algorithm. Unfortunately, SIS suffers from efficiency issues when used for filtering problems. We explain this deficiency in a simple example. It is possible to resolve the issues by extending SIS to the so-called *Sequential Monte Carlo* (SMC) algorithm which we discuss in Sect. 4.2.

Remark 2 The sampling procedures presented in this section approximate posterior measures $(\mu_t)_{t=1}^{\infty}$. In addition, it is also possible to approximate integrals of the form $(\mu_t(f))_{t=1}^{\infty}$, where $f : X \to \mathbb{R}$ is a μ_t-integrable *quantity of interest*. To simplify the presentation we focus on the approximation of integrals rather than on the approximation of measures for the remaining part of this tutorial. This can be done without loss of generality within the framework of weak representations of measures. We outline this approach and give some examples for quantities of interest in Appendix 1.

4.1 Sequential Importance Sampling

We are now interested in constructing an algorithm to approximate the sequence of posterior measures $(\mu_t)_{t=1}^{\infty}$ and thereby efficiently discretise the update rule of the learning process in Fig. 1. To this end we consider *particle-based* approximations. These consist of a collection of M weighted *particles* (or *samples*) $\{X_i, W_i\}_{i=1}^{M}$ where $X_i \in X$, $W_i \geq 0$ for $i = 1, \ldots, M$, and $\sum_{i=1}^{M} W_i = 1$. Additionally, the particles should be constructed in such a way that the *random measure*

$$\mu_t^M := \sum_{i=1}^{M} W_i \delta_{X_i}$$

converges *weakly with probability one* to μ_t as $M \to \infty$. In particular, for any bounded and continuous function $f : X \to \mathbb{R}$ (or bounded and Lipschitz-continuous function; see Prop. 2 in Appendix 1),

$$\mu_t^M(f) := \sum_{i=1}^{M} W_i f(X_i) \xrightarrow{M\to\infty} \mathbb{E}[f(\widehat{\theta}_t)] \qquad \text{almost surely.}$$

Note that $(\mu_t^M)_{M\geq 1}$ is a sequence of measure-valued random variables; that is a sequence of random variables defined on the space of measures. Almost every realisation of this sequence of measures converges weakly to a common deterministic measure as $M \to \infty$. More precisely, it holds that

$$\mathbb{P}(\mu_t^M \to \mu_t, \text{ weakly, as } M \to \infty) = 1.$$

For simple distributions μ where direct sampling is possible an estimator with the desired properties above can be constructed by choosing independent random variables $X_1, \ldots, X_M$ distributed according to μ and by using uniform weights $W_i = 1/M$ for $i = 1, \ldots, M$. We denote this construction by the operator S^M and call $\mu^M = S^M \mu$ the *standard Monte Carlo* estimate of μ. Since S^M creates an empirical measure by using random variables the operator S^M maps probability measures to random probability measures and is thus a non-deterministic operator.

However, it is typically impossible to sample from the posterior measures $(\mu_t)_{t=1}^{\infty}$ in the Bayesian filtering problem. This precludes the construction of Monte Carlo estimates. Alternatively, we can use the *Importance Sampling* (IS) method. Let μ, the *target measure*, denote the probability measure to be estimated. Let ν denote a probability measure from which it is possible to sample. The measure ν is called *importance measure*. Let μ be *absolutely continuous* with respect to ν. This means that $\nu(A) = 0$ implies $\mu(A) = 0$ for any measurable set $A \subseteq X$. Let furthermore f denote a measurable, bounded function. Then there exists a non-negative function $w : X \to \mathbb{R}$ such that

$$\mu(f) = \frac{\nu(f \cdot w)}{\nu(w)}. \tag{4}$$

The intuition behind importance sampling is the following. If the importance measure ν is close to the target measure μ, then sampling from ν should be approximately equivalent to sampling from μ.

The IS estimate of μ is constructed by creating a Monte Carlo estimate of ν. Then, the Monte Carlo samples are reweighed using the expression in (4). This is necessary since we wish to obtain samples distributed according to the target measure μ, and not samples distributed according to the importance measure ν. We arrive at

$$\mu^M(f) = \frac{\nu^M(f \cdot w)}{\nu^M(w)} = \frac{\frac{1}{M}\sum_{i=1}^M w(X_i)f(X_i)}{\frac{1}{M}\sum_{j=1}^M w(X_j)} := \sum_{i=1}^M W_i f(X_i).$$

In summary, importance sampling maps the Monte Carlo estimate ν^M to an updated estimate μ^M by adjusting the weights of the particles $X_1, \ldots, X_M$ according to the formula

$$W_i = \frac{w(X_i)}{\sum_{j=1}^M w(X_j)}.$$

Now we apply the IS approximation in the context of the Bayesian filtering problem. The prior measure μ_0 is often tractable and direct sampling algorithms are available. It is then possible to create an initial Monte Carlo estimate of the prior measure. Afterwards we can iteratively update the particles to incorporate the new knowledge from the observation $y_1, y_2, \ldots, y_T$ through importance sampling. This follows the learning process described in Sect. 2. Observe that Bayes' formula in (2) tells us that each measure μ_{t+1} is absolutely continuous with respect to the previous measure. We use this relation to define the nonlinear operator IS_t for any probability measure μ over X as follows:

$$(\mathrm{IS}_{t+1}\mu)(f) = \frac{\mu(L_{t+1}(y_{t+1}|\cdot) \cdot f)}{\mu(L_{t+1}(y_{t+1}|\cdot))}.$$

Since $Z(y_{t+1}) = \mu_t(L_{t+1}(y_{t+1}|\cdot))$ for every $t \geq 0$ these operators can be used to describe reweighing in an importance sampling estimate with target measure μ_{t+1} and importance measure μ_t. In particular, if for some $t \geq 0$ a particle approximation μ_t^M of μ_t is given by the particles $\{X_i^{(t)}, W_i^{(t)}\}$, the operator IS_{t+1} can be used to define the following approximations:

$$Z^M(y_{t+1}) := \mu_t^M(L_{t+1}(y_{t+1}|\cdot)) = \sum_{i=1}^{M} W_i^{(t)} L_{t+1}(y_{t+1}|X_i),$$

$$\begin{aligned}\mu_{t+1}^M(f) &:= (\mathrm{IS}_{t+1}\mu_t^M)(f) \\ &= \frac{1}{Z^M(y_{t+1})} \sum_{i=1}^{M} W_i^{(t)} L_{t+1}(y_{t+1}|X_i^{(t)}) f(X_i^{(t)}) \\ &= \sum_{i=1}^{M} W_i^{(t+1)} f(X_i^{(t)}),\end{aligned}$$

where the particle weights $W_i^{(t+1)}$ are given by

$$W_i^{(t+1)} = \frac{W_i^{(t)} L_{t+1}(y_{t+1}|X_i^{(t)})}{\sum_{j=1}^{M} W_j^{(t)} L_{t+1}(y_{t+1}|X_j^{(t)})}.$$

Note that the IS_{t+1} update requires $\sum_{j=1}^{M} W_j^{(t)} L_{t+1}(y_{t+1}|X_j^{(t)}) > 0$. If the likelihood $L_{t+1}(y_{t+1}|\cdot)$ is strictly positive, then this condition is always satisfied. We observe that the IS update formula only changes the weights

$$(W_i^{(t)})_{i=1}^{M} \mapsto (W_i^{(t+1)})_{i=1}^{M}.$$

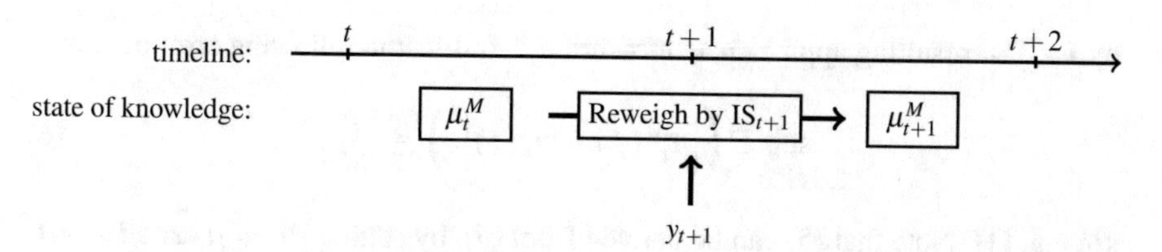

Fig. 4 Plot of the SIS approximation of the learning procedure. The starting point is the approximation μ_t^M of the current state of knowledge at timepoint t. At the timepoint $t+1$ we observe y_{t+1} and use it to construct the importance sampling operator IS_{t+1}. The new state of knowledge is approximated by μ_{t+1}^M

The positions $(X_i^{(t)})_{i=1}^M = (X_i^{(t+1)})_{i=1}^M$ of the particles remain unchanged. Based on the recursive update formula, we construct an approximation to the sequence $(\mu_t)_{t=0}^\infty$ as follows:

$$\mu_0^M = S^M \mu_0,$$
$$\mu_{t+1}^M = \text{IS}_{t+1}(\mu_t^M)$$

see also Fig. 4 for a graphical representation. The sequential importance sampling algorithm is a natural and asymptotically correct approximation of the Bayesian filtering process, see e.g. [5, 7] for up to a finite number of observations. Now we discuss the accuracy of (sequential) importance sampling measured in terms of the so-called *effective sample size (ESS)*. Let $t \geq 0$. First, we define the constant $\rho_t > 0$ by

$$\rho_t := \frac{\mu_0(\mathbf{L}_t^2)}{\mu_0(\mathbf{L}_t)^2},$$

where $\mathbf{L}_t(\theta) = \prod_{i=1}^t L(y_t|\theta)$ is the joint likelihood of the dataset $y_{1:t}$. One can show that the accuracy of (sequential) importance sampling up to time t is equivalent to the accuracy of a standard Monte Carlo approximation with M/ρ_t samples (see Remark 3). The fraction M/ρ_t is the *effective sample size*.

Remark 3 We sketch the derivation of the ESS in SIS for step $t \geq 0$. We assume that we can sample M' times independently from μ_t. Using these samples, we approximate μ_t by $S^{M'}(\mu_t)$, the standard Monte Carlo estimator. In this case, one can show that

$$\sup_{|f|\leq 1} \mathbb{E}\left(\left|S^{M'}(\mu_t)(f) - \mu_t(f)\right|^2\right) \leq \frac{4}{M'}. \tag{5}$$

This is an upper bound on the expected squared error between the integrals of any bounded test function f. Hence, we can now use the *sample size* M' in equation (5) as an indicator for the accuracy of the standard Monte Carlo approximation $S^{M'}(\mu_t)$ of μ_t. In the SIS algorithm, M samples are drawn from μ_0 and are then reweighed by $\mathbf{L}_t$. The resulting approximate measure μ_t^M fulfills the following error bound:

$$\sup_{|f|\leq 1} \mathbb{E}\left(\left|\mu_t^M(f) - \mu_t(f)\right|^2\right) \leq \frac{4\rho_t}{M}, \tag{6}$$

see e.g. [1]. Note that (5) can be derived from (6), by setting $\mu_0 = \mu_t$ and $\mathbf{L}_t \equiv 1$, hence $\rho_t = 1$. To make the upper bounds in (5) and (6) comparable, we replace the sample size M' in (5) by the *effective sample size* M/ρ_t in (6).

In conclusion, we can interpret the ESS as follows:

- If $M/\rho_t \approx M$, then the estimation is nearly as accurate as sampling directly from the correct measure. Hence a value $\rho_t \approx 1$ is desirable.
- If $M/\rho_t \ll M$, then the estimation is only as accurate as a Monte Carlo approximation with a very small number of particles.

Unfortunately, during the course of a filtering procedure, it may happen that ρ_t explodes. We illustrate this by a simple example.

Example 4 (Degeneracy) Consider the problem of estimating the mean $m^\dagger$ of a normal distribution $\mathrm{N}(m^\dagger, 1)$ using $t \in \mathbb{N}$ samples $y_1, \ldots, y_t$ of the distribution. By choosing the conjugate prior $\mathrm{N}(0, 1)$ it is easy to see that the posterior distribution of the unknown parameter m is equal to $\mathrm{N}\left(\frac{S_t}{t+1}, (1+t)^{-1}\right)$, where $S_t = \sum_{i=1}^t y_i$. It is possible to compute ρ_t analytically,

$$\rho_t = \frac{t+1}{\sqrt{2t+1}} \exp\left(\frac{S_t^2}{(2t+1)(t+1)}\right).$$

Hence, $\rho_t = O(t^{1/2}; t \to \infty)$ grows unboundedly as t increases. This implies that the effective sample size M/ρ_t converges to 0 as $t \to \infty$.

In a filtering problem where $M/\rho_t \to 0$ as $t \to \infty$ we cannot use SIS since the estimation accuracy deteriorates over time. It is possible, however, to estimate the effective sample size, and to use this estimate to improve SIS.

Remark 4 The SIS algorithm presented in Sect. 4.1 is basic in the sense that the importance measure for μ_{t+1} at time point $t+1$ is simply the measure μ_t at time point t. It is possible to construct alternative importance measures by using Markov kernels in the SIS algorithm. We refer to the generic framework in [12, §2] for more details.

4.2 Sequential Monte Carlo

The effective sample size is a good indicator of the impoverishment of the particle estimate for μ_t due to the discrepancy between the prior and target distributions. In practice, we approximate M/ρ_t by

$$M_{\text{eff}} := M \frac{\mu_0^M(\mathbf{L}_t)^2}{\mu_0^M(\mathbf{L}_t^2)} = \frac{\left(\sum_{i=1}^M \mathbf{L}_t(X_i)\right)^2}{\sum_{i=1}^M \mathbf{L}_t(X_i)^2}. \tag{7}$$

Note that M_{eff} is a consistent estimator of M/ρ_t. Throughout the rest of this tutorial we will also refer to M_{eff} as *effective sample size*.

If M_{eff} is small it is possible to discard particles situated in regions of the parameter space with low probability. We can then replace these particles with particles that are more representative of the target distribution in the sense that they carry a larger weight. This is done by introducing a *resampling step* after adjusting the weights of the particles in the IS estimate. Precisely, we consider an approximation μ_t^M given by the particles $\{X_i^{(t)}, W_i^{(t)}\}_{i=1}^M$ at time $t \geq 0$. The weights of the approximation are updated by the IS_{t+1} operator, and M_{eff} is computed using the updated set of weights $(W_i^{(t+1)})_{i=1}^M$. If M_{eff} is larger than a pre-defined threshold $M_{\text{thresh}} \in (0, M]$, then the positions of the particles remain unchanged in the step from μ_t^M to μ_{t+1}^M, giving $X_i^{(t+1)} = X_i^{(t)}, i = 1, \ldots, M$. If, on the other hand, $M_{\text{eff}} \leq M_{\text{thresh}}$, then a new set of particles is sampled according to the updated weights. The particle estimate is then given by $\mu_{t+1}^M = S^M(\text{IS}_{t+1}\mu_t^M)$ where we use the new particle set in the Monte Carlo estimate.

Remark 5 The choice of the threshold parameter M_{thresh} is highly problem dependent. Doucet and Johansen [14] mention that $M_{\text{thresh}} = M/2$ is a typical choice. On the other hand, Beskos et al. [5] use $M_{\text{thresh}} = M$, that is, the resampling step is always carried out. Empirical tests targeting a small variance of the posterior mean estimator may also be helpful to define a suitable threshold.

The resampling procedure can be performed as follows. For $i = 1, \ldots, M$ sample $U_i \sim \text{Unif}[0, 1]$, a uniform random variable between 0 and 1. Then, define

$$X_i^{(t+1)} := X_j^{(t)}, \quad j = \min\left\{k \in \{1, \ldots, M\} : \sum_{n=1}^{k} W_n^{(t+1)} \geq U_i\right\},$$

$$W_i^{(t+1)} := 1/M.$$

Alternative ways of resampling are possible (see [16]). Note that the resampling step can successfully eliminate particles in low density areas of the parameter space. However, it still fails to reduce the particle degeneracy, since several particles may occur in the same position. Moreover, even with resampling the particles cannot fully explore the parameter space, since their position remains fixed at all times.

Due to the resampling in case of a small value M_{eff}, or if M_{eff} is large, it is reasonable to assume that the remaining particles are approximately μ_{t+1}-distributed. The idea is now to scatter the particles in a way such that they explore the parameter space to reduce degeneracy without destroying the approximate μ_{t+1}-distribution. This can be achieved by a scattering with a μ_{t+1}-invariant dynamic.

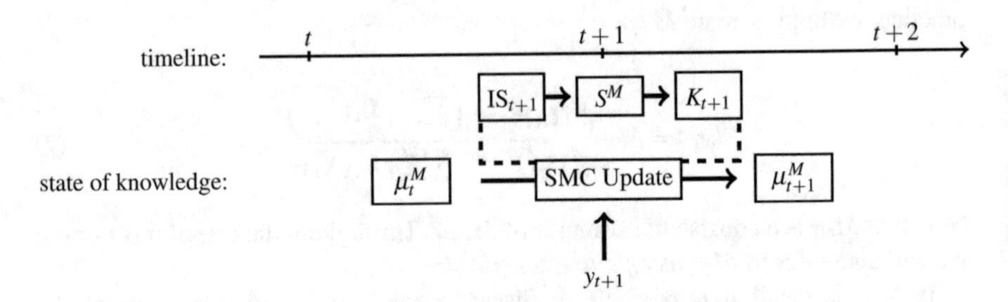

Fig. 5 Plot of the SMC approximation of the learning procedure. The approximation is constructed in three steps: *reweighing* to identify representative particles, *resampling* to discard particles in low probability regions, and a *correction* to adjust the position of the particles to the new state of knowledge

Such a dynamic is given by an ergodic Markov kernel $K_{t+1} : X \times \mathscr{B}(X) \to [0, 1]$ that has μ_{t+1} as stationary distribution. This means, if $X \sim \mu_{t+1}$ and $X' \sim K_{t+1}(X, \cdot)$, then $X' \sim \mu_{t+1}$. Applying the Markov kernel repeatedly to a single particle will asymptotically produce particles that are μ_{t+1} distributed independently of the initial value (see [31, Chapter 6]). Since we assumed that the particles are approximately μ_{t+1}-distributed, we typically rely only weakly on this asymptotic result.

The reader might find it not easy to construct an ergodic Markov kernel K_t that has μ_t as a stationary measure, $t \geq 1$. However, this is a standard task in so-called Markov Chain Monte Carlo methods (MCMC). The literature on MCMC offers a large variety of suitable Markov kernels. We mention Gibbs samplers, Hamiltonian Monte Carlo, Metropolis–Hastings, Random–Walk–Metropolis, Slice samplers (see e.g. [26, 31]). In summary, we add a final step to the approximation, and apply once or several times the transition kernel K_t to each of the particles to obtain a better coverage of the probability density of the posterior at time t. The resulting algorithm is the *Sequential Monte Carlo* (SMC) method. Its approximation of the learning process is depicted in Fig. 5.

5 Particle Approximation of the Bayesian Pendulum Filtering

We are now ready to compute an approximate solution to the Bayesian filtering problem in Sect. 3.2. To this end we consider the likelihoods in (3) and use the set of measurements given in Table 1. Note that the distance between τ_6 and τ_7 is very short; we suspect that the timer has been operated twice instead of once.

Table 1 Time measurements corresponding to the angle $x(\tau_t) = 0$ of a simple pendulum

Measurement	t	1	2	3	4	5	6	7	8	9	10
Time (in s)	τ_t	1.51	4.06	7.06	9.90	12.66	15.40	15.58	18.56	21.38	24.36

The pendulum length is $\ell = 7.4m$. The initial angle $x_0 = 5^\circ = \pi/36$, and the initial speed $v_0 = 0$. Since x_0 is small and $v_0 = 0$, it would be possible to consider the linearised IVP with an analytical solution (see Remark 1). In tests not reported here the filtering solution obtained with the analytical solution of the linearised IVP did not differ from the numerical solution of the nonlinear IVP.

```
import numpy as np
from scipy.stats import norm,truncnorm
from scipy.integrate import odeint
from particle_approximation import ParticleApproximation

measurements = np.array([...]) # see Table 1.

# Define the probabilistic model.
prior = truncnorm(-10, 10, loc=10, scale=1)
error_model = norm(loc=0, scale=0.05)

# The IVP is defined by a RHS and initial values.
theta0 = [5*np.pi/180, 0]
def pendulum_rhs(theta, t, g):
    return [theta[1],-g/7.4 * np.sin(theta[0])]

# The forward response operator is discretized.
mesh = np.append(0, measurements)
def forward_response(g, n):
    sol = odeint(pendulum_rhs, theta0, mesh[:n+1], args=(g,))
    return sol[1:,0]

# The potential at time n defined using n data points.
def potential(g, n):
    if n == 0: return prior.logpdf(g)
    return error_model.logpdf(forward_response(g, n)).sum()
vec_potential = np.vectorize(potential)

# Define the proposal kernel for the correction steps.
def gaussian_proposal(x): return norm(loc=x, scale=.25).rvs()

# Approximate the sequence of posteriors.
approximation = ParticleApproximation(2500, prior)
for n in range(measurements.size):
    importance_potential = lambda x: vec_potential(x, n)
    target_potential = lambda x: vec_potential(x, n+1)
    approximation.smc_update(importance_potential, target_potential,
        ↪ gaussian_proposal, correction_steps=5, ess_ratio=1.3)
    mean = approximation.integrate(lambda x: x)
    var = approximation.integrate(np.square) - mean**2
    print("(posterior #%d) mean=%f, var=%f" % (n+1,mean,var))
```

Display 1 Python file smc_approx.py

The numerical results presented in the following sections have been computed with PYTHON. The associated Python code is available online, see https://github.com/BayesianLearning/PenduSMC. A part of the code is printed in Display 1. It can be combined with the data set of time measurements in Table 1 to reproduce our results, and to carry out further experiments. The code in Display 1 also summarises the complete modelling cycle of the pendulum filtering problem. The class `ParticleApproximation` can be found in Appendix 2.

5.1 Sequential Importance Sampling

We first construct approximations of the learning process using the SIS algorithm as illustrated in Fig. 4. We employ different numbers of particles, $M = 2^4, 2^5, \ldots, 2^{12}$, and perform 50 simulations for each value of M. In Fig. 6 we present the results of the SIS approximations of the posterior distribution at time $t = 10$ depending on the number of particles M. The left display in this figure shows that the variance of the posterior mean estimate is reduced as the number of particles M is increased. Moreover, we observe in the right display of the figure that the convergence rate of the variance of the posterior mean estimate is $O(1/M)$. As expected this coincides with the convergence rate of a standard Monte Carlo approximation (see Remark 3).

Next, we fix the number of particles $M = 2500$, and investigate the accuracy in the SIS approximation by studying the effective sample size. Recall that in SIS only samples from the prior distribution are used throughout the filtering procedure. Anticipated by the plot of the densities in Fig. 3 we expect that these prior samples lead to a poor approximation of the sequence of posterior distributions constructed during the learning process. This intuition is confirmed by numerical experiments

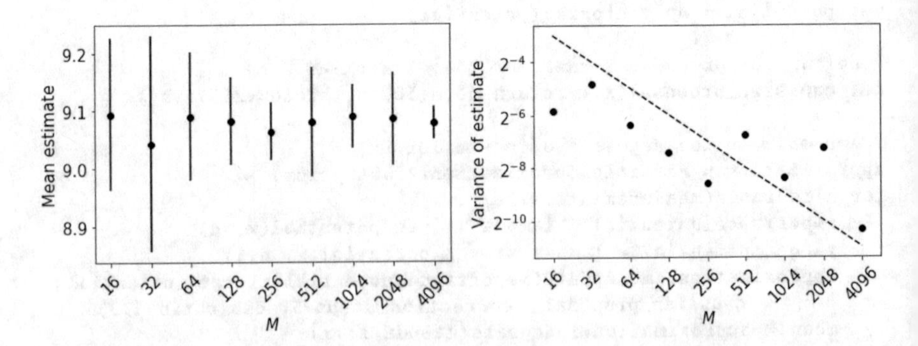

Fig. 6 Accuracy of the SIS approximation. In the plots we use data from runs with a variable number of particles indicated on the horizontal axis. We show results averaged over 50 runs per setting. Left: Plot displaying the convergence of the estimated mean of the posterior distribution. Each black dot represents the sample average over 50 runs, and the bars represent the standard deviation of the mean within the runs. Right: Plot displaying the asymptotic convergence of the variance of the posterior mean estimate

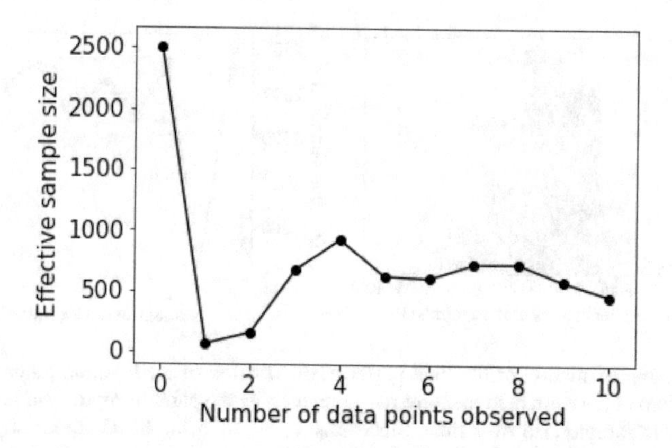

Fig. 7 Plot of the effective sample size for the SIS approximation of the learning process

shown in Fig. 7. We see that the effective sample size is reduced dramatically in the first step of SIS with $t = 1$. In the final step with $t = 10$ the effective sample size is only about 20% of the initial sample size. Unfortunately, a reduced effective sample size implies a loss of estimator accuracy.

Recall that the mean square error of a Monte Carlo estimator with $20\% \cdot M$ samples is 5 times larger compared to the mean square error with M samples. In the simple pendulum setting, where the parameter space space is one-dimensional and compact, a sample size of $20\% \cdot M$ is likely still sufficient to obtain a useful posterior estimate. In real-world applications, however, we typically encounter high-dimensional and unbounded parameter spaces which require the exploration with a large number of representative samples. In this case, the decrease of the effective sample size in SIS is a serious drawback.

5.2 *Sequential Monte Carlo*

We proceed by constructing an approximation of the learning process using the SMC algorithm. The Markov kernel K_t is given by a Random-Walk-Metropolis algorithm with target distribution μ_t and a Gaussian random walk with standard deviation 0.25 as proposal distribution. Each Markov kernel K_t is applied 5 times to correct the approximated distribution. We choose a minimal effective sample size corresponding to 75% of the total population M. This means that we resample if $M_{\text{eff}} < M_{\text{thresh}} := 75\% \cdot M$.

We present a typical run of the algorithm in Fig. 8. Again $M = 2500$ particles are used to approximate the learning process. In the right panel of Fig. 8 we see that four resampling steps have been performed. This maintains an ESS well over 1500 and thus improves the accuracy of the posterior estimate with SMC compared to the estimate with SIS.

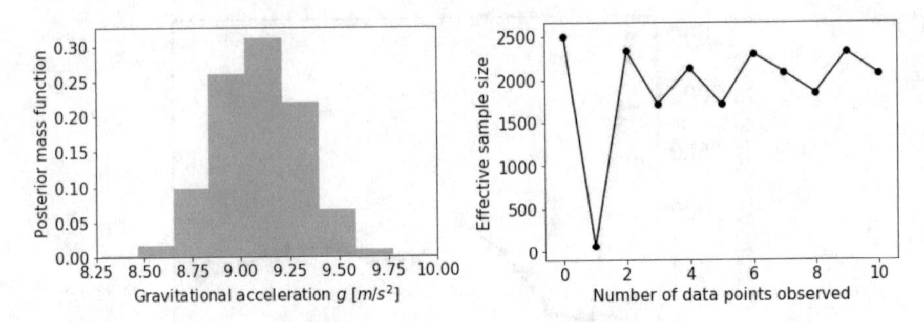

Fig. 8 SMC approximation of the final posterior distribution of the learning process at $t = 10$. Left: Histogram of the estimated measure μ_{10} computed by the particle approximation. Right: Plot of the effective sample size over time. Since $M_{\text{thresh}} = 1875$ the SMC algorithm performed 4 resampling steps to maintain a large effective sample size

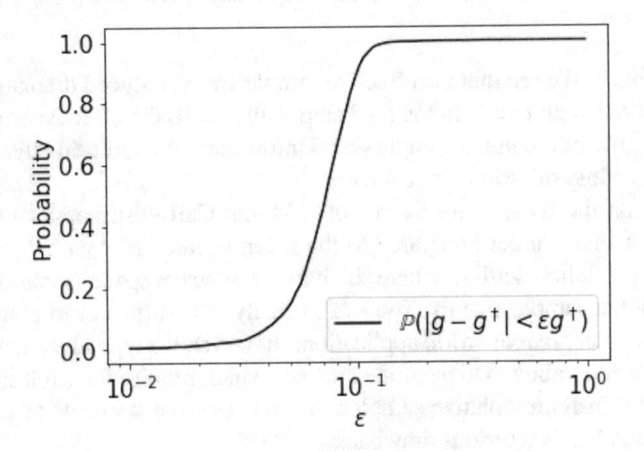

Fig. 9 Comparison of the posterior measure μ^y with the true parameter $g^\dagger$. We plot $\mu_{10}(|\cdot - g^\dagger| < \varepsilon g^\dagger)$ for various values of ε. This value shows how likely the posterior measures sees the uncertain parameter g to be ε-close to the true value $g^\dagger$. The posterior probabilities are computed with SMC based on 2500 particles

Moreover, in the left panel of Fig. 8 we see that the SMC approximation to the posterior measure for $t = 10$ is centered around the value $g \approx 9.12m/s^2$. To compare the estimate with the true value $g^\dagger = 9.808\,\text{m/s}^2$ we compute the posterior probability of the event $\{|g - g^\dagger| < \varepsilon g^\dagger\}$. This describes our posterior expectations about the closeness of the estimated parameter to the truth. We plot the results in Fig. 9. The posterior μ_{10} considers values close to the true parameter $g^\dagger$ unlikely. Aside from the first digit we would not be able to identify $g^\dagger$. However, the estimate is not bad given the very simple experimental setup. The relative error associated with $g = 9.12\,\text{m/s}^2$ is 7%; a similar result was obtained in [2].

We suspect that the estimate for g can be improved by using a more sophisticated physical model for the pendulum dynamics. Moreover, the noise model could be improved. In tests not reported here we observed that the measurement error

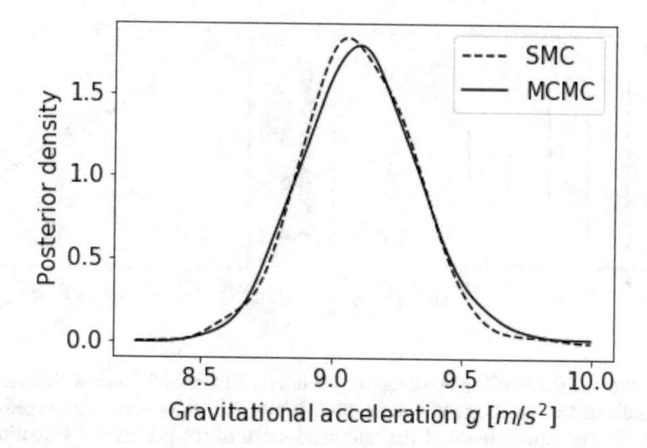

Fig. 10 Kernel density estimates of μ_{10} using the MCMC reference solution and the SMC solution. Each estimate is based on 2500 samples

increases as time increases. This is plausible since the timer has been operated manually and not automatically by a sensor. Finally, we did not include the uncertainty in the initial condition x_0.

To validate the SMC posterior estimate of μ_{10} with $M = 2500$ particles, we have performed a reference run using a Markov chain Monte Carlo sampler. The MCMC sampler was run with 3000 posterior samples where the first 500 samples have been discarded to mitigate the burn-in effect. Using the MCMC and SMC samples we perform a kernel density estimation for the posterior pdf. We present the estimation results in Fig. 10. The posterior approximations do not differ significantly which leads us to conclude that the SMC approximation is consistent with the MCMC approximation.

Moreover, we again investigate the convergence of SMC based on simulations with different numbers of particles $M = 2^4, 2^5, \ldots, 2^{12}$. In each of these settings we consider the results of 50 simulations. We present the test results in Fig. 11. We plot the posterior mean approximation at time $t = 10$ and the variance of the estimators within 50 simulations.

Comparing Fig. 6 with Fig. 11 we see that the posterior mean estimates obtained with SMC differ only slightly from the estimates obtained with SIS. Moreover, we observe again that the variance of the SMC estimate decreases with the rate $O(1/M)$ as M increases (cf. Remark 3). However, we see a much higher variability in the posterior mean estimates in SIS. In particular, we need a smaller number of particles in SMC compared to SIS to reach a certain variance of the estimator. Recall, however, that the SMC algorithm involves parameters which need to be selected by the user; we mention the threshold sample size M_{thresh}, and the Markov kernels K_t together with the number of MCMC steps. In contrast, the SIS algorithm does not require parameter tuning, and is simpler to implement. Hence SIS could be used if the forward response operators $(\mathcal{G}_t)_{t\geq 1}$ can be evaluated cheaply; this allows more evaluations within a given computational budget to reach a desired accuracy.

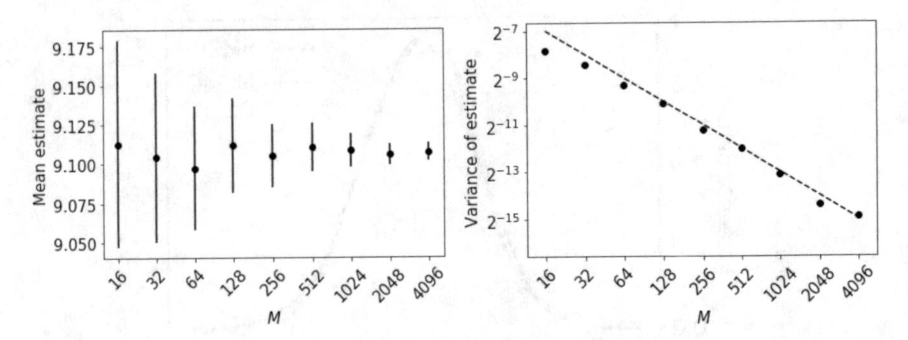

Fig. 11 Accuracy of the SMC approximation. We use data from SMC runs with a variable number of particles indicated on the horizontal axis of each plot. We show results averaged over 50 runs per setting. Left: The convergence of the estimated mean of the posterior distribution. The black dots represent the average over 50 runs and the bars represent the standard deviation of the mean within the runs. Right: The convergence of the variance of the posterior mean estimate

SMC should be preferred if the evaluation of the forward response operators is computationally expensive, and if a small number of MCMC steps is sufficient to mix the particles efficiently in the parameter space.

5.3 Continuing to Learn: Data Sets from Further Experiments

The filtering problem we discussed so far is kept simple for demonstration purposes. However, in reality one would perhaps not update the posterior after the collection of every single measurement (so every couple of seconds), and one would try to use more data sets, possibly from different sources. Indeed, the data set in Table 1 is only one of several data sets that were collected in independent experiments, each carried out by different individuals. A more realistic filtering problem is to update the posterior measure for g after each round of measurements has been completed by an individual. Then, the posterior measure reflects the knowledge obtained from a fixed number of independently performed experiments.

Notably, the SIS and SMC algorithm can be used in this situation as well. It is possible to process the measurements individually or in batches, with both SIS and SMC, without changes to the implementation. Moreover, there is in principle no limit for the number of measurements that can be used in an update. However, since we expect that the posterior measures associated with a larger number of measurements will be more concentrated, it is likely that the degeneracy of the effective sample size of SIS will become more pronounced. For this reason we use SMC to include data sets from further pendulum experiments.

In Fig. 12 we plot the posterior density associated with a single experiment with 10 measurements (see the set-up in Sect. 5.2) along with the posterior density for six experiments with a total number of 50 measurements; this includes the

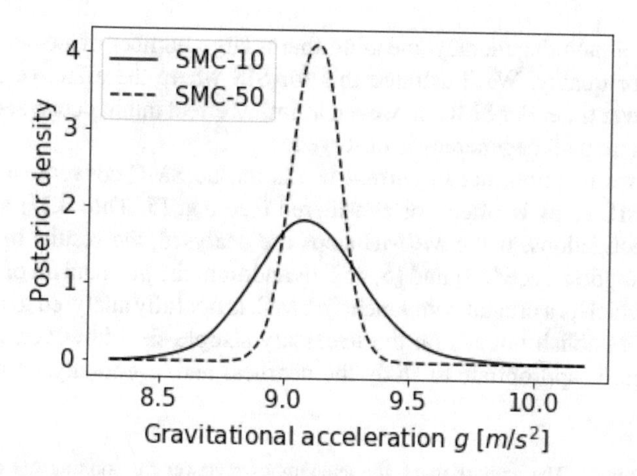

Fig. 12 Posterior density based on the 10 measurements given in Table 1 (SMC-10) and based on 40 further measurements (SMC-50)

10 measurements in the single experiment setting. We observe that the posterior measure after six experiments is more concentrated (informed) than the posterior after one experiment. Thus the uncertainty in the parameter g is reduced after seeing data from different experiments. This is consistent with our intuition about learning.

6 Discussion

We finish with some comments on current research directions.

For Bayesian filtering problems with finite-dimensional data and parameter spaces together with a simple mathematical model it is straightforward to prove the well-posedness of the Bayesian filtering problem (i.e. the existence, uniqueness, and stability of the posterior measure). In general, we will have to consider infinite-dimensional parameter spaces (e.g. random fields), infinite-dimensional data spaces (e.g. the image of a tumor), and complex mathematical models. The theoretical framework for the well-posedness proof has been established in [35], however, the conditions therein need to be verified on a case-by-case basis.

A related problem is the inheritance of certain properties of measures (e.g. Gaussianity, convexity, tail behavior) within the sequence of posterior measures. This is important since in a (time-dependent) filtering problem the posterior at time t becomes the prior at time $t + 1$. Hence it is not sufficient to establish the existence of a posterior without studying its properties as well.

It is often possible to prove that the Bayesian learning process converges to a measure concentrated in a small neighbourhood around the true parameter in the large data limit (Bernstein-von-Mises theorem, Doob's consistency theorem, see [37, §10]). However, the particle filters that approximate the learning process often suffer from a path degeneracy, meaning that a large number of updates decreases the estimator quality. We illustrated this for SIS where the effective sample size decreased over time. For SMC, however, a similar effect might occur, see [3] or [23, §5.1.2.], where path degeneracy is observed.

Finally, we mention that in current works on the SMC convergence the effect of the MCMC steps is often not considered (see e.g. [5, Thm 3.1]) and requires further investigations. If the MCMC steps are analysed, the results require rather strong assumptions, see [38] and [5, §5]. In addition, the mechanism of importance sampling, which is a crucial component in SMC, is not fully analysed to date. Recent works [32] establish bounds on the necessary sample size, however, it is unclear which metric is appropriate to study the nearness and, eventually, convergence of measures.

Acknowledgments The authors thank the anonymous reviewer for pointing out an error in an earlier version of the implementation of SIS, and for several helpful comments. JL and EU gratefully acknowledge the support by the German Research Foundation (DFG) and the Technische Universität München through the International Graduate School of Science and Engineering within project 10.02 BAYES. The measurements were collected by the authors and with the much appreciated help by Elizabeth Bismut, Ionuţ-Gabriel Farcaş, Mario Teixeira Parente, and Laura Scarabosio during the Open House Day on 21 October 2017 at the Forschungszentrum Garching, Germany.

Appendix 1: Quantities of Interest and Weak Representations of Measures

For a sequence of measures $(\mu_t)_{t=1}^{\infty}$ we are typically interested in computing the expected value of a measurable and μ_t-integrable ($t \geq 1$) *quantity of interest* $f : X \to \mathbb{R}$ with respect to a measure in the sequence, that is

$$\mu_t(f) := \int f \mathrm{d}\mu_t, \quad t \geq 1.$$

Example 5 (Quantities of Interest) Let $t \geq 1$.

- Let $f(x) := x_i$ denote the canonical projection onto the coordinate x_i of x. Then, $\mu_t(f)$ is the *mean* of the ith marginal density of μ_t. Moreover, $\mu_t(f^k)$ is the kth moment ($k \in \mathbb{N}$) of the same marginal.
- Let $A \subseteq X$ be measurable, and let f be given by

$$f(x) = \begin{cases} 1, & \text{if } x \in A, \\ 0, & \text{otherwise.} \end{cases}$$

Then, $\mu_t(f) =: \mu_t(A)$ is the *probability* that the parameter takes on values in A.

The quantities of interest in Example 5 are interesting in their own right. More importantly, it is possible to use integrals with respect to certain functions to fully represent a measure.

Proposition 1 *Let μ, ν be two measures on $(X, \mathscr{B}X) := (\mathbb{R}^N, \mathscr{B}\mathbb{R}^N)$. Then, the identity $\mu = \nu$ holds, if one of the following conditions is satisfied.*

1. *$\mu(A) = \nu(A)$ for all $A \in \mathscr{B}X$,*
2. *$\mu(f) = \nu(f)$ for all bounded, measurable functions $f : X \to \mathbb{R}$,*
3. *$\mu(f) = \nu(f)$ for all bounded, continuous functions $f : X \to \mathbb{R}$,*
4. *$\mu(f) = \nu(f)$ for all bounded, Lipschitz-continuous functions $f : X \to \mathbb{R}$.*

Proof Condition (1.) is the definition of $\mu = \nu$. Condition (2.) implies (1.) by setting

$$f(x) := \begin{cases} 1, & \text{if } x \in A \\ 0, & \text{otherwise,} \end{cases}$$

for any $A \in \mathscr{B}X$. Conditions (3.) and (4.) imply equivalence of the characteristic functions of μ, ν which implies that $\mu = \nu$ [21, Thm. 13.16, Thm. 15.8].

Moreover, we can use the criteria in Proposition 1 to investigate the convergence of a sequence of measures.

Proposition 2 *Let $(\mu^M)_{M=1}^{\infty}$ be a sequence of measures and let ν denote a further measure on $(X, \mathscr{B}X) := (\mathbb{R}^N, \mathscr{B}\mathbb{R}^N)$. Then $\mu^M \to \nu$ as $M \to \infty$*

1. *in total variation, if one of the following conditions holds:*
   a. *$\mu^M(A) \to \nu(A)$ as $M \to \infty$ for all $A \subseteq X$ measurable,*
   b. *$\mu^M(f) \to \nu(f)$ as $M \to \infty$ for all bounded, measurable functions $f : X \to \mathbb{R}$,*
2. *weakly, if it converges in total variation, or if one of the following conditions holds:*
   a. *$\mu^M(f) \to \nu(f)$ as $M \to \infty$ for all bounded, continuous functions $f : X \to \mathbb{R}$,*
   b. *$\mu^M(f) \to \nu(f)$ as $M \to \infty$ for all bounded, Lipschitz-continuous functions $f : X \to \mathbb{R}$.*

Proof Convergence in total variation holds if

$$\lim_{M \to \infty} \sup_{A \in \mathscr{B}X} |\mu^M(A) - \nu(A)| = 0.$$

This follows directly by condition (1.a) or (1.b). (2.a) is the definition of weak convergence, (2.b) is implied by the Portmanteau Theorem [21, Thm. 13.16].

Thus, instead of investigating the properties of measures directly, it is possible to study integrals of functions with respect to these measures.

Appendix 2: Source Code of the Particle Approximation Class

```
import numpy as np
import matplotlib.pyplot as plt

class ParticleApproximation:
    def __init__(self, num_particles, prior, init=True):
        self.num_particles = num_particles
        self.prior = prior

        # create an initial Monte Carlo approximation of the prior
        if init:
            self.particles = prior.rvs(size=num_particles)
            self.weights = np.full(num_particles, 1.0/num_particles)

    @staticmethod
    def load(filename, prior):
        data = np.load(filename)
        approximation = ParticleApproximation(data['particles'].size,
            ↪ prior, init=False)
        approximation.particles = data['particles']
        approximation.weights = data['weights']
        return approximation

    def save(self, filename):
        np.savez(filename, particles=self.particles, weights=self.
            ↪ weights)

    def hist(self, **kwargs):
        plt.hist(self.particles, weights=self.weights, **kwargs)

    def integrate(self, f):
        fv = np.vectorize(f)
        return np.dot(fv(self.particles), self.weights)

    def sample(self, size):
        return np.random.choice(self.particles, size=size, p=self.
            ↪ weights, replace=True)
```

```
    def resample(self):
        self.particles = np.random.choice(self.particles, size=self.
            ↪ num_particles, p=self.weights, replace=True)
        self.weights = np.full(self.num_particles, 1/self.
            ↪ num_particles)

    def effective_sample_size(self):
        return 1 / np.dot(self.weights, self.weights)

    def reweight(self, importance_potential, target_potential):
        # Update is done in log-scale
        self.weights = np.log(self.weights)

        # Compute log-importance-weights and update current weights
        importance_weights = target_potential(self.particles) -
            ↪ importance_potential(self.particles)
        self.weights += importance_weights

        # Return to linear-scale to normalize weights
        self.weights = np.exp(self.weights)
        self.weights /= self.weights.sum()

    def mh_correction(self, target_potential, proposal_kernel,
        ↪ n_steps):
        total_accepted = 0

        # Sample from the proposal kernel, conditioned on currect
            ↪ particles
        proposals = proposal_kernel(self.particles)

        proposal_potentials = target_potential(proposals)
        current_potentials = target_potential(self.particles)

        for i in range(n_steps):
            # Compute the log acceptance ratio
            potential_ratio = proposal_potentials -
                ↪ current_potentials
            prior_ratio = self.prior.logpdf(proposals) - self.prior.
                ↪ logpdf(self.particles)
            acceptance_ratio = np.exp(potential_ratio + prior_ratio)

            # Randomly accept the transitions based on the log
                ↪ acceptance ratio
```

```
            accepted = np.random.uniform(size=self.num_particles) <
                ↪ acceptance_ratio
            self.particles[accepted] = proposals[accepted]

            total_accepted += np.sum(accepted)

            # Recompute necessary potentials for next step
            if i < n_steps - 1:
                # Update current potentials
                current_potentials[accepted] = proposal_potentials[
                    ↪ accepted]

                # Sample new proposals
                proposals = proposal_kernel(self.particles)
                proposal_potentials = target_potential(proposals)

        return total_accepted / (n_steps * self.num_particles)

    def smc_update(self, importance_potential, target_potential,
        ↪ proposal_kernel, correction_steps, ess_ratio):
        self.reweight(importance_potential, target_potential)
        acceptance_ratio = self.mh_correction(target_potential,
            ↪ proposal_kernel, n_steps=correction_steps)

        ess = self.effective_sample_size()
        if ess < self.num_particles/ess_ratio:
            self.resample()

        return (acceptance_ratio, ess)
```

Display 2 Python file `particle_approximation.py`

References

1. Agapiou, S., Papaspiliopoulos, O., Sanz-Alonso, D., Stuart, A.M.: Importance sampling: intrinsic dimension and computational cost. Stat. Sci. **32**(3), 405–431 (2017)
2. Allmaras, M., Bangerth, W., Linhart, J.M., Polanco, J., Wang, F., Wang, K., Webster, J., Zedler, S.: Estimating parameters in physical models through Bayesian inversion: a complete example. SIAM Rev. **55**(1), 149–167 (2013)
3. Andrieu, C., De Freitas, N., Doucet, A.: Sequential MCMC for Bayesian model selection. In: Proceedings of the IEEE Workshop HOS (1999)
4. Beskos, A., Jasra, A., Kantas, N., Thiery, A.: On the convergence of adaptive sequential Monte Carlo methods. Ann. Appl. Probab. **26**(2), 1111–1146 (2016)

5. Beskos, A., Jasra, A., Muzaffer, E.A., Stuart, A.M.: Sequential Monte Carlo methods for Bayesian elliptic inverse problems. Stat. Comput. **25**(4), 727–737 (2015)
6. Beskos, A., Jasra, A., Law, K.J.H., Tempone, R., Zhou, Y.: Multilevel sequential Monte Carlo samplers. Stoch. Proc. Appl. **127**(5), 1417–1440 (2017)
7. Bulté, M.: Sequential Monte Carlo for time-dependent Bayesian inverse problems. Bachelor's thesis, Technische Universität München (2018)
8. Collis, J., Connor, A.J., Paczkowski, M., Kannan, P., Pitt-Francis, J., Byrne, H.M., Hubbard, M.E.: Bayesian calibration, validation and uncertainty quantification for predictive modelling of tumour growth: a tutorial. Bull. Math. Biol. **79**(4), 939–974 (2017)
9. Dashti, M., Stuart, A.M.: The Bayesian approach to inverse problems. In: Ghanem, R., Higdon, D., Owhadi, H. (eds.) Handbook of Uncertainty Quantification, pp. 311–428. Springer, Cham (2017)
10. Del Moral, P.: Feynman-Kac Formulae - Genealogical and Interacting Particle Systems with Applications. Springer, New York (2004)
11. Del Moral, P.: Mean Field Simulation for Monte Carlo Integration. Chapman and Hall/CRC, Boca Raton (2013)
12. Del Moral, P., Doucet, A., Jasra, A.: Sequential Monte Carlo samplers. J. R. Stat. Soc. B **68**(3), 411–436 (2006)
13. Doucet, A.: Sequential Monte Carlo methods & particle filters resources. http://www.stats.ox.ac.uk/~doucet/smc_resources.html. Accessed 20 July 2018
14. Doucet, A., Johansen, A.M.: A tutorial on particle filtering and smoothing: fifteen years later. In: The Oxford Handbook of Nonlinear Filtering, pp. 656–704. Oxford University Press, Oxford (2011)
15. Evensen, G.: The Ensemble Kalman Filter: theoretical formulation and practical implementation. Ocean Dynam. **53**(4), 343–367 (2003)
16. Gerber, M., Chopin, N., Whiteley, N.: Negative association, ordering and convergence of resampling methods (2017). ArXiv e-prints
17. Humpherys, J., Redd, P., West, J.: A Fresh Look at the Kalman Filter. SIAM Rev. **54**(4), 801–823 (2012)
18. Kahle, C., Lam, K.F., Latz, J., Ullmann, E.: Bayesian parameter identification in Cahn-Hilliard models for biological growth (2018). ArXiv e-prints
19. Kaipio, J., Somersalo, E.: Statistical and Computational Inverse Problems. Springer, New York (2005)
20. Kantas, N., Beskos, A., Jasra, A.: Sequential Monte Carlo methods for high-dimensional inverse problems: a case study for the Navier-Stokes equations. SIAM/ASA J. Uncertain. Quant. **2**(1), 464–489 (2014)
21. Klenke, A.: Probability Theory: A Comprehensive Course. Springer, London (2014)
22. Latz, J.: Bayes Linear Methods for Inverse Problems. Master's thesis, University of Warwick (2016)
23. Latz, J., Papaioannou, I., Ullmann, E.: Multilevel Sequential^2 Monte Carlo for Bayesian inverse problems. J. Comput. Phys. **368**, 154–178 (2018)
24. Law, K., Stuart, A., Zygalakis, K.: Data assimilation. A mathematical introduction, In: Texts in Applied Mathematics, vol. 62. Springer, Cham (2015)
25. Lima, E.A.B.F., Oden, J.T., Wohlmuth, B., Shahmoradi, A., Hormuth II, D.A., Yankeelov, T.E., Scarabosio, L., Horger, T.: Selection and validation of predictive models of radiation effects on tumor growth based on noninvasive imaging data. Comput. Methods Appl. Mech. Eng. **327**, 277–305 (2017)
26. Liu, J.S.: Monte Carlo Strategies in Scientific Computing. Springer, New York (2004)
27. Nakamura, G., Potthast, R.: Inverse Modeling. 2053–2563. IOP Publishing, Bristol (2015)
28. Neal, R.M.: Annealed importance sampling. Stat. Comp. **11**(2), 125–139 (2001)
29. Nocedal, J., Wright, S.J.: Numerical Optimization, 2nd edn. Springer, New York (2006)
30. Robert, C.P.: The Bayesian Choice, 2nd edn. Springer, New York (2007)
31. Robert, C.P., Casella, G.: Monte Carlo Statistical Methods. Springer, Berlin (2004)

32. Sanz-Alonso, D.: Importance sampling and necessary sample size: an information theory approach. SIAM/ASA J. Uncertain. Quant. **6**(2), 867–879 (2018)
33. Särkkä, S.: Bayesian filtering and smoothing, In: *Institute of Mathematical Statistics Textbooks*, vol. 3. Cambridge University Press, Cambridge (2013)
34. Schillings, C., Stuart, A.M.: Analysis of the ensemble Kalman filter for inverse problems. SIAM J. Numer. Anal. **55**(3), 1264–1290 (2017)
35. Stuart, A.M.: Inverse problems: A Bayesian perspective. In: Acta Numerica, vol. 19, pp. 451–559. Cambridge University Press, Cambridge (2010)
36. Thorpe, S., Fize, D., Marlot, C.: Speed of processing in the human visual system. Nature **381**(6582), 520 (1996)
37. van der Vaart, A.W.: Asymptotic Statistics. Cambridge Series in Statistical and Probabilistic Mathematics. Cambridge University Press, Cambridge (1998)
38. Whiteley, N.: Sequential Monte Carlo samplers: error bounds and insensitivity to initial conditions. Stoch. Anal. Appl. **30**(5), 774–798 (2013)

Editorial Policy

1. Volumes in the following three categories will be published in LNCSE:

i) Research monographs
ii) Tutorials
iii) Conference proceedings

Those considering a book which might be suitable for the series are strongly advised to contact the publisher or the series editors at an early stage.

2. Categories i) and ii). Tutorials are lecture notes typically arising via summer schools or similar events, which are used to teach graduate students. These categories will be emphasized by Lecture Notes in Computational Science and Engineering. **Submissions by interdisciplinary teams of authors are encouraged.** The goal is to report new developments – quickly, informally, and in a way that will make them accessible to non-specialists. In the evaluation of submissions timeliness of the work is an important criterion. Texts should be well-rounded, well-written and reasonably self-contained. In most cases the work will contain results of others as well as those of the author(s). In each case the author(s) should provide sufficient motivation, examples, and applications. In this respect, Ph.D. theses will usually be deemed unsuitable for the Lecture Notes series. Proposals for volumes in these categories should be submitted either to one of the series editors or to Springer-Verlag, Heidelberg, and will be refereed. A provisional judgement on the acceptability of a project can be based on partial information about the work: a detailed outline describing the contents of each chapter, the estimated length, a bibliography, and one or two sample chapters – or a first draft. A final decision whether to accept will rest on an evaluation of the completed work which should include

- at least 100 pages of text;
- a table of contents;
- an informative introduction perhaps with some historical remarks which should be accessible to readers unfamiliar with the topic treated;
- a subject index.

3. Category iii). Conference proceedings will be considered for publication provided that they are both of exceptional interest and devoted to a single topic. One (or more) expert participants will act as the scientific editor(s) of the volume. They select the papers which are suitable for inclusion and have them individually refereed as for a journal. Papers not closely related to the central topic are to be excluded. Organizers should contact the Editor for CSE at Springer at the planning stage, see *Addresses* below.

In exceptional cases some other multi-author-volumes may be considered in this category.

4. Only works in English will be considered. For evaluation purposes, manuscripts may be submitted in print or electronic form, in the latter case, preferably as pdf- or zipped ps-files. Authors are requested to use the LaTeX style files available from Springer at http://www.springer.com/gp/authors-editors/book-authors-editors/manuscript-preparation/5636 (Click on LaTeX Template → monographs or contributed books).

For categories ii) and iii) we strongly recommend that all contributions in a volume be written in the same LaTeX version, preferably LaTeX2e. Electronic material can be included if appropriate. Please contact the publisher.

Careful preparation of the manuscripts will help keep production time short besides ensuring satisfactory appearance of the finished book in print and online.

5. The following terms and conditions hold. Categories i), ii) and iii):

Authors receive 50 free copies of their book. No royalty is paid.
Volume editors receive a total of 50 free copies of their volume to be shared with authors, but no royalties.

Authors and volume editors are entitled to a discount of 40 % on the price of Springer books purchased for their personal use, if ordering directly from Springer.

6. Springer secures the copyright for each volume.

Addresses:

Timothy J. Barth
NASA Ames Research Center
NAS Division
Moffett Field, CA 94035, USA
barth@nas.nasa.gov

Michael Griebel
Institut für Numerische Simulation
der Universität Bonn
Wegelerstr. 6
53115 Bonn, Germany
griebel@ins.uni-bonn.de

David E. Keyes
Mathematical and Computer Sciences
and Engineering
King Abdullah University of Science
and Technology
P.O. Box 55455
Jeddah 21534, Saudi Arabia
david.keyes@kaust.edu.sa
and

Department of Applied Physics
and Applied Mathematics
Columbia University
500 W. 120 th Street
New York, NY 10027, USA
kd2112@columbia.edu

Risto M. Nieminen
Department of Applied Physics
Aalto University School of Science
and Technology
00076 Aalto, Finland
risto.nieminen@aalto.fi

Dirk Roose
Department of Computer Science
Katholieke Universiteit Leuven
Celestijnenlaan 200A
3001 Leuven-Heverlee, Belgium
dirk.roose@cs.kuleuven.be

Tamar Schlick
Department of Chemistry
and Courant Institute
of Mathematical Sciences
New York University
251 Mercer Street
New York, NY 10012, USA
schlick@nyu.edu

Editor for Computational Science
and Engineering at Springer:

Martin Peters
Springer-Verlag
Mathematics Editorial IV
Tiergartenstrasse 17
69121 Heidelberg, Germany
martin.peters@springer.com

Lecture Notes in Computational Science and Engineering

1. D. Funaro, *Spectral Elements for Transport-Dominated Equations.*
2. H.P. Langtangen, *Computational Partial Differential Equations.* Numerical Methods and Diffpack Programming.
3. W. Hackbusch, G. Wittum (eds.), *Multigrid Methods V.*
4. P. Deuflhard, J. Hermans, B. Leimkuhler, A.E. Mark, S. Reich, R.D. Skeel (eds.), *Computational Molecular Dynamics: Challenges, Methods, Ideas.*
5. D. Kröner, M. Ohlberger, C. Rohde (eds.), *An Introduction to Recent Developments in Theory and Numerics for Conservation Laws.*
6. S. Turek, *Efficient Solvers for Incompressible Flow Problems.* An Algorithmic and Computational Approach.
7. R. von Schwerin, ***M****ulti* ***B****ody* ***S****ystem* ***SIM****ulation.* Numerical Methods, Algorithms, and Software.
8. H.-J. Bungartz, F. Durst, C. Zenger (eds.), *High Performance Scientific and Engineering Computing.*
9. T.J. Barth, H. Deconinck (eds.), *High-Order Methods for Computational Physics.*
10. H.P. Langtangen, A.M. Bruaset, E. Quak (eds.), *Advances in Software Tools for Scientific Computing.*
11. B. Cockburn, G.E. Karniadakis, C.-W. Shu (eds.), *Discontinuous Galerkin Methods.* Theory, Computation and Applications.
12. U. van Rienen, *Numerical Methods in Computational Electrodynamics.* Linear Systems in Practical Applications.
13. B. Engquist, L. Johnsson, M. Hammill, F. Short (eds.), *Simulation and Visualization on the Grid.*
14. E. Dick, K. Riemslagh, J. Vierendeels (eds.), *Multigrid Methods VI.*
15. A. Frommer, T. Lippert, B. Medeke, K. Schilling (eds.), *Numerical Challenges in Lattice Quantum Chromodynamics.*
16. J. Lang, *Adaptive Multilevel Solution of Nonlinear Parabolic PDE Systems.* Theory, Algorithm, and Applications.
17. B.I. Wohlmuth, *Discretization Methods and Iterative Solvers Based on Domain Decomposition.*
18. U. van Rienen, M. Günther, D. Hecht (eds.), *Scientific Computing in Electrical Engineering.*
19. I. Babuška, P.G. Ciarlet, T. Miyoshi (eds.), *Mathematical Modeling and Numerical Simulation in Continuum Mechanics.*
20. T.J. Barth, T. Chan, R. Haimes (eds.), *Multiscale and Multiresolution Methods.* Theory and Applications.
21. M. Breuer, F. Durst, C. Zenger (eds.), *High Performance Scientific and Engineering Computing.*
22. K. Urban, *Wavelets in Numerical Simulation.* Problem Adapted Construction and Applications.
23. L.F. Pavarino, A. Toselli (eds.), *Recent Developments in Domain Decomposition Methods.*

24. T. Schlick, H.H. Gan (eds.), *Computational Methods for Macromolecules: Challenges and Applications.*

25. T.J. Barth, H. Deconinck (eds.), *Error Estimation and Adaptive Discretization Methods in Computational Fluid Dynamics.*

26. M. Griebel, M.A. Schweitzer (eds.), *Meshfree Methods for Partial Differential Equations.*

27. S. Müller, *Adaptive Multiscale Schemes for Conservation Laws.*

28. C. Carstensen, S. Funken, W. Hackbusch, R.H.W. Hoppe, P. Monk (eds.), *Computational Electromagnetics.*

29. M.A. Schweitzer, *A Parallel Multilevel Partition of Unity Method for Elliptic Partial Differential Equations.*

30. T. Biegler, O. Ghattas, M. Heinkenschloss, B. van Bloemen Waanders (eds.), *Large-Scale PDE-Constrained Optimization.*

31. M. Ainsworth, P. Davies, D. Duncan, P. Martin, B. Rynne (eds.), *Topics in Computational Wave Propagation.* Direct and Inverse Problems.

32. H. Emmerich, B. Nestler, M. Schreckenberg (eds.), *Interface and Transport Dynamics.* Computational Modelling.

33. H.P. Langtangen, A. Tveito (eds.), *Advanced Topics in Computational Partial Differential Equations.* Numerical Methods and Diffpack Programming.

34. V. John, *Large Eddy Simulation of Turbulent Incompressible Flows.* Analytical and Numerical Results for a Class of LES Models.

35. E. Bänsch (ed.), *Challenges in Scientific Computing - CISC 2002.*

36. B.N. Khoromskij, G. Wittum, *Numerical Solution of Elliptic Differential Equations by Reduction to the Interface.*

37. A. Iske, *Multiresolution Methods in Scattered Data Modelling.*

38. S.-I. Niculescu, K. Gu (eds.), *Advances in Time-Delay Systems.*

39. S. Attinger, P. Koumoutsakos (eds.), *Multiscale Modelling and Simulation.*

40. R. Kornhuber, R. Hoppe, J. Périaux, O. Pironneau, O. Wildlund, J. Xu (eds.), *Domain Decomposition Methods in Science and Engineering.*

41. T. Plewa, T. Linde, V.G. Weirs (eds.), *Adaptive Mesh Refinement – Theory and Applications.*

42. A. Schmidt, K.G. Siebert, *Design of Adaptive Finite Element Software.* The Finite Element Toolbox ALBERTA.

43. M. Griebel, M.A. Schweitzer (eds.), *Meshfree Methods for Partial Differential Equations II.*

44. B. Engquist, P. Lötstedt, O. Runborg (eds.), *Multiscale Methods in Science and Engineering.*

45. P. Benner, V. Mehrmann, D.C. Sorensen (eds.), *Dimension Reduction of Large-Scale Systems.*

46. D. Kressner, *Numerical Methods for General and Structured Eigenvalue Problems.*

47. A. Boriçi, A. Frommer, B. Joó, A. Kennedy, B. Pendleton (eds.), *QCD and Numerical Analysis III.*

48. F. Graziani (ed.), *Computational Methods in Transport.*

49. B. Leimkuhler, C. Chipot, R. Elber, A. Laaksonen, A. Mark, T. Schlick, C. Schütte, R. Skeel (eds.), *New Algorithms for Macromolecular Simulation.*

50. M. Bücker, G. Corliss, P. Hovland, U. Naumann, B. Norris (eds.), *Automatic Differentiation: Applications, Theory, and Implementations.*

51. A.M. Bruaset, A. Tveito (eds.), *Numerical Solution of Partial Differential Equations on Parallel Computers.*

52. K.H. Hoffmann, A. Meyer (eds.), *Parallel Algorithms and Cluster Computing.*

53. H.-J. Bungartz, M. Schäfer (eds.), *Fluid-Structure Interaction.*

54. J. Behrens, *Adaptive Atmospheric Modeling.*

55. O. Widlund, D. Keyes (eds.), *Domain Decomposition Methods in Science and Engineering XVI.*

56. S. Kassinos, C. Langer, G. Iaccarino, P. Moin (eds.), *Complex Effects in Large Eddy Simulations.*

57. M. Griebel, M.A Schweitzer (eds.), *Meshfree Methods for Partial Differential Equations III.*

58. A.N. Gorban, B. Kégl, D.C. Wunsch, A. Zinovyev (eds.), *Principal Manifolds for Data Visualization and Dimension Reduction.*

59. H. Ammari (ed.), *Modeling and Computations in Electromagnetics: A Volume Dedicated to Jean-Claude Nédélec.*

60. U. Langer, M. Discacciati, D. Keyes, O. Widlund, W. Zulehner (eds.), *Domain Decomposition Methods in Science and Engineering XVII.*

61. T. Mathew, *Domain Decomposition Methods for the Numerical Solution of Partial Differential Equations.*

62. F. Graziani (ed.), *Computational Methods in Transport: Verification and Validation.*

63. M. Bebendorf, *Hierarchical Matrices.* A Means to Efficiently Solve Elliptic Boundary Value Problems.

64. C.H. Bischof, H.M. Bücker, P. Hovland, U. Naumann, J. Utke (eds.), *Advances in Automatic Differentiation.*

65. M. Griebel, M.A. Schweitzer (eds.), *Meshfree Methods for Partial Differential Equations IV.*

66. B. Engquist, P. Lötstedt, O. Runborg (eds.), *Multiscale Modeling and Simulation in Science.*

67. I.H. Tuncer, Ü. Gülcat, D.R. Emerson, K. Matsuno (eds.), *Parallel Computational Fluid Dynamics 2007.*

68. S. Yip, T. Diaz de la Rubia (eds.), *Scientific Modeling and Simulations.*

69. A. Hegarty, N. Kopteva, E. O'Riordan, M. Stynes (eds.), *BAIL 2008 – Boundary and Interior Layers.*

70. M. Bercovier, M.J. Gander, R. Kornhuber, O. Widlund (eds.), *Domain Decomposition Methods in Science and Engineering XVIII.*

71. B. Koren, C. Vuik (eds.), *Advanced Computational Methods in Science and Engineering.*

72. M. Peters (ed.), *Computational Fluid Dynamics for Sport Simulation.*

73. H.-J. Bungartz, M. Mehl, M. Schäfer (eds.), *Fluid Structure Interaction II - Modelling, Simulation, Optimization.*

74. D. Tromeur-Dervout, G. Brenner, D.R. Emerson, J. Erhel (eds.), *Parallel Computational Fluid Dynamics 2008.*

75. A.N. Gorban, D. Roose (eds.), *Coping with Complexity: Model Reduction and Data Analysis.*

76. J.S. Hesthaven, E.M. Rønquist (eds.), *Spectral and High Order Methods for Partial Differential Equations.*
77. M. Holtz, *Sparse Grid Quadrature in High Dimensions with Applications in Finance and Insurance.*
78. Y. Huang, R. Kornhuber, O.Widlund, J. Xu (eds.), *Domain Decomposition Methods in Science and Engineering XIX.*
79. M. Griebel, M.A. Schweitzer (eds.), *Meshfree Methods for Partial Differential Equations V.*
80. P.H. Lauritzen, C. Jablonowski, M.A. Taylor, R.D. Nair (eds.), *Numerical Techniques for Global Atmospheric Models.*
81. C. Clavero, J.L. Gracia, F.J. Lisbona (eds.), *BAIL 2010 – Boundary and Interior Layers, Computational and Asymptotic Methods.*
82. B. Engquist, O. Runborg, Y.R. Tsai (eds.), *Numerical Analysis and Multiscale Computations.*
83. I.G. Graham, T.Y. Hou, O. Lakkis, R. Scheichl (eds.), *Numerical Analysis of Multiscale Problems.*
84. A. Logg, K.-A. Mardal, G. Wells (eds.), *Automated Solution of Differential Equations by the Finite Element Method.*
85. J. Blowey, M. Jensen (eds.), *Frontiers in Numerical Analysis - Durham 2010.*
86. O. Kolditz, U.-J. Gorke, H. Shao, W. Wang (eds.), *Thermo-Hydro-Mechanical-Chemical Processes in Fractured Porous Media - Benchmarks and Examples.*
87. S. Forth, P. Hovland, E. Phipps, J. Utke, A. Walther (eds.), *Recent Advances in Algorithmic Differentiation.*
88. J. Garcke, M. Griebel (eds.), *Sparse Grids and Applications.*
89. M. Griebel, M.A. Schweitzer (eds.), *Meshfree Methods for Partial Differential Equations VI.*
90. C. Pechstein, *Finite and Boundary Element Tearing and Interconnecting Solvers for Multiscale Problems.*
91. R. Bank, M. Holst, O. Widlund, J. Xu (eds.), *Domain Decomposition Methods in Science and Engineering XX.*
92. H. Bijl, D. Lucor, S. Mishra, C. Schwab (eds.), *Uncertainty Quantification in Computational Fluid Dynamics.*
93. M. Bader, H.-J. Bungartz, T. Weinzierl (eds.), *Advanced Computing.*
94. M. Ehrhardt, T. Koprucki (eds.), *Advanced Mathematical Models and Numerical Techniques for Multi-Band Effective Mass Approximations.*
95. M. Azaïez, H. El Fekih, J.S. Hesthaven (eds.), *Spectral and High Order Methods for Partial Differential Equations ICOSAHOM 2012.*
96. F. Graziani, M.P. Desjarlais, R. Redmer, S.B. Trickey (eds.), *Frontiers and Challenges in Warm Dense Matter.*
97. J. Garcke, D. Pflüger (eds.), *Sparse Grids and Applications – Munich 2012.*
98. J. Erhel, M. Gander, L. Halpern, G. Pichot, T. Sassi, O. Widlund (eds.), *Domain Decomposition Methods in Science and Engineering XXI.*
99. R. Abgrall, H. Beaugendre, P.M. Congedo, C. Dobrzynski, V. Perrier, M. Ricchiuto (eds.), *High Order Nonlinear Numerical Methods for Evolutionary PDEs - HONOM 2013.*
100. M. Griebel, M.A. Schweitzer (eds.), *Meshfree Methods for Partial Differential Equations VII.*

101. R. Hoppe (ed.), *Optimization with PDE Constraints - OPTPDE 2014.*

102. S. Dahlke, W. Dahmen, M. Griebel, W. Hackbusch, K. Ritter, R. Schneider, C. Schwab, H. Yserentant (eds.), *Extraction of Quantifiable Information from Complex Systems.*

103. A. Abdulle, S. Deparis, D. Kressner, F. Nobile, M. Picasso (eds.), *Numerical Mathematics and Advanced Applications - ENUMATH 2013.*

104. T. Dickopf, M.J. Gander, L. Halpern, R. Krause, L.F. Pavarino (eds.), *Domain Decomposition Methods in Science and Engineering XXII.*

105. M. Mehl, M. Bischoff, M. Schäfer (eds.), *Recent Trends in Computational Engineering - CE2014.* Optimization, Uncertainty, Parallel Algorithms, Coupled and Complex Problems.

106. R.M. Kirby, M. Berzins, J.S. Hesthaven (eds.), *Spectral and High Order Methods for Partial Differential Equations - ICOSAHOM'14.*

107. B. Jüttler, B. Simeon (eds.), *Isogeometric Analysis and Applications 2014.*

108. P. Knobloch (ed.), *Boundary and Interior Layers, Computational and Asymptotic Methods – BAIL 2014.*

109. J. Garcke, D. Pflüger (eds.), *Sparse Grids and Applications – Stuttgart 2014.*

110. H. P. Langtangen, *Finite Difference Computing with Exponential Decay Models.*

111. A. Tveito, G.T. Lines, *Computing Characterizations of Drugs for Ion Channels and Receptors Using Markov Models.*

112. B. Karazösen, M. Manguoğlu, M. Tezer-Sezgin, S. Göktepe, Ö. Uğur (eds.), *Numerical Mathematics and Advanced Applications - ENUMATH 2015.*

113. H.-J. Bungartz, P. Neumann, W.E. Nagel (eds.), *Software for Exascale Computing - SPPEXA 2013-2015.*

114. G.R. Barrenechea, F. Brezzi, A. Cangiani, E.H. Georgoulis (eds.), *Building Bridges: Connections and Challenges in Modern Approaches to Numerical Partial Differential Equations.*

115. M. Griebel, M.A. Schweitzer (eds.), *Meshfree Methods for Partial Differential Equations VIII.*

116. C.-O. Lee, X.-C. Cai, D.E. Keyes, H.H. Kim, A. Klawonn, E.-J. Park, O.B. Widlund (eds.), *Domain Decomposition Methods in Science and Engineering XXIII.*

117. T. Sakurai, S.-L. Zhang, T. Imamura, Y. Yamamoto, Y. Kuramashi, T. Hoshi (eds.), *Eigenvalue Problems: Algorithms, Software and Applications in Petascale Computing*. EPASA 2015, Tsukuba, Japan, September 2015.

118. T. Richter (ed.), *Fluid-structure Interactions*. Models, Analysis and Finite Elements.

119. M.L. Bittencourt, N.A. Dumont, J.S. Hesthaven (eds.), *Spectral and High Order Methods for Partial Differential Equations ICOSAHOM 2016.* Selected Papers from the ICOSAHOM Conference, June 27-July 1, 2016, Rio de Janeiro, Brazil.

120. Z. Huang, M. Stynes, Z. Zhang (eds.), *Boundary and Interior Layers, Computational and Asymptotic Methods BAIL 2016.*

121. S.P.A. Bordas, E.N. Burman, M.G. Larson, M.A. Olshanskii (eds.), *Geometrically Unfitted Finite Element Methods and Applications*. Proceedings of the UCL Workshop 2016.

122. A. Gerisch, R. Penta, J. Lang (eds.), *Multiscale Models in Mechano and Tumor Biology*. Modeling, Homogenization, and Applications.

123. J. Garcke, D. Pflüger, C.G. Webster, G. Zhang (eds.), *Sparse Grids and Applications - Miami 2016*.

124. M. Schäfer, M. Behr, M. Mehl, B. Wohlmuth (eds.), *Recent Advances in Computational Engineering*. Proceedings of the 4th International Conference on Computational Engineering (ICCE 2017) in Darmstadt.

125. P.E. Bjørstad, S.C. Brenner, L. Halpern, R. Kornhuber, H.H. Kim, T. Rahman, O.B. Widlund (eds.), *Domain Decomposition Methods in Science and Engineering XXIV*. 24th International Conference on Domain Decomposition Methods, Svalbard, Norway, February 6–10, 2017.

126. F.A. Radu, K. Kumar, I. Berre, J.M. Nordbotten, I.S. Pop (eds.), *Numerical Mathematics and Advanced Applications – ENUMATH 2017*.

127. X. Roca, A. Loseille (eds.), *27th International Meshing Roundtable*.

128. Th. Apel, U. Langer, A. Meyer, O. Steinbach (eds.), *Advanced Finite Element Methods with Applications*. Selected Papers from the 30th Chemnitz Finite Element Symposium 2017.

129. M. Griebel, M.A. Schweitzer (eds.), *Meshfree Methods for Partial Differencial Equations IX*.

130. S. Weißer, BEM-based Finite Element *Approaches on Polytopal Meshes*.

131. V.A. Garanzha, L. Kamenski, H. Si (eds.), *Numerical Geometry, Grid Generation and Scientific Computing*. Proceedings of the 9th International Conference, NUMGRID2018/Voronoi 150, Celebrating the 150th Anniversary of G. F. Voronoi, Moscow, Russia, December 2018.

132. H. van Brummelen, A. Corsini, S. Perotto, G. Rozza (eds.), *Numerical Methods for Flows*.

133. H. van Brummelen, C. Vuik, M. Möller, C. Verhoosel, B. Simeon, B. Jüttler (eds.), *Isogeometric Analysis and Applications 2018*.

134. S.J. Sherwin, D. Moxey, J. Peiro, P.E. Vincent, C. Schwab (eds.), *Spectral and High Order Methods for Partial Differential Equations ICOSAHOM 2018*.

135. G.R. Barrenechea, J. Mackenzie (eds.), *Boundary and Interior Layers, Computational and Asymptotic Methods BAIL 2018*.

136. H.-J. Bungartz, S. Reiz, B. Uekermann, P. Neumann, W.E. Nagel (eds.), *Software for Exascale Computing - SPPEXA 2016–2019*.

137. M. D'Elia, M. Gunzburger, G. Rozza (eds.), *Quantification of Uncertainty: Improving Efficiency and Technology*.

138. R. Haynes, S. MacLachlan, X.-C. Cai, L. Halpern, H.H. Kim, A. Klawonn, O. Widlund (eds.), *Domain Decomposition Methods in Science and Engineering XXV*.

139. F.J. Vermolen, C. Vuik (eds.), *Numerical Mathematics and Advanced Applications ENUMATH 2019*.

For further information on these books please have a look at our mathematics catalogue at the following URL: www.springer.com/series/3527

Monographs in Computational Science and Engineering

1. J. Sundnes, G.T. Lines, X. Cai, B.F. Nielsen, K.-A. Mardal, A. Tveito, *Computing the Electrical Activity in the Heart.*

For further information on this book, please have a look at our mathematics catalogue at the following URL: www.springer.com/series/7417

Texts in Computational Science and Engineering

1. H. P. Langtangen, *Computational Partial Differential Equations.* Numerical Methods and Diffpack Programming. 2nd Edition
2. A. Quarteroni, F. Saleri, P. Gervasio, *Scientific Computing with MATLAB and Octave.* 4th Edition
3. H. P. Langtangen, *Python Scripting for Computational Science.* 3rd Edition
4. H. Gardner, G. Manduchi, *Design Patterns for e-Science.*
5. M. Griebel, S. Knapek, G. Zumbusch, *Numerical Simulation in Molecular Dynamics.*
6. H. P. Langtangen, *A Primer on Scientific Programming with Python.* 5th Edition
7. A. Tveito, H. P. Langtangen, B. F. Nielsen, X. Cai, *Elements of Scientific Computing.*
8. B. Gustafsson, *Fundamentals of Scientific Computing.*
9. M. Bader, *Space-Filling Curves.*
10. M. Larson, F. Bengzon, *The Finite Element Method: Theory, Implementation and Applications.*
11. W. Gander, M. Gander, F. Kwok, *Scientific Computing: An Introduction using Maple and MATLAB.*
12. P. Deuflhard, S. Röblitz, *A Guide to Numerical Modelling in Systems Biology.*
13. M. H. Holmes, *Introduction to Scientific Computing and Data Analysis.*
14. S. Linge, H. P. Langtangen, *Programming for Computations* - A Gentle Introduction to Numerical Simulations with MATLAB/Octave.
15. S. Linge, H. P. Langtangen, *Programming for Computations* - A Gentle Introduction to Numerical Simulations with Python.
16. H.P. Langtangen, S. Linge, *Finite Difference Computing with PDEs* - A Modern Software Approach.
17. B. Gustafsson, *Scientific Computing from a Historical Perspective.*
18. J. A. Trangenstein, *Scientific Computing.* Volume I - Linear and Nonlinear Equations.

19. J. A. Trangenstein, *Scientific Computing*. Volume II - Eigenvalues and Optimization.

20. J. A. Trangenstein, *Scientific Computing*. Volume III - Approximation and Integration.

21. H. P. Langtangen, K.-A. Mardal, *Introduction to Numerical Methods for Variational Problems.*

22. T. Lyche, *Numerical Linear Algebra and Matrix Factorizations.*

For further information on these books please have a look at our mathematics catalogue at the following URL: `www.springer.com/series/5151`

Printed in the United States
by Baker & Taylor Publisher Services

Printed in the United States
by Baker & Taylor Publisher Services